About Island Press

Island Press is the only nonprofit organization in the United States whose principal purpose is the publication of books on environmental issues and natural resource management. We provide solutions-oriented information to professionals, public officials, business and community leaders, and concerned citizens who are shaping responses to environmental problems.

In 2000, Island Press celebrates its sixteenth anniversary as the leading provider of timely and practical books that take a multidisciplinary approach to critical environmental concerns. Our growing list of titles reflects our commitment to bringing the best of an expanding body of literature to the environmental community throughout North America and the world.

Support for Island Press is provided by The Jenifer Altman Foundation, The Bullitt Foundation, The Mary Flagler Cary Charitable Trust, The Nathan Cummings Foundation, The Geraldine R. Dodge Foundation, The Charles Engelhard Foundation, The Ford Foundation, The German Marshall Fund of the United States, The George Gund Foundation, The Vira I. Heinz Endowment, The William and Flora Hewlett Foundation, The W. Alton Jones Foundation, The John D. and Catherine T. MacArthur Foundation, The Andrew W. Mellon Foundation, The Charles Stewart Mott Foundation, The Curtis and Edith Munson Foundation, The National Fish and Wildlife Foundation, The New-Land Foundation, The Oak Foundation, The Overbrook Foundation, The David and Lucile Packard Foundation, The Pew Charitable Trusts, The Rockefeller Brothers Fund, Rockefeller Financial Services, The Winslow Foundation, and individual donors.

About the Pacific Institute for Studies in Development, Environment, and Security

The Pacific Institute for Studies in Development, Environment, and Security, in Oakland, California, is an independent, nonprofit organization created in 1987 to conduct research and policy analysis in the areas of environmental protection, sustainable development, and international security. Underlying all of the Institute's work is the recognition that the urgent problems of environmental degradation, regional and global poverty, and political tension and conflict are fundamentally interrelated, and that long-term solutions dictate an interdisciplinary approach. Since 1987, we have produced more than 60 research studies, organized roundtable discussions, and held widespread briefings for policymakers and the public. The Institute has formulated a new vision for long-term water planning in California and internationally, developed a new approach for valuing well-being in the Sierra Nevada, worked on transborder environment and trade issues in North America, analyzed ISO 14000's role in global environmental protection, clarified key concepts and criteria for sustainable water use in the lower Colorado basin, offered recommendations for reducing conflicts over water in the Middle East and elsewhere, assessed the impacts of global warming on freshwater resources, and created programs to address environmental justice concerns in poor communities and communities of color.

For detailed information about the Institute's activities, visit www.pacinst.org, www.worldwater.org, and www.globalchange.org.

THE WORLD'S WATER
2000-2001

THE WORLD'S WATER

2000–2001

The Biennial Report on Freshwater Resources

Peter H. Gleick

Pacific Institute for Studies in Development,
Environment, and Security
Oakland, California

ISLAND PRESS

Washington, D.C. • Covelo, California

ISLAND PRESS is a trademark of the Center for Resource Economics.

Library of Congress Card Catalog Number: 98-24877

ISBN 1-55963-792-7

ISSN 1528-7165

Printed on recycled, acid-free paper ∞ ✪

Manufactured in the United States of America

10 9 8 7 6 5 4 3 2 1

To all those working to solve the world's water problems

Contents

Foreword

The twentieth century was one of tremendous political, demographic, economic, and technological change. Politically, the concept of world war twice came into play, and for nearly fifty years we endured the Cold War. At century's end, regional and ethnic conflicts continued to bring home the divisive nature of politics and ideology. Demographically, rapidly growing human populations and dramatic increases in consumption of resources transformed human activity into one of the most profound natural forces on Earth. The emergence of a gaping seasonal hole in the Earth's protective ozone layer demonstrated to all the interconnected nature of the world in which we now live.

Indeed, the end of the twentieth century saw the emergence of interdependence—the idea that certain problems transcend borders and link the fates of all the world's people. Accordingly, international organizations were created to begin addressing a wide range of problems, ranging from military security to legal justice to ecological and human health. Economic systems became similarly intertwined—a phenomenon that has come to be known as globalization. Finally, advances in communications, information, and transportation drew the world together as never before, transforming the roles of governments, the private sector, and nongovernmental organizations.

The world of the twenty-first century will no doubt be equally complex and surprising. But as we enter this new century, it is important that the unresolved problems and unmet challenges of the twentieth century not be forgotten. While there are many issues that must be addressed by the world community, few are as fundamental as those that surround human needs and uses of water.

I can think of no one who is better able to identify both our water problems and their solutions than Peter Gleick, the author of this invaluable series. I first met Gleick while I was serving the state of Colorado in the United States Senate and working on the issues of global climate change and its potential effects on U.S. water resources. Peter Gleick was then and is now the expert who is widely regarded and recognized for his work in this field. His ability to blend science and policy—all in plain English—made him an outstanding witness in hearings I helped chair for the U.S. Senate Committee on Energy and Natural Resources in the late 1980s. Not surprisingly, he has been asked to brief such policymakers as the secretary of the interior, the secretary of state, and the vice president on both domestic and international water issues.

During subsequent work on climate change as undersecretary of state for global affairs, I too found myself calling on Gleick to help sort through the potential ramifications of current and future climatic and human influences on water and its growing relationship to the potential for dispute and conflict within and among nations.

This, the second volume of *The World's Water,* serves as a fine complement to its predecessor, which was published in 1998. This new book includes updates on such important topics as the fascinating chronology of water-related conflicts. Gleick also continues to argue cogently for more attention to be paid to one of the most basic water problems: the need to provide clean drinking water and sanitation services to the billion people who still lack them.

This theme of poverty alleviation and meeting basic human needs runs throughout Gleick's writing. In 1996 he defined and quantified a basic water requirement for all humans and argued that the provision of this water should be the top priority for international, national, and regional water policymakers. In 1998, he described the terrible human health implications of the failure to meet these basic water needs and the high economic costs of that failure. And in the very first chapter of this new book he makes a compelling argument that there is a legal and institutional human "right" to water, further raising the stakes for the water community.

But this second volume traverses other new terrain, such as the complex and critical relationship between water and food production. One of the most important concerns we face in the new century is whether or not we will be able to produce and distribute enough food to meet the needs of a growing global population. Directly related to this are the questions of whether there will be enough water to grow this food, where that water will come from, and what the social and environmental implications of that water use will be. Gleick addresses these issues perceptively and does not shy away from such controversial issues as whether certain dams ought to be removed. He describes and analyzes the trend toward the removal of dams that have outlived their usefulness or have caused so much ecological damage as to warrant their decommissioning. He also brings a clear eye to the question of the potential for desalination and water reclamation and reuse to solve water-supply problems and offers an unbiased look at the possibilities of and limitations to these solutions. Gleick concludes that both approaches have great potential and can be a part of water systems in many places, but they must be considered as only pieces of the water puzzle and not as panaceas for our problems.

One of the most valuable attributes of this series of water books is the vast amount of information and data it contains. Each of Gleick's books includes separate tables and figures with clear descriptions of the data, their limitations, and their sources. I am aware of no other single source of information about water that has the breadth and depth of information that these books have. The interested reader can find data on dams removed in the United States and elsewhere, the capacity of desalination plants around the world, the new registry of international river basins, water supply and demand by country, irrigated area, the distribution of reservoirs, and much more.

Water runs as a thread through all of our lives, connecting us to the food we eat, the ecosystems on which we depend, our climate, and to our neighbors, often determining whether we live in peace or in conflict. *The World's Water* helps to trace that thread and weaves it into a clear tapestry that informs and guides us on this critical aspect of our global future.

TIMOTHY E. WIRTH
President, United Nations Foundation

Acknowledgments

This book has benefited from numerous conversations, reviews, critiques, and suggestions. Many people also either steered me toward important sources of data or provided those data outright. For all this assistance, I would like to thank Joe Alcamo, Kent Anderson, Jim Birkett, Margaret Bowman, Wil Burns, Beth Chalecki, Catterina Ferreccio, Beth Gleick, Donen Gleick, Pamela Hyde, Richard Jolly, George Kent, Peter MacLaggan, Elizabeth Maclin, Steve McCaffrey, Ruth Meinzen-Dick, Lisa Owens-Viani, Rajul Pandya-Lorch, Sandra Postel, Kevin Price, Jerry delli Priscoli, Frank Rijsberman, David Seckler, Igor Shiklomanov, Klaus Wangnick, Brian Ward, Paul Ward, Jim Wescoat, Aaron Wolf, Arlene Wong, and probably many others I've forgotten. Any errors are, of course, my own.

Special thanks are also due to Todd Baldwin, my editor at Island Press. Todd somehow manages to tread the fine line between letting me do my work and letting me know that he is anxiously awaiting the next draft. He also has an uncanny ability to find the holes in my arguments or writing and the diplomatic skills needed to point them out to me in a way that makes me want to fix them, rather than throw the whole thing in the round file under my desk.

Support for *The World's Water 2000–2001* was provided by grants to the Pacific Institute for Studies in Development, Environment, and Security from the Flora Family, Horace W. Goldsmith, William and Flora Hewlett, Henry P. Kendall, and John D. and Catherine T. MacArthur foundations. I thank them for their foresight in identifying water as a central issue in the environment, development, and security arena and their willingness to support the research and policy work these books represent.

Finally, as always, I can't thank Nicki Norman enough for all she is and does.

Introduction

Large or round numbers seem to hold a special significance. We celebrate major birthdays and anniversaries and keep our eye out for numbers with lots of zeros or a repeating digit on an odometer. It was front-page news when the Earth's population reached 6 billion. Most of us remember calculating how old we would be in the year 2000. And when both the century and the millennium numbers turned over, we threw ourselves a global party.

We are living in the midst of an information revolution where numbers rule. But this revolution is a selective one, reaching only certain people with only limited information. Some numbers, and their meaning, are beyond our understanding or have such depressing implications that we tend to ignore them. Thus, while many people know Bill Gates's net worth, who knows the worth of the billion poorest people on the planet? People in richer countries know how much bottled water costs, but have no idea that it costs hundreds or even thousands of times more than pure water from their taps. We hear about the growing numbers of people with access to the Internet, but who hears about how many people suffer from lack of access to clean drinking water? Every day we hear about the NASDAQ, S&P 500, Nikkei, or Hang Seng Indexes, but why don't we hear more about the Human Development Index—a much better indicator of overall human well-being? We know immediately when people die in airplane crashes and train derailments, but who knows how many tens of thousands of children die each day from easily prevented water-related diseases? My children and their peers know by heart the names and characteristics of more than 150 make-believe creatures called Pokéman, but who knows the names of the many real creatures driven to extinction by human actions?

Water runs like a river through our lives, touching everything from our health and the health of ecosystems around us to farmers' fields and the production of the goods we consume. The story of water is a complex one, told partly by numbers, partly by real human stories, and partly by intangible, immeasurable things. *The World's Water* is an effort to tell some of this story—to offer pieces of the puzzle that will let us understand the role of water in the human equation and, conversely, the effects of humans on the hydrologic cycle that sustains us. These pieces, however, should be viewed as only part of a rapidly changing story, which is one reason why

The World's Water was designed to be produced every two years—to publish new information and data on water problems and concerns, and pose new solutions to be tried, discarded, or improved upon.

The first two reports, 1998–1999 and 2000–2001, are not meant to be viewed separately, but rather as complementary to each other. *The World's Water 1998–1999* included chapters on the changing nature of water management and the connections between water and human health, an update on large dams, reviews of the links between water and international conflict and water and climatic changes, and developments in international organizations and structures working on water problems. Extensive data on water were presented in tables not easily accessible elsewhere.

A major focus of the first report was the changing nature of water management, development, and planning—I described it as "the changing water paradigm." There are many components to this change: a shift away from sole, or even primary, reliance on finding new sources of supply to address perceived new demands; a growing emphasis on incorporating ecological values into water policy; a reemphasis on meeting basic needs for water services; and a conscious breaking of the ties between economic growth and water use. Much has happened in the world of water in the subsequent two years. I believe the evidence for a true change in the way we think about water continues to accumulate.

The World's Water 2000–2001 builds on this idea. In an analysis stimulated in part by the fiftieth anniversary of the Universal Declaration of Human Rights, the legal, moral, and institutional implications of recognizing a human right to water are presented in the first chapter. A human right to water appears to be firmly rooted in international law and the norms of expected state behavior. Acknowledging that right is an important new step in meeting unmet basic water needs for billions of people and in forcing a re-evaluation of water priorities and policies.

New information is now available on the world's stocks and flows and the first new analysis of international river basins in over 20 years is described and analyzed. In the last 20 years, the number of international rivers basins has increased from 214 to 261—just one indicator of the increasingly political nature of this vital natural resource. The first book explicitly discussed water-related conflicts; this one expands further on the history and nature of such conflicts.

The connections between water and food are addressed in detail here as concerns of food experts begin to encompass the realities of water availability. In some ways, thinking about food is undergoing a revolution similar to that in the water world. It is becoming harder and harder to bring new lands into production and to maintain the historically large annual increases in crop yields. Yet there is great potential for improving the "efficiency" with which we produce food, by changing cropping patterns, by reducing wasteful applications of resources, by cutting losses between the field and the plate, and by altering diets and the manner in which international markets function. Each of these approaches has a parallel in the debate over meeting water needs. But more importantly, the question of whether we can produce enough food to feed a burgeoning population—and get it to where it is needed—is intricately connected to the question of where and when fresh water is available. Decisions made today about water policy will affect whether people continue to be undernourished in the coming decades.

The first book reviewed the state of the world's dams and noted that the old paradigm of relying on ever larger numbers of dams to capture ever larger fractions of freshwater runoff is beginning to fail for environmental, economic, and social reasons. This book focuses on a related trend to take out or decommission dams that either no longer serve a useful purpose or have caused such egregious ecological impacts as to warrant removal. Nearly 500 dams in the United States and elsewhere have already been removed, and the movement toward river restoration is accelerating. Within a few months of the removal of the Edwards Dam in Maine in mid-1999, salmon, striped bass, alewives, and other affected fish returned to waters from which they had been absent for 162 years. Several other case studies are described in this volume, together with some ambitious proposals to remove some of the world's largest dams.

As traditional approaches to the supply of water become less appropriate or more expensive, unconventional methods are receiving more attention. The concept and practice of transporting fresh water in large oceangoing plastic bags was described in the first book. Several additional approaches are described here, including large- and small-scale desalination technology, water reclamation and reuse, and techniques such as fog collection. More and more cities are discovering that wastewater can be a resource, not a liability, for purposes ranging from irrigation to drinking. Matching water demands with available waters of different quality can reduce water-supply constraints, increase system reliability, and solve costly wastewater disposal problems. Water-quality issues are also addressed in the context of the discovery of the widespread contamination of groundwater in Bangladesh and West Bengal, India, with arsenic. The laudable successes in the 1970s and 1980s in meeting the water-supply requirements of millions of Bangladeshis are now threatened by the failure to detect arsenic and to protect public health.

In one of the greatest technological embarrassments of the twentieth century, which certainly had its share of technological embarrassments, errors caused by the use of incompatible units of measure by two different groups involved in spacecraft navigation led to the destruction of the Mars Climate Orbiter spacecraft just as it reached the red planet—a $125 million dollar goof-up. The water world has its share of strange units of measure. In a modest effort to encourage students of water problems to check their work and avoid such expensive mistakes, I've included a comprehensive set of water units and conversions at the end of the Data Section. Now readers should be readily able to find the appropriate meaning (and conversions) for leagues, dekameters, feddans, acre-feet, Imperial gallons, morgen-feet, miner's inches, quinaria, and more.

The Data Section in the present volume also provides new, updated, and expanded data sets. When no new data were available, data tables from the first edition were not reproduced. For example, new data on cholera, access to clean drinking water and sanitation services, and hydroelectric capacity and production are either not available or vary little from the 1998 edition. New tables, however, have been added on water and agriculture, international river basins, basic stocks and flows of the world's water, and more. Downloadable selections from both sets of tables are posted on the Web site associated with this book: http://www.worldwater.org.

Some of the other information found in the first edition of *The World's Water* is also updated here. The chronology on water-related conflicts has been modified and expanded and now appears in the Water Briefs section. As the information available on the Internet has grown, the section on water-related Internet sites has been enlarged and updated. This list, too, is available at the above Web site.

A few readers of the first book noted the lack of a detailed discussion of food and agriculture, or flood control, or some other important issue. No doubt some readers of this edition will note the lack of a detailed discussion of water and ecosystems, or privatization, or something else. But I repeat a comment made in the introduction to the first edition: No single publication can adequately address all of the issues of interest to water experts, students, and the public. I urge readers to use these publications as stepping stones into the large, turbulent world of water, with its many different streams and pools.

PETER H. GLEICK
OAKLAND, CALIFORNIA

The Human Right to Water

If the misery of our poor be caused not by the laws of nature, but by our institutions, great is our sin.
CHARLES ĐARWIN

The test of our progress is not whether we add more to the abundance of those who have much; it is whether we provide enough for those who have little.
FRANKLIN ĐELANO ROOSEVELT

Universal access to basic water services is one of the most fundamental conditions of human development. Yet as we enter the twenty-first century, billions of people lack such access. The numbers are stark: more than 1 billion people in the developing world do not have safe drinking water, and nearly 3 billion people live without access to adequate sanitation systems necessary for reducing exposure to water-related diseases. The failure of the international aid community, nations, and local organizations to satisfy these basic human needs has led to substantial, unnecessary, and preventable human suffering. An estimated 14 to 30 thousand people, mostly young children and the elderly, die every day from water-related diseases. At any given moment, approximately one-half of the people in the developing world suffer from disease caused by drinking contaminated water or eating contaminated food (United Nations 1997b). This chapter argues that access to a basic water requirement is a fundamental human right implicitly supported by international law, declarations, and state practice. This right to water could also be considered even more basic and vital than some of the more explicit human rights already acknowledged by the international community. And a transition is underway making a right to water explicit.

As we enter the twenty-first century, governments, international aid agencies, nongovernmental organizations, and local communities must work to provide all

Portions of this chapter were published in the article "A Human Right to Water" *Water Policy*, Vol. 1, No. 5, pp. 487–503.

humans with a basic water requirement and to guarantee that water as a human right. By acknowledging a human right to water and expressing the willingness to meet this right for those currently deprived of it, the water community would have a useful tool for addressing one of the most fundamental failures of twentieth-century development.

Is There a Human Right to Water?

The question of what qualifies as a human right has generated a substantial body of literature as well as many organizations and conferences. The term "right" is here used in the sense of genuine rights under international law, where states have a duty to protect and promote those rights. The initial reason for the international community to develop human rights agreements was to address violations of moral values and standards related to violence and loss of freedoms. Over the past few decades, however, the international community has increasingly expanded rights laws and agreements to encompass a broader set of concerns related to human well-being, including rights associated with environmental and social conditions and access to resources. Box 1.1 lists some of the formal definitions commonly found in discussions of human rights and international law.

BOX 1.1

Human Rights and International Law: Some Common Definitions

Convention
Agreement concluded among states; synonym of treaty. It is legally binding on the states that ratify it.

Protocol
Agreement that completes an international treaty or convention, and which has the same legal force as the initial document.

Covenant
Synonym of treaty, convention; legally binding agreement among states.

Ratification
Approval of a treaty, convention, or other document by a country's competent bodies, thereby securing that country's commitment to it.

Declaration
A document whose signatories express their agreement with a set of objectives and principles. It is not legally binding but carries moral weight.

Resolution
Text adopted by a deliberative body or an international organization (for example, by the United Nations General Assembly or the General Conference of UNESCO).

Until recently the question of whether individuals or groups have a legal right to a minimum set of resources, specifically water, and whether there is an obligation for states or other parties to provide those resources when they are lacking, has not been adequately addressed. Several of the major references and bibliographies related to the issue of human rights have no entries or citations related to water (Lawson 1991, United Nations 1993, Steiner and Alston 1996). The new edition of Lawson (1996) adds a single water entry, but in the context of human health, not human rights. Even the current index of the Web site of the UN high commissioner for human rights has no entry for water (http://www.unhchr.ch/index.htm). In 1992 McCaffrey tackled the legal background from the perspective of the UN (and related international law) human rights framework in a comprehensive and perceptive assessment. His initial conclusion was that there is a right at least to sufficient water to sustain life and that a state has the "due diligence obligation to safeguard these rights" as a priority (McCaffrey 1992). Gleick (1999) expands upon that analysis and concludes that international law, international agreements, and evidence from the practice of states strongly and broadly support the human right to a basic water requirement.

What is the purpose or value of explicitly acknowledging a human right to water, as the international community has explicitly acknowledged a human right to food and to life? After all, despite the declaration of a formal right to food, nearly 850 million people remain undernourished (see Chapter 4). One reason is to encourage the international community and individual governments to renew their efforts to meet the basic water needs of their populations. Efforts are underway in this area now through the Vision 21 process of the Water Supply and Sanitation Collaborative Council (WSSCC). International discussion of the necessity of meeting this basic need for all humans is extremely important—it raises issues that are global but often ignored on the national or regional level. Secondly, by acknowledging such a right, pressures to translate that right into specific national and international legal obligations and responsibilities are more likely. As Richard Jolly of the UNDP notes:

> To emphasize the human right of access to drinking water does more than emphasize its importance. It grounds the priority on the bedrock of social and economic rights, it emphasizes the obligations of states parties to ensure access, and it identifies the obligations of states parties to provide support internationally as well as nationally. (Jolly 1998)

A third reason is to spotlight the deplorable state of water management in many parts of the world. Even where progress is being made in meeting basic water needs, such as in South Africa, systems are not always being put in place to ensure their long-term sustainability (FTGWR 1999). A fourth reason is to help focus attention on the need to more widely address international watershed disputes and to help resolve conflicts over the use of shared water by identifying minimum water requirements and allocations for all basin parties. Finally, explicitly acknowledging a human right to water can help set specific priorities for water policy: meeting a basic water requirement for all humans to satisfy this right should take precedence over other water allocations and investment decisions.

Existing Human Rights Laws, Covenants, and Declarations

There are many covenants and international agreements formally identifying and declaring a range of human rights (see Box 1.2). Among the most important of these are the 1948 Universal Declaration of Human Rights (UDHR), the 1950 European Convention on Human Rights, the 1966 International Covenant on Economic, Social and Cultural Rights (ICESCR), the 1966 International Covenant on Civil and Political Rights (ICCPR), the 1969 American Convention on Human Rights, the 1986 Declaration on the Right to Development (DRD), and the 1989 Convention of the Rights of the Child (CRC). Among the rights explicitly protected by these various declarations and covenants are the right to life, to the enjoyment of a standard of living adequate for health and well-being, to protection from disease, and to adequate food. While access to clean water would seem to be a precondition for many of these rights, water is explicitly mentioned only in the Convention of the Rights of the Child. Can a right to water be considered a "derivative" right—that is to say, is a comparable human right to water implied by these declarations—or can it be inferred from the debate over, and background materials from, the existing covenants? Is water so fundamental a resource, like air, that it was thought unnecessary to explicitly include it at the time these agreements were forged? Or could the

---

BOX 1.2

Selected Human Rights Laws, Covenants, and International Agreements

- 1948 Universal Declaration of Human Rights (UDHR).
  http://www.unhchr.ch/udhr/index.htm
- 1950 European Convention on Human Rights ("Convention for the Protection of Human Rights and Fundamental Freedoms").
  http://www.coe.fr/eng/legaltxt/5e.htm
- 1966 International Covenant on Economic, Social and Cultural Rights (ICESCR).
  http://www.unhchr.ch/html/menu3/b/a_cescr.htm
- 1966 International Covenant on Civil and Political Rights (ICCPR).
  http://www.unhchr.ch/html/menu3/b/a_ccpr.htm
- 1969 American Convention on Human Rights.
  http://www.oas.org/EN/PROG/ichr/enbas3.htm
- 1986 Declaration on the Right to Development (DRD).
  http://www.unhchr.ch/html/menu3/b/74.htm
- 1989 Convention of the Rights of the Child (CRC).
  http://www.unhchr.ch/html/menu3/b/k2crc.htm

framers of these agreements have actually intended to exclude access to water as a right, while including access to food and other conditions or qualities of life?

A detailed review of international legal and institutional agreements relevant to these questions supports the conclusion that the drafters implicitly considered water to be a fundamental resource. Moreover, several of the rights explicitly protected by international rights conventions and agreements, specifically those guaranteeing food, human health, and development, cannot be satisfied without also guaranteeing access to basic clean water. These conclusions are discussed in the following text.

The Right to Water as an Implicit Part of the Earliest Human Rights Agreements

At the United Nations Conference on International Organization, held in San Francisco in 1945, it was suggested that the United Nations General Assembly develop a bill of rights. The subsequent UN Charter requires the Economic and Social Council to set up a commission for the promotion of human rights—the only commission specifically named in the Charter. The Commission on Human Rights (the Commission) held its first meetings in 1947 and agreed to prepare for the General Assembly both a declaration and a convention on human rights. A "declaration" is a statement of basic principles of inalienable human rights and imposes moral, not legal, weight. Such declarations, however, often either express already existing norms of customary international law (human rights or otherwise), or, as in the case of the UDHR, may over time crystallize into customary norms. The "convention" or "covenant," on the other hand, is a treaty legally binding on ratifying parties (United Nations 1949, p. 524 ff.).

During late 1947 and early 1948, a draft declaration was developed and debated by the Commission. In mid-1948, the Commission presented a draft declaration to the Economic and Social Council. Article 22 of the draft stated:

> Everyone has the right to a standard of living, including food, clothing, housing and medical care, and to social services, adequate for the health and well-being of himself and his family. (United Nations 1948, p. 576)

In the final debate over this document, the emphasis was refocused from providing a general standard of living to a more encompassing right to health and well-being. Why was water not included in this list? The debate around the wording makes clear that the specific provisions for food, clothing, housing, and so on were not meant to be all-inclusive, but representative or indicative of the "*component elements of an adequate standard of living*" (emphasis added, United Nations 1956, p. 216). In 1948 the United Nations General Assembly approved the UDHR by 48 votes, with 8 abstentions. The reworded Article 22, now Article 25 of the declaration, was adopted unanimously and states:

> Everyone has the right to a standard of living adequate for the health and well-being of himself and of his family, including food, clothing, housing. . . .
> (United Nations General Assembly 1948)

Satisfying the standards of Article 25 cannot be done without water of a sufficient quantity and quality to maintain human health and well-being. Meeting a standard of living adequate for the health and well-being of individuals requires the availability of a minimum amount of clean water. Some basic amount of clean water is necessary to prevent death from dehydration, to reduce the risk of water-related diseases, and to provide for basic cooking and hygienic requirements. This fact has long been recognized by the World Health Organization and other UN and international aid agencies that specify basic water standards for quantity and quality. Logic further suggests that the framers of the UDHR considered water to be one of the "component elements"—as fundamental as air.

The 1948 declaration also includes rights that must be considered less fundamental than a right to water, such as the right to work, to protection against unemployment, to form and join trade unions, and to rest and leisure (Articles 23 and 24). This further supports the conclusion that Article 25 was intended to implicitly support the right to a basic water requirement.

The Universal Declaration also implies a need for water to grow sufficient food for an adequate standard of living. But an important distinction can be made between water for food and the much smaller amount of water required to support the health and well-being of individuals. In particular, the food necessary to meet the rights described in Article 25 can be produced in distant locations and moved to the point of demand. It can thus be argued that the provision of adequate food to satisfy Article 25 does not require local provision of water. This issue has been discussed more completely in the work of Allan (1995) and a final background document to the Comprehensive Assessment of the Freshwater Resources of the World prepared for the United Nations in 1997 (Lundqvist and Gleick 1997).

As a resolution of the UN General Assembly, the 1948 UDHR is not binding on states. Many of the provisions of the Declaration, however, are now considered to be customary international law, and the broad human rights found there have since been reasserted in many other international documents.

In the 20 years following the UDHR, work continued at the United Nations on the more binding convention, which became two separate covenants in 1966: the International Covenant on Economic, Social and Cultural Rights (ICESCR) and the International Covenant on Civil and Political Rights (ICCPR). These came into force in 1976. As of January 1999 there were nearly 140 parties to the ICESCR and the ICCPR (Churchill 1996, Danieli et al. 1999). Under these covenants, each state would undertake to ensure to all individuals within its jurisdiction certain human rights and would adopt "the necessary legislative or other measures to give them practical effect" (United Nations 1949, p. 538). Article 2(1) of the ICESCR requires that each party to the covenant

> undertakes to take steps, individually and through international assistance and co-operation, especially economic and technical, to the maximum of its available resources, with a view to achieving progressively the full realisation of the rights recognised by the present Covenant by all appropriate means including particularly the adoption of legislative measures.

Work was not completed on the entire agreement until 1966, 18 years after the initial draft was presented for debate (United Nations 1966). But 10 years earlier, in 1956, Articles 11 and 12 of the ICESCR addressing the right to an adequate standard of living and human health were both adopted without any dissenting votes (United Nations 1963). Article 11 formalizes the right to food and some minimum quality of life:

> The States Parties to the present Covenant recognize the right of everyone to an adequate standard of living for himself and his family, including adequate food, clothing and housing, and to the continuous improvement of living conditions.

Article 12 continues:

> The States Parties to the present Covenant recognize the right of everyone to the enjoyment of the highest attainable standard of physical and mental health. The steps to be taken . . . to achieve the full realization of this right shall include those necessary for . . . [t]he prevention, treatment and control of epidemic, endemic, occupational and other diseases.

As with the UDHR, access to water can be inferred as a derivative right necessary to meet the explicit rights to health and an adequate standard of life. In their review of major human rights progress over the past 50 years, Danieli et al. support the right to water as implicit in the rights guaranteed by the ICESCR:

> There is nothing ill-defined or fuzzy about being deprived of the basic human rights to food and clean water, clothing, housing, medical care, and some hope for security in old age. As for legal toughness, the simple fact is that the 138 governments which have ratified the International Covenant on Economic, Social, and Cultural Rights have a legal obligation to ensure that their citizens enjoy these rights. (Danieli et al. 1999)

The International Covenant for Civil and Political Rights (ICCPR) was debated and developed at the same time as the ICESCR. Article 6 of the ICCPR states:

> Every human being has the inherent right to life. This right shall be protected by law. No one shall be arbitrarily deprived of his life.

Water once again is not explicitly mentioned in the final document of the Covenant, but the right to life implies the right to the fundamental conditions necessary to support life. Furthermore, an analysis of the accompanying history and interpretation of the negotiations and discussions surrounding the preparation of the Covenant reveals that the Human Rights Committee (HRC) established by the ICCPR took a broad interpretation of the right to life. In particular, in a critical observation, the HRC called for an inclusive interpretation of the provision that requires states to take positive action to provide the "appropriate means of subsistence" necessary to support life:

> [T]he right to life has been too often narrowly interpreted. The expression "inherent right to life" cannot properly be understood in a restricted

manner, and the protection of this right requires that States adopt positive measures. (United Nations 1989a)

The American Convention on Human Rights and the European Convention on Human Rights also support the requirement that states take positive, proactive steps to support the right to life. Article 2 of the European Convention requires that states have an obligation "not only to refrain from taking life 'intentionally' but further, to take appropriate steps to safeguard life" (DRECHR 1979, Churchill 1996). Even narrow definitions of Article 6 of the ICCPR interpret it as guaranteeing protection against arbitrary and intentional denial of access to sustenance (Dinstein 1981, McCaffrey 1992). This must include water.

At a minimum, therefore, the explicit right to life and the broader rights to health and well-being include the right to sufficient water, of appropriate quality, to sustain life. To assume the contrary would mean that there is no right to the single most important resource necessary to satisfy the human rights more explicitly guaranteed by the world's primary rights declarations and covenants.

Explicit Support for the Human Right to Water: International Statements, Agreements, and State Practice

Other international agreements and examples of state practice offer further evidence of the transition toward an explicit right to water. Beginning in the 1970s, a series of international environmental or water conferences have taken on the issue of access to basic resource needs and rights to water. While the statements and agreements coming from these meetings are not legal documents with the same standing as the covenants described earlier, they offer strong evidence of international intent and policy.

One of the earliest water conferences was held in 1977 at Mar del Plata. The conference statement issued at the close of the meeting explicitly recognized the right of access to water for basic needs:

> [A]ll peoples, whatever their stage of development and their social and economic conditions, have the right to have access to drinking water in quantities and of a quality equal to their basic needs. (United Nations 1977)

More recently, the question of "development" has become central to overall actions and priorities of the United Nations and international organizations. The right to development has increasingly come to be considered as a "universal and inalienable right and an integral part of fundamental human rights" (Article I[10] of the Vienna Declaration, Principle 3 of the Cairo Programme of Action, Commitment 1[n] of the Copenhagen Declaration, and Article 213 of the Beijing Platform of Action, cited in UNDP 1998).

In 1986, the United Nations General Assembly adopted the Declaration on the Right to Development (DRD) (United Nations 1986). According to Article 8 of the Declaration,

> States should undertake, at the national level, all necessary measures for the realization of the right to development and shall ensure, *inter alia,* equality of opportunity for all in their access to basic resources.

In interpreting Article 8 of the DRD, the United Nations explicitly includes water as a basic resource when it states that the persistent conditions of underdevelopment in which millions of humans are "denied access to such essentials as food, water, clothing, housing and medicine in adequate measure" represent a clear and flagrant "mass violation of human rights" (United Nations 1995). At a minimum, this implies that nations should implement continued and strong efforts to progressively meet these needs to the extent of their available resources, as required by the ICESCR. As noted later, resource limitations should not constrain these efforts in the case of water.

Explicit recognition of water continued with the 1989 Convention of the Rights of the Child (CRC). Article 24 of the CRC, paralleling Article 25 of the Universal Declaration of Human Rights, provides that a child has the right to enjoy the highest attainable standard of health. Among the actions states are to take to secure this right are measures to

> combat disease and malnutrition . . . through, *inter alia,* . . . the provision of adequate nutritious foods and clean drinking water. (United Nations 1989b)

Here for the first time is explicit recognition of the connections between water resources, the health of the environment, and human health. This Convention has been signed and ratified by more nations than any other international agreement. By November 1999, over 190 nations had ratified the CRC; only the United States and Somalia have failed to do so (http://www.unicef.org/crc/status.htm).

Regional and national conventions and constitutions are also increasingly making the right to basic resources a part of accepted state practice. For example, Article 11 of the American Convention on Human Rights in the Area of Economic, Social, and Cultural Rights of 1988 provides that "Everyone shall have the right to live in a healthy environment and to have access to basic public services" (OASTS 1988). Although few states have made formal commitments recognizing a right to water, more and more of the newer national constitutions discuss either water or the right to a healthy environment. South Africa has moved strongly in this direction. The Bill of Rights of the new Constitution of South Africa, adopted in 1994, offers a clear example of state practice relevant to an explicit human right to water. Section 27(1)(b) states: "Everyone has the right to have access to sufficient food and water." Water policies to implement this right in South Africa are now being developed (Gleick 1998).

Defining and Meeting a Human Right to Water

What are the social and institutional implications of a human right to water? A right to water cannot imply a right to an unlimited amount of water. Resource limitations, ecological constraints, and economic and political factors must limit human water availability and use. Given such constraints, how much water is necessary to satisfy this right? Enough solely to sustain a life? Enough to grow all food sufficient to sustain a life? Enough to maintain a certain economic standard of living?

Answers to these questions come from international discussions over development, analysis of the human rights literature, and an understanding of human needs and uses of water. These lead to the conclusion here that a human right to water should only apply to "basic needs" for drinking, cooking, and fundamental domestic uses, such as those described in Gleick (1996).

Both the 1977 Mar del Plata statement and the 1986 UN Right to Development set a goal of meeting "basic" needs. The concept of meeting basic water needs was further strongly reaffirmed during the 1992 Earth Summit in Rio de Janeiro and expanded to include water for ecosystem functions:

> In developing and using water resources, priority has to be given to the satisfaction of basic needs and the safeguarding of ecosystems. (United Nations 1992)

In 1997, the Comprehensive Assessment of the Freshwater Resources of the World prepared for the Commission on Sustainable Development of the UN stated:

> All people require access to adequate amounts of clean water, for such basic needs as drinking, sanitation and hygiene (p. 3). [It is essential to] develop sustainable water strategies that address basic human needs, as well as the preservation of ecosystems (p. 29). [I]t is essential that water planning secure basic human and environmental needs for water (p. 25). (United Nations 1997b)

The UN Convention on the Law of the Non-navigational Uses of International Watercourses, approved by the General Assembly on May 21, 1997, also explicitly addresses this question of water for basic human needs, including food. Article 10 states that in the event of a conflict between uses of water in an international watercourse, special regard shall be given "to the requirements of vital human needs." The states negotiating the Convention included in the accompanying Statement of Understanding an explicit definition that

> In determining "vital human needs," special attention is to be paid to providing sufficient water to sustain human life, including both drinking water and water required for production of food in order to prevent starvation. (United Nations 1997a)

Article 10 is obligatory. In interpreting Article 10, priority allocation of water in the event of conflicting demands goes to water for fundamental human needs.

Implicit in the concept of vital or basic needs is the idea of minimum requirements for certain human and ecological functions and the allocation of sufficient resources to meet those needs. A true minimum human need for water can only be defined as the amount needed to maintain human survival, approximately 3 to 5 liters of clean water per day. Setting a minimum at this level would have little meaning: except in accidental rare circumstances, no one dies from a lack of water. Studies also show that improvements in human health can be realized by increasing amounts of clean water up to around 20 liters per person per day (lpcd) (Esrey and Habicht 1986).

Various international organizations have made recommendations over the years for basic drinking water and sanitation requirements. The U.S. Agency for International Development, the World Bank, and the World Health Organization have rec-

TABLE 1.1　A Recommended Basic Water
Requirement for Human Domestic Needs

Purpose	Liters per Person per Day
Drinking water[a]	5
Sanitation services	20
Bathing	15
Food preparation[b]	10

Source: Gleick 1996.

[a]This is a true minimum to sustain life in moderate climatic conditions and average activity levels.

[b]Excluding water required to grow food. A rough estimate of the water required to grow the average daily food needs of an individual is 2,700 liters, but this varies enormously depending on the diet, climate, and agricultural practices of a region. A current average North American diet requires 5,000 liters per day, while the average diet in sub-Saharan Africa may require only 1,800 liters per day. These issues are discussed more fully in Chapter 4.

ommended between 20 and 40 lpcd, excluding water for cooking, bathing, and basic cleaning. This is also in line with recommended standards from the UN International Drinking Water Supply and Sanitation Decade and Agenda 21 of the Earth Summit.

Adopting a standard of 5 liters of clean water per person per day for drinking water and 20 lpcd for sanitation and hygiene, Gleick (1996, 1998) recommended a basic water requirement of 25 lpcd to meet the most basic of human needs with an additional 15 lpcd for bathing and 10 lpcd for cooking. International organizations and water providers should adopt an overall basic water requirement (BWR) for meeting these four domestic basic needs, independent of climate, technology, and culture (see Table 1.1). The recommendation of 50 liters (around 13 gallons) per person per day is justifiable and appropriate, but the specific number is less important than the principle of setting a goal and implementing actions to reach that goal. Table 1.1 also notes the much larger volume of water necessary for growing food, but as others have noted, the right to food is already addressed in the human rights literature, and international trade in food can permit this right to be met with water from other regions.

Billions of people lack access to even a basic water requirement of 50 lpcd, though not because of inadequate water availability. Table 1.2 shows those countries where the average domestic (reported) water use falls below 100 lpcd. Nearly 2,200 million people—more than a third of the world's population—live in the 62 countries that report average domestic water use below 50 lpcd. Yet absolute water *availability* is not the problem: only 12 countries have less than 1,000 lpcd of water available on average and only Kuwait reports having a natural renewable freshwater supply of less than 100 liters per person per day. Despite its limited natural endowment, Kuwait provides more than the recommended BWR to its population by supplementing its natural supplies with desalinated water (see Chapter 5).

There are, of course, problems with the data. Average water-use figures by country are known to be unreliable or old. (See Table 2 in the Data Section, page 203.)

TABLE 1.2 Countries with Reported Per-Capita Domestic Water Use Below 100 Liters per Person per Day (lpcd) for the Year 2000

Country	2000 Population (millions)	2000 Estimated Domestic lpcd	Country	2000 Population (millions)	2000 Estimated Domestic lpcd
Gambia	1.24	3	Honduras	6.49	26
Haiti	7.82	3	Guinea	7.86	26
Djibouti	0.69	4	Indonesia	212.57	28
Somalia	11.53	6	Afghanistan	25.59	28
Mali	12.56	6	Cote D'Ivoire	15.14	28
Cambodia	11.21	6	Swaziland	0.98	29
Mozambique	19.56	7	Liberia	3.26	30
Uganda	22.46	8	El Salvador	6.32	30
Tanzania	33.69	8	India	1006.77	31
Ethiopia (and Eritrea)	69.99	9	Yemen	18.12	31
Albania	3.49	9	Paraguay	5.50	32
Bhutan	2.03	10	Uruguay	3.27	33
Chad	7.27	11	Togo	4.68	33
Central African Republic	3.64	11	Cameroon	15.13	33
Congo, DR (formerly Zaire)	51.75	11	Kenya	30.34	36
Nepal	24.35	12	Zimbabwe	12.42	38
Rwanda	7.67	13	Laos	5.69	38
Lesotho	2.29	13	Costa Rica	3.80	39
Burundi	6.97	14	Bolivia	8.33	41
Angola	12.80	14	Guyana	0.87	46
Bangladesh	128.31	14	Dominican Republic	8.50	48
Ghana	19.93	14	Equatorial Guinea	0.45	49
Benin	6.20	15	Cyprus	0.79	51
Sierra Leone	4.87	15	Morocco	28.98	51
Guatemala	12.22	15	Pakistan	156.01	55
Myanmar	49.34	15	Thailand	60.50	58
Papua New Guinea	4.81	17	China	1276.30	59
Burkina Faso	12.06	17	Mongolia	2.74	61
Cape Verde	0.44	17	Botswana	1.62	61
Sri Lanka	18.82	18	Oman	2.72	62
Fiji	0.85	19	Singapore	3.59	65
Senegal	9.50	20	Netherlands	15.87	67
Niger	10.81	20	Tunisia	9.84	73
Congo	2.98	23	Sudan	29.82	73
Belize	0.24	23	Zambia	9.13	81
Guinea-Bissau	1.18	23	Trinidad and Tobago	1.34	83
Malawi	10.98	24	Ecuador	12.65	84
Jamaica	2.59	24	Jordan	6.33	94
Nigeria	128.79	24	Gabon	1.24	96
Madagascar	17.40	26	Algeria	31.60	97
			Syria	16.13	98

Source: These data come from reported domestic water use for various years (from Gleick 1998) and the United Nations medium 2000 population projections.

Notes: Improvements are needed in collection of water-use data (see text). Some of the countries listed here are relatively well endowed with water and it is likely that domestic water use is higher, perhaps substantially higher, than reported. In other countries, however, where reported average per capita domestic water use is higher than 100 lpcd, many people may receive less than the recommended 50 lpcd.

Several large countries, such as India and China, report that their average domestic water use is very close to 50 liters per person per day. In these countries large segments of populations no doubt receive less than the average, while wealthier portions of the population receive more. There are many countries in Table 1.2 that are relatively water-rich, suggesting that official data on water withdrawals may miss substantial domestic water use that is self-supplied (i.e., does not come from publicly operated systems). There are few data to indicate the typical quality of the water received. Poor quality of domestic water is a severe and widespread problem, and it is likely that many people who may receive more than the recommended BWR are getting contaminated water. Finally, in most of the countries listed, populations are growing faster than improvements to water availability. Improving the scope, quality, and extent of water-use data—a subject on which this author has often written—is vitally important. Notwithstanding these data problems, however, we must conclude that meeting a basic water requirement for all people is constrained by institutional and management failures, not by basic water availability.

Translating the Right to Water into Specific Legal Obligations

If there is a human right to water, to what extent does a state have an obligation to provide that water to its citizens? While the many international declarations and formal conference statements supporting a right to water do not directly require states to meet individuals' water requirements, Article 2(1) of the ICESCR obligates states to provide the institutional, economic, and social environment necessary to help individuals to progressively realize those rights. In certain extreme circumstances, moreover, when individuals are unable to meet basic needs for reasons beyond their control, including disaster, discrimination, economic impoverishment, age, or disability, states must provide for basic needs (Gleick 1996). Meeting these minimum needs should take precedence over other allocations of spending for economic development. This will require a redirection of current priorities at international and local levels, and will likely require that new resources be invested as well.

The overall economic and social benefits of meeting basic water requirements far outweigh any reasonable assessment of the costs of providing for these needs. One early estimate was that water-related diseases cost society on the order of $125 billion per year (in late-1970 dollars) just in direct medical expenses and lost work time (Pearce and Warford 1993). Even this estimate excluded costs associated with social disruptions caused by disease, lost educational opportunities for families, long-term debilitation of children, or any other poorly quantified or hidden costs. Yet the cost of providing new infrastructure needs for all major urban water sectors has been estimated at around $25 to $50 billion per year (Christmas and de Rooy 1991, Rogers 1997, Jolly 1998).

While these costs are far below the costs of failing to meet these needs, they are two to three times the average rate of spending for water during the 1980s and 1990s (United Nations 1997b). The WSSCC estimates that 80 percent of water investments in the 1980s represented expenditures to meet the needs of a relatively small number of affluent urban dwellers (WSSCC 1997). Studies on investment alternatives reveal that 80 percent of the unserved can be reached for only 30 percent of the costs of providing the highest level of service to all. The WSSCC, for example, estimates that 35,000 rural people could be provided with basic sanitation services for the same cost of providing 1,000 urban residents with a centralized sewerage system.

McCaffrey (1992), who supports the conclusion that "in some form, the right [to water] may be inferred under the basic instruments of international human rights law," argues that the devastating consequences of being denied such water require that relevant provisions of existing human rights instruments "ought to be interpreted broadly, so as to facilitate the implementation of the right to water as quickly and comprehensively as possible."

McCaffrey also raised the concern that defining a basic human right to water might have the unintentional effect of causing disputes between neighboring countries that share water. Would such a human right require that one state has the right to receive water from another to meet this basic need? The final statement from the 1997 Convention appears to resolve this question affirmatively: in the unusual case in which a basic water requirement cannot be met solely from a state's internal water resources, neighboring states do not have the right to deny a co-riparian sufficient water to meet those needs on the grounds that the upstream nation needs the water for economic development. A country is thus not permitted to exploit a shared water resource in a manner that deprives individuals in a neighboring country of access to their basic human needs. While this interpretation raises the specter of water-related conflicts over basic needs, such conflicts seem unlikely to arise in practice: in almost all regions of the world absolute water availability is no constraint to meeting minimal basic needs.

Consequences of the Failure to Meet Basic Needs for Water

Many international organizations work to meet the unmet water needs of human populations, including the United Nations, the Water Supply and Sanitation Collaborative Council, the World Bank, international aid organizations such as USAID, the Swedish International Development Agency, the Canadian International Development Agency, and nongovernmental organizations such as Water Aid and Water for People. These efforts have made significant progress in increasing access to basic water needs for hundreds of millions of people.

Yet, despite these efforts, many water-related problems have worsened. The incidence of cholera soared in the 1990s and expanded in geographic extent. The populations in urban areas without access to clean water and sanitation actually increased between 1980 and 1990, despite great efforts to meet these needs (WHO 1996). Even more distressing has been the apparent difficulty the world water community has had in setting new targets and goals for meeting basic needs.

Similar kinds of problems beset the world food community, which has set and continually revised action plans for reducing hunger. The World Food Council met in 1989 in Cairo to propose a Programme of Cooperative Action. In that same year, a meeting of food experts in Bellagio, Italy, set nutritional goals for the year 2000, which were reaffirmed at the 1990 UN World Summit for Children. The 1992 UN International Conference on Nutrition laid out a World Declaration and Plan of Action for Nutrition. While huge populations remain undernourished, even less success has been achieved in setting and meeting water-related goals.

It seems likely that an appropriate mix of economic, political, and social strategies can be developed to reliably provide for basic needs. And despite a growing emphasis on markets, if a "market" system is unable to provide a basic water requirement, states have responsibilities to meet these needs under the human rights agreements discussed above. Unless international organizations, national and local governments, and water providers adopt and work to meet a basic water requirement standard, large-scale human misery and suffering will continue, contributing to impoverishment, ill-health, and the risk of social and military conflict. Ultimately, decisions about defining and applying a basic water requirement will depend on political and institutional will.

CONCLUSIONS

All human beings have an inherent right to water in quantities and of a quality necessary to meet their basic needs. This right should be protected by law. The right to water is satisfied when every person has physical and economic access to a basic water requirement at all times.

A communications and computer revolution is sweeping the globe. International financial markets and industries are increasingly integrated and connected. Efforts are being made to ensure regional and global security. In this context, our inability to meet the most basic water requirements of billions of people has resulted in enormous human suffering and tragedy and is one of the twentieth century's greatest failures.

International law, declarations of governments and international organizations, and state practices support the conclusion that access to a basic water requirement must be considered a fundamental human right. The major human rights treaties, statements, and formal covenants contain implicit and explicit evidence that reinforces the application of rights law in this area. If the framers of early human rights language had foreseen that reliable provision of a resource as fundamental as clean water would be so problematic, they would certainly have included a right to water in the basic rights documents.

Will the recognition of the human right to water actually improve conditions worldwide? Perhaps not. The challenge of meeting human rights obligations in all areas is a difficult one, which has been inadequately and incompletely addressed. But the imperatives to meet basic human water needs are more than just moral, they are rooted in justice and law and the responsibilities of governments. It is time for the international community to reexamine its fundamental development goals. A first step toward meeting a human right to water would be for governments, water providers, and international organizations to guarantee all humans the most fundamental of basic water needs and to develop the necessary institutional, economic, and management strategies necessary for meeting them.

REFERENCES

Allan, J.A. 1995. "Water in the Middle East and in Israel-Palestine: Some local and global issues." In M. Haddad and E. Feitelson, (eds.), *Joint Management of Shared Aquifers.* Palestine Consultancy Group and the Truman Research Institute of Hebrew University, Jerusalem, pp. 31–44.

Christmas, J., and C. de Rooy. 1991. "The decade and beyond: At a glance." *Water International,* Vol. 16, pp. 127–134.

Churchill, R. R. 1996. "Environmental rights in existing human rights treaties." In A.E. Boyle and M.R. Anderson (eds.), *Human Rights Approaches to Environmental Protection.* Clarendon Press, Oxford, United Kingdom, pp. 89–108.

Danieli, Y., E. Stamatopoulou, C.J. Diaz. 1999. *The Universal Declaration of Human Rights: Fifty Years and Beyond.* Baywood Publishing Company, Inc., Amityville, New York.

DRECHR, Decisions and Reports of the European Commission on Human Rights. 1979. Association X v. United Kingdom, *Application 7154/75,* Vol. 14, pp. 31–32.

Dinstein, Y. 1981. "The right to life, physical integrity, and liberty." In L. Henkin (ed.), *The International Bill of Rights: The Covenant on Civil And Political Rights,* Columbia University Press, New York.

Esrey, S.A., and J.P. Habicht. 1986. "Epidemiological evidence for health benefits from improved water and sanitation in developing countries." *Epidemiological Reviews,* Vol. 8, pp. 117–128.

Financial Times Global Water Report (FTGWR). 1999. "South Africa: Mbeki's task." *Financial Times Global Water Report,* Issue 74, June 12, pp. 1–2.

Gleick, P.H. 1996. "Basic water requirements for human activities: Meeting basic needs." *Water International,* Vol. 21, pp. 83–92.

Gleick, P.H. 1998. *The World's Water 1998–1999: The Biennial Report on Freshwater Resources.* Island Press, Washington D.C.

Gleick, P.H. 1999. "A human right to water." *Water Policy,* Vol. 1, No. 5, pp. 487–503.

Jolly, R. 1998. "Water and human rights: Challenges for the twenty-first century." Address at the Conference of the Belgian Royal Academy of Overseas Sciences, March 23, Brussels.

Lawson, E. 1991 and 1996. *Encyclopedia of Human Rights.* First and second editions. Taylor and Francis, Inc., New York.

Lundqvist, J., and P.H. Gleick. 1997. *Sustaining Our Waters into the Twenty-first Century.* Comprehensive Assessment of the Freshwater Resources of the World, Stockholm Environment Institute, for the United Nations, Stockholm, Sweden.

McCaffrey, S.C. 1992. "A human right to water: Domestic and international implications." *Georgetown International Environmental Law Review,* Vol. 5, No. 1, pp. 1–24.

Organization of American States Treaty Series (OASTS). 1988. *American Convention on Human Rights in the Area of Economic, Social, and Cultural Rights.* 69, OEA/Ser. L.V/II.92, Doc. 31, Rev. 3.

Pearce, D.W., and J.J. Warford. 1993. *World without End: Economics, Environment, and Sustainable Development.* Oxford University Press, New York.

Rogers, P. 1997. "Water for big cities: Big problems, easy solutions?" Draft paper, Harvard University, Cambridge, Massachusetts.

Steiner, H.J., and P. Alston. 1996. *International Human Rights in Context.* Clarendon Press, Oxford, United Kingdom.

United Nations. 1948. *Yearbook of the United Nations 1947–48.* Department of Public Information, United Nations, Lake Success, New York.

United Nations. 1949. *Yearbook of the United Nations 1948–49.* Columbia University Press/ United Nations Publications, New York.

United Nations. 1956. *Yearbook of the United Nations 1956.* Columbia University Press/ United Nations Publications, New York.

United Nations. 1963. *Yearbook of the United Nations 1963.* Columbia University Press/ United Nations Publications, New York.

United Nations. 1966. *Yearbook of the United Nations 1966.* Columbia University Press/ United Nations Publications, New York.

United Nations. 1977. *Report of the United Nations Water Conference, Mar del Plata.* March 14–25, 1997. No. E.77.II.A.12, United Nations Publications, New York.

United Nations. 1986. *Declaration on the Right to Development,* adopted by General Assembly resolution 41/128, 4 December 1986. (available at http://www.unhchr.ch/html/menu3/b/74.htm).

United Nations. 1989a. *General Comments of the Human Rights Committee of the International Covenant on Civil and Political Rights.* UN Doc. CCPR/C/21/ Rev. 1 (May). United Nations Publications, New York.

United Nations. 1989b. *Convention on the Rights of the Child.* General Assembly resolution 44/25 of 20 November 1989; entry into force 2 September 1990 (available at http://www.unhchr.ch/html/menu3/b/k2crc.htm).

United Nations. 1992. "Protection of the quality and supply of freshwater resources: Application of integrated approaches to the development, management and use of water resources." *Agenda 21,* Chapter 18, United Nations Publications, New York.

United Nations. 1993. *Human Rights Bibliography 1980–1990.* Centre for Human Rights, United Nations Documents and Publications, New York.

United Nations. 1995. *The United Nations and Human Rights 1945–1995.* United Nations Blue Book Series, Vol. VII, Department of Public Information, United Nations Publications, New York.

United Nations. 1997a. *Convention on the Law of the Non-navigational Uses of International Watercourses.* UN General Assembly Doc. A/51/869 (April 11). United Nations Publications, New York.

United Nations. 1997b. *Comprehensive Assessment of the Freshwater Resources of the World.* Commission on Sustainable Development. United Nations, New York. Printed by the World Meteorological Organization for the Stockholm Environment Institute.

United Nations Development Programme (UNDP). 1998. *Integrating Human Rights with Sustainable Development.* United Nations Development Programme. United Nations, New York (January).

United Nations General Assembly. 1948. *Universal Declaration of Human Rights.* Resolution 217 UN Doc. A/64. Final authorized text available at gopher://gopher.undp.org/00/unearth/rights.txt

Water Supply and Sanitation Collaborative Council (WSSCC). 1997. "Vision 21: Water, sanitation, and global well-being." A Statement from the Water Supply and Sanitation Collaborative Council (Draft), United Nations (December 22, 1997).

World Health Organization (WHO). 1996. *Water Supply and Sanitation Sector Monitoring Report 1996 (Sector Status as of 1994).* Water Supply and Sanitation Collaborative Council and the United Nations Children's Fund, UNICEF, New York.

How Much Water Is There and Whose Is It?

The World's Stocks and Flows of Water and International River Basins

The availability and quality of freshwater resources around the world are of growing concern to academics, policymakers, and the international community. Understanding the stocks and flows of water through the world's hydrologic cycle is essential to any discussion about the world's water problems. Human well-being, ecosystem health and functions, even economics and politics all depend on how much, when, and where water is available. Surprisingly, great uncertainties still remain about water availability and the natural variations over time in the hydrologic cycle. Despite our increasingly accurate remote-sensing technology, mapping capabilities, and computer models, even the total amount of water on the planet is not known to a high degree of accuracy. On top of these uncertainties, constantly changing political borders make assessments of regional and national water availability or rights rapidly outdated and contentious.

In the last decade, the water community has launched several major water-related assessments, including the Comprehensive Assessment of the Freshwater Resources of the World (United Nations 1997a), the new UN Food and Agriculture Organization's AQUASTAT program (see www.fao.org), an expansion of the International Hydrological Programme of UNESCO (see, for example IHP 1998), and the Vision activities preparing for the major conference in The Hague in March 2000. Analysts have also started to explore the complex connections between the hydrologic cycle, water availability and use, and regional or international security (see, for example, Gleick 1993, 1998; Wolf 1998). The nexus between the science of the world's water resources and the political and social implications of water availability and use has become one of the most exciting and complex areas for research, education, and water policy. This is especially true for the field of water and international politics, because of the growing importance of water in regions like the Middle East, southern Africa, and Asia, and even many industrialized nations of Europe.

Despite a lively semantic debate over definitions of international "security" and conflict, it is indisputable that the world's limited freshwater resources cross many political borders and that the ownership and management of some of these resources are disputed, sometimes in violent ways. In addition, there is a long history of the use of water and water systems as weapons, tools, or targets during war (see Gleick 1998 for a chronology of water-related conflicts in the Middle East from

3000 B.C. to 300 B.C., and the Water Briefs section of this volume for an updated chronology of more current water-related conflicts). The fact that the geophysical characteristics of freshwater rarely coincide with the geopolitical ones means that growing populations and levels of economic development are likely to increase the political and economic competition for water where absolute (the total amount of renewable freshwater) or relative (such as per capita) availability of water is limited. Such competition may be particularly intense where water is shared by disputing political entities.

In recent years there have been notable improvements in collecting information about freshwater availability and use, applying new technology for monitoring the hydrologic cycle, analyzing geographical information, and assessing water-related concerns. In particular, satellite monitoring of water flows, declassified high-resolution digital elevation data, and computers and mapping systems capable of analyzing and presenting these data, are now available. This chapter presents the results of the most up-to-date assessment of the world's stocks and flows (Shiklomanov 1998) and a new registry of international watersheds (Wolf et al. 1999). It discusses the data, analyzes changes over the past couple of decades, and raises some ongoing policy questions about reducing the risks of conflicts over shared freshwater resources.

How Much Water Is There?
The Basic Hydrologic Cycle

Water is found in many places and forms, and it is in continuous and rapid transformation from one form and stock to another. Fresh water is a renewable resource made available on an ongoing basis by the flow of solar energy reaching the earth from the sun. This energy evaporates fresh water into the atmosphere from the oceans and land surfaces and redistributes it around the world in what is known as the hydrologic cycle—the stocks, flows, and interactions of water as ice, liquid, and vapor. The quantity of water in the atmosphere is reduced by precipitation and replenished by evaporation. Less water falls on the oceans as precipitation than leaves through evaporation; thus there is a continuous transfer of water to the land, which runs off in rivers and streams or is stored in lakes, soils, and groundwater aquifers.

Because of the heterogeneous nature of our atmosphere, land surfaces, and energy fluxes, the distribution of fresh water around the world is also heterogeneous in both space and time. This uneven distribution of water determines the nature of many of the problems related to freshwater management and use. Some places receive enormous quantities of water regularly; others are extremely dry and arid. Seasonal cycles of rainfall and evaporation are the rule, not the exception. And variability from one period to another can be large.

The hydrologic cycle is constantly operating and producing renewable supplies of water. The rates of renewal, however, vary greatly. The average length of time a water molecule stays in the atmosphere before cycling out (typically called the residence time) is about eight days. The residence time of water in deep groundwater aquifers, or large glaciers, may be measured in hundreds, thousands, or even hundreds of thousands of years. When humans use water at rates that exceed natural renewal rates, that use is unsustainable. A sophisticated science of hydrology and water management has been developing throughout the history of civilization to help water planners and managers deal with these factors.

Stocks of Fresh Water

A clear understanding of the natural hydrological cycle requires reliable estimates of the stocks and flows of freshwater resources. Despite vast improvements in monitoring technology, data storage, and computer modeling, information on the amount of fresh water on earth is still neither reliable nor accurate—any estimates of water stocks and flows thus remain approximations. This is partly the result of inadequate efforts to actually monitor water stocks and partly the result of the impossibility of accurately measuring—and integrating multiple measurements of—factors like soil moisture, water in wetlands and groundwater aquifers, and the vastly complex and chaotic flows of water vapor and liquids. Even the quantity of water in lakes, glaciers, polar ice, and other more well-defined stocks can only be approximated.

Table 2.1 presents a recent estimate of the major water stocks on earth, separated by salt- and freshwater stocks. The total volume of water on earth is approximately 1.4 billion cubic kilometers (km^3) and only 2.5 percent of it, or about 35 million km^3, is fresh water. The vast majority of fresh water is in the form of permanent ice or snow, locked up in Antarctica and Greenland, or in deep groundwater aquifers. The principal sources of water for human use are lakes, rivers, soil moisture, and relatively shallow groundwater basins. The usable portion of these sources is estimated to be only about 200,000 km^3 of water—less than 1 percent of all fresh water on

TABLE 2.1   Major Stocks of Water on Earth (thousand cubic kilometers)

	Volume (1000 km^3)	Percentage of Total Water	Percentage of Total Fresh Water
Salt water stocks			
Oceans	1,338,000	96.54	
Saline/brackish groundwater	12,870	0.93	
Saltwater lakes	85	0.006	
Freshwater stocks			
Glaciers, permanent snowcover	24,064	1.74	68.70
Fresh groundwater	10,530	0.76	30.06
Ground ice, permafrost	300	0.022	0.86
Freshwater lakes	91	0.007	0.26
Soil moisture	16.5	0.001	0.05
Atmospheric water vapor	12.9	0.001	0.04
Marshes, wetlands[a]	11.5	0.001	0.03
Rivers	2.12	0.0002	0.006
Incorporated in biota	1.12	0.0001	0.003
Total Water on Earth (1000 km^3)	1,386,000	100	100
Total Fresh Water on Earth (1000 km^3)	35,029		

Source: Shiklomanov 1993.

Note: Totals may not add due to rounding.

[a]Marshes, wetlands, and water incorporated in biota are often mixed salt and fresh water.

earth and only one one-hundredth of a percent (0.01%) of all water on the planet. And much of this water is located far from human populations.

Flows of Fresh Water

The water bodies described above serve as major stocks of water. The hydrologic cycle, however, consists of continuous flows of water into and out of every stock under the influence of solar energy, and the dynamics of the atmosphere, gravity, and human activities. The rates of flux of water vary enormously from stock to stock. These cycling rates are of great importance when evaluating the possible uses of water for human activities and the impacts of those uses on natural systems.

The major source of fresh water is evaporation off the surface of the oceans. Approximately 505,000 km³ a year, or a layer 1,400 mm thick, evaporates from the oceans. Another 72,000 km³ evaporates from land surfaces annually. Approximately 80 percent of all precipitation, or about 458,000 km³ falls on the oceans; the remaining 119,000 km³ of precipitation falls over land. The difference between precipitation onto land surfaces and evaporation from those surfaces (119,000 km³ annually minus 72,000 km³ annually) is runoff and groundwater recharge—approximately 47,000 km³ per year.

These global averages hide considerable variation in both the spatial and the temporal distribution of water. Precipitation and evaporation vary on every time scale, ranging from interannual variations to sharp differences in the intensity of storm events. Nearly 80 percent of all runoff in Asia occurs between May and October; three-quarters of runoff in Africa occurs between January and June; in Australia as much as 30 percent of runoff may occur in the single month of March. Another way to think about these variations over time is to realize that nearly 7,000 km³ more water is stored on land in snow, soil moisture, and lakes in March than in September, and that 600 km³ more water is stored in the atmosphere in September than in March (van Hylckama 1970). A few spots on earth receive essentially no regular rainfall for long periods, while others are deluged, receiving as much as 10 meters in the span of a few months.

Similar variability is evident in the regional and annual distribution of runoff. Table 2.2 shows one estimate of the average annual water balance of major continental areas, including precipitation, evaporation, and runoff.

TABLE 2.2   Surface Water Balances, by Continent

Continent	Precipitation (km³/yr)	Evaporation (km³/yr)	Runoff (km³/yr)
Europe	8,290	5,320	2,970
Asia	32,200	18,100	14,100
Africa	22,300	17,700	4,600
North America	18,300	10,100	8,180
South America	28,400	16,200	12,200
Australia/Oceania	7,080	4,570	2,510
Antarctica	2,310	0	2,310
Total Land Area	118,880	71,990	46,870

Source: Shiklomanov 1993.

Note: Runoff includes flows to groundwater, inland basins, and ice flows of Antarctica. These data can be compared with Table 2.3, which excludes non-renewable water flows and Antarctic ice flows.

Table 2.3 presents a comparable data set that shows average runoff by continent together with maximum and minimum flows, excluding flows to long-term groundwater and Antarctic ice flows. Global runoff is very unevenly distributed, even accounting for differences in area. More than half of all runoff occurs in Asia and South America. An unusually large fraction of all runoff occurs in just a single river, the Amazon, which carries more than 6,000 km³ of water a year. The water available in Brazil exceeds that available in Russia and is more than is available in Canada and the United States combined, which have the third- and fourth-largest average annual renewable water availability (Shiklomanov 1998). Figure 2.1 shows average annual water availability for selected countries.

TABLE 2.3   Continental Average, Maximum, and Minimum Runoff (km³ per year)

	Maximum	Average	Minimum
Europe	3,410	2,900	2,254
North America	8,917	7,890	6,895
Africa	5,082	4,050	3,073
Asia	15,008	13,510	11,800
South America	14,350	12,030	10,320
Australia and Oceania	2,843	2,360	1,850

Source: Shiklomanov 1998.

Note: Excluding nonrenewable water flows and Antarctica ice flows. For the period 1921–1985.

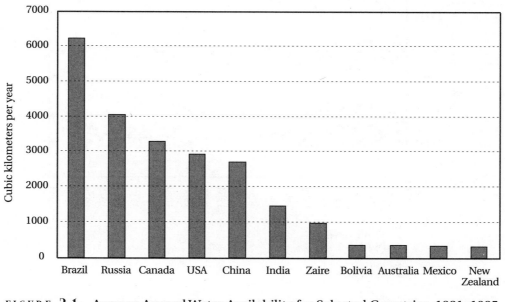

FIGURE 2.1   Average Annual Water Availability for Selected Countries, 1921–1985

The average amount of fresh water available for various countries is shown here, in cubic kilometers per year, as measured over the period 1921 to 1985. This figure shows the vast differences in the natural distribution of fresh water among different regions.
Source: Shiklomanov 1998.

Figure 2.2 shows how runoff has varied from year to year (1921 to 1985) for major continental areas. Periods of high and low flows can be seen in every region. While one may be tempted to try to discern trends in these annual figures (particularly readers interested in the issue of global climate change), I strongly urge the reader to resist this temptation. First of all, the data themselves only go to 1985, likely prior to any discernible climate change effect on runoff. More importantly, however, these data were collected with an enormous range of instruments, measurement techniques, and reliability. No effort has been made here to correct for biases, changing human withdrawals over time, and regional heterogeneities.

The total volume of stocks and flows provides only a single measure of water availability. From a human perspective, *relative* or *specific* (see Box 2.1) water availability per person (per capita) or per unit area can be more enlightening and useful for public policy. Figure 2.3 shows total average annual water availability by continent. Figure 2.4 shows per capita availability by continent. Is Asia water-rich or water-poor? Total volumes of water availability are greatest in Asia. Using per capita measures, however, Asia has the lowest per capita water availability because of its large population. Australia and Oceania have the lowest total volume of water of these continental regions, but by far the greatest per capita availability.

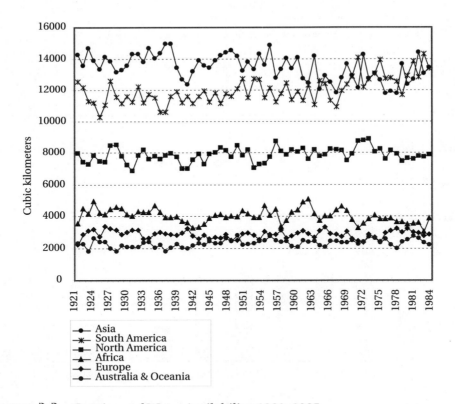

FIGURE 2.2   Continental Water Availability, 1921–1985

The annual availability of fresh water by continent is shown here in cubic kilometers per year. Note the natural variability from year to year. These data only go through 1985. The reader is urged to avoid any temptation to discern trends or patterns over time. *Source:* Shiklomanov 1998.

BOX 2.1
Quantitative Measures of Water Availability or Use

To estimate the quantitative characteristics of water resources, many different measures or indicators are used. Among the most common are *absolute* indicators that measure total volumes of water and *relative* or *specific* indicators that normalize water volumes to some other standard such as population, area, or economic values. A list of common (and some uncommon) measures of units (physical and water-related) is included in the Data Section (see Table 25). This list can also be found at www.worldwater.org.

The most common absolute measure for a stock of water is a volume. The most common measure for a flow of water is a volume per unit time. Thus a stock of water in a lake or groundwater aquifer can be measured in km³ or million gallons or thousand acre-feet. River flow might be measured in km³ per year or cubic feet per second or acre-feet per year or one of any number of similar units. When flows are described, it is important to know the period over which the flow is presented. Typical river flows are described as "average annual," meaning an average of a number of annual measurements. But the average annual flow of a river measured over five years may be different than that measured over twenty-five years. Or the average flow in two different five-year periods may vary considerably.

Common relative or specific measures are an amount or flow of water per unit area or per person (per capita). Examples are km³/km² or m³/person/year. These terms are "relative" because they depend on the size of another factor. This factor may or may not be constant. Areal measures are typically unchanging; population measures are constantly changing. Per capita global average runoff in 1900 was around 30,000 m³/person. In 2000, this same measure was around 7,000 m³/person because of the large increase in the world's population, not because of any change in the hydrologic cycle.

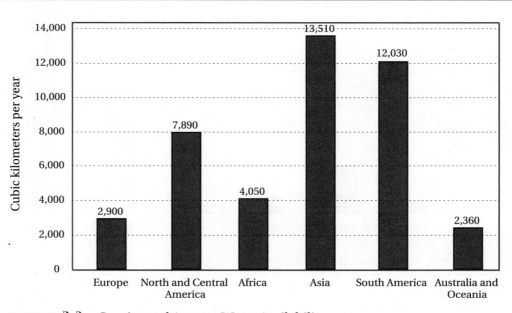

FIGURE 2.3   Continental Average Water Availability

The long-term average annual availability of fresh water by continent in cubic kilometers per year. Note the substantial total volume of water available in Asia, North America, and South America. *Source:* Shiklomanov 1998.

Water availability also has a strong political component to it. The world is not divided along watershed boundaries but along political boundaries that are defined by tradition, culture, conquest, economics, and myriad other factors. Occasionally a political border will coincide with a river; very rarely do political borders coincide with watershed or aquifer boundaries. How does one measure the water availability for a person living in the Nile River basin? One literal measure would be the total average runoff for the basin divided by the total population of the basin—approximately 85 km³/year divided by around 145 million people (in the 1990s)—or about 600 m³/p/yr. But this number is far different than the theoretical water availability of the average Egyptian, or Sudanese, which are 925 m³/p/yr and 3,150 m³/p/yr, respectively. And the theoretical water availability rarely represents the actual water available to any particular person, which depends on economic factors, legal water rights, technical ability to capture, store, and move water from place to place, political agreements with neighboring countries, and so on. On paper, the Sudan has a vast amount of water available on average, but it is compelled by a treaty signed with Egypt to pass on much of the water it receives in the Nile from upstream nations. In recent years, internal turmoil and civil war have prevented the Sudan from using even its legal share from the Nile treaty. Looking at purely hydrologic measures of water availability thus paints a misleading picture of how much water may actually be accessible, usable, or even used by a particular population or region. For this, we must look at the intersection of water and politics. The next section reviews the latest information on international river basin definitions and

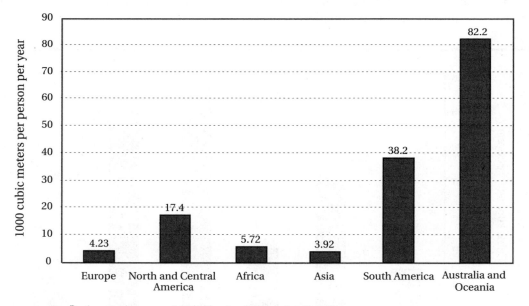

FIGURE 2.4   Continental Per Capita Water Availability

The long-term average annual per-capita availability of fresh water by continent in thousand cubic meters per person per year. Comparing this figure with the previous shows the importance of population in determining overall water availability. While Asia has the largest total water availability, it has the lowest per-capita availability because of its large population. Data on population are for 1995. *Source:* Shiklomanov 1998.

areas, which policymakers will find vital for addressing water-related disputes and conflicts.

International River Basins: A New Assessment

A critical factor in assessing the world's stocks and flows of water is the role of political borders in dividing those waters. Many rivers, lakes, and groundwater aquifers are shared by two or more nations and most of the renewably available freshwater of the earth crosses political borders. In 1958, the United Nations published the first comprehensive collection of information on shared international rivers of the world (United Nations 1958). This early assessment identified 166 major international river basins. In 1978, the United Nations published an updated assessment (United Nations 1978) identifying 214 such basins (see Box 2.2 for definitions of "international rivers"). For over 20 years, this assessment guided policy and analysis. By today's standards, the mapping of these river basins was crude and subject to large errors. Measurements were based on regional maps and often made by hand with a planimeter—a tool today's generation of digital mapmakers has never used.

Two major changes have occurred since the 1978 assessment: dramatic improvements in our abilities to precisely measure topography, identify geographical characteristics in flat terrain, and accurately map watersheds; and widespread geopolitical changes redefining nations and borders. As a result, the last assessment has been seriously out of date for a long time. In 1998, the World Resources and

---

BOX 2.2

Limitations, Definitions, and Caveats

What is an "international river basin?" A river basin is considered international if any tributary crosses the political boundaries of two or more nations. Wolf et al. (1999) make the further distinction that the tributary must be perennial (i.e., must flow all the time). A river that flows intermittently and is shared by two nations is therefore not included in their registry.

In the 1978 assessment, only "first order" basins communicating directly with the final recipient of the water (an ocean or a closed inland sea or lake) were included to distinguish them from tributary basins. This approach is still used today, even though some second- or even third-order tributaries of major rivers may be substantially larger in size than many first-order basins. Many tributary basins may also be more valuable politically and economically. Thus the scale of analysis is of vital importance, and one should not presume that river basins excluded from international registries are unimportant or irrelevant for regional or even international politics. For example, the Cauvery River basin is entirely contained within one nation, India, and hence is not included in international registries. Yet the Cauvery River has been the source of intense interstate rivalry, and even violent conflict, between the Indian states of Karnataka and Tamil Nadu (Gleick 1993).

Worldwatch institutes presented new basin areas for major watersheds using new digital elevation map sets (Revenga et al. 1998). These were determined from five-minute gridded elevation data—a relatively coarse resolution data set—further edited using a 1 km digital elevation model.

In 1999, Aaron Wolf and several colleagues (Wolf et al. 1999) released a new comprehensive analysis taking advantage of global digital coverage and a detailed compilation of changes in political borders. The new registry was compiled using digital elevation models at spatial resolutions of 30 arc seconds with cross checking of maps at scales of 1:20,000 to 1:1,000,000. The watershed maps thus created (Figures 2.5 to 2.9) were then superimposed over complete coverage of current political boundaries. The new registry now identifies 261 major international river basins, covering 45 percent of the land surface of the earth, excluding Antarctica. Altogether, 145 nations include territory within international river basins and 33 countries have more than 95 percent of their total land area in such basins (Wolf et al. 1999). The increase in the number of basins since the 1978 survey reflects both

FIGURE 2.5   International River Basins, Africa

Source: Wolf et al. 1999.

changes in the political landscape over the past two decades and changes that resulted from improvements in mapping technology.

Technology is not the only thing to have changed over the past two decades: the political landscape has also been radically altered. The most important of these changes has been the disintegration of the Soviet Union—once the largest single county in the world—into 15 separate nations. Many of the world's largest rivers flow in the territories of these nations and the breakup of the USSR has resulted in many new international rivers. The breakup of Yugoslavia and division of Czechoslovakia have also greatly increased the numbers of nations sharing the rivers of Europe. At the same time, the consolidation of Germany has eliminated one formerly international river, as has the confederation of North and South Yemen.

Even the highly sophisticated computer mapping methods now available have not produced agreement about basin boundaries and areas. The watershed areas from the World Resources Institute/Worldwatch report do not agree with those of the new registry, in part because of differences in the resolution of the mapping data used and differences in the algorithms for allocating basin areas to watersheds. Table 2.4 shows area estimates for several major river basins from the 1978 UN

FIGURE 2.6 International River Basins, Asia

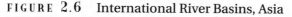

Source: Wolf et al. 1999.

FIGURE 2.7   International River Basins, Europe

Source: Wolf et al. 1999.

registry, the WRI/WWI report, and the new register. Note that the number of significant figures used in the latter two assessments have increased from 1978, though the size of the disagreements among them remain very large. This is an example of our growing ability to measure things with extreme *precision*, without necessarily improving the actual *accuracy* of the results. In recognition of the potential errors involved in this process, Wolf et al. (1999) rounded the last two significant figures in basins 100 km² or larger.

Table 2.5 compares the number of international basins, by continent, for the 1978 register and the updated register of Wolf et al. (1999). The largest differences are in Asia and Europe, reflecting the political changes in those regions. Europe now has 71 international river basins, up from 48 in 1978, while Asia has 53, up from 40. Even the Americas, however, show an increase in the total number of shared rivers, which reflects a new analysis of watershed boundaries rather than changes in political boundaries. Figures 2.5 (Africa), 2.6 (Asia), 2.7 (Europe), 2.8 (North and Central America), and 2.9 (South America) show the international river basins of the world by continental area.

FIGURE 2.8 International River Basins, North and Central America

Source: Wolf et al. 1999.

Some watersheds are highly divided. Nineteen watersheds are shared by five or more states (see Table 2.6). In some cases "states" with unresolved formal national status in 1999 are included in this register, such as the West Bank for the Jordan River. Topping the list is the Danube, with 17 political entities, up from only 12 in 1978. This change is the result of changes in the political landscape of Eastern Europe. Second on the list are the Congo and Niger rivers shared by 11 nations each. Changes in these basins since 1978 are primarily the result of better geographical information and watershed mapping, which identify more nations as part of the watersheds. Many of the new additions are nations that only share a tiny portion of the basins.

FIGURE 2.9   International River Basins, South America

Source: Wolf et al. 1999.

There are 17 international basins exceeding 1 million square kilometers in area (see Table 2.4 and Table 7 in the Data Section). This subset of large river basins encompasses more than 40 million square kilometers in area, over 60 percent of the total area of international basins. The largest international river basin measured by area is the Amazon, shared by eight nations. The Amazon is also the world's largest river measured by total annual flow. Three other basins, the Congo, Mississippi, and Nile, exceed 3 million square kilometers in total area. The Congo and Nile also have very large numbers of states sharing the watershed. The Mississippi, on the other hand, is contained almost entirely within the United States, with only 1.5 percent of its area falling within Canada.

Some of the changes since the 1978 assessment are surprising and may complicate international politics in particular regions. Egypt, for example, is now listed as a riparian member of the Jordan basin—a basin with already difficult politics. In the new register, the area of the Golan Heights is separated out for the Jordan basin, but ultimately this area will be part of Syria or Israel, or will be split between them. Similarly, Saudi Arabia is now considered a riparian of the Tigris-Euphrates system, and

TABLE 2.4   Comparison of Watershed Area Estimates for Major International River Basins (square kilometers)

Watershed	United Nations Registry (1978)	Revenga et al. (1998)	Wolf et al. (1999)
Amazon	5,870,000	6,144,727	5,866,100
Congo	3,720,000	3,730,474	3,699,100
Mississippi	3,250,000	3,202,230	3,226,300
Nile	3,030,700	3,254,555	3,038,100
La Plata	3,200,000	Not reported separately	2,966,900
Ob	3,010,000	2,972,497	2,734,800
Yenisey	2,530,000	2,554,482	2,497,600
Lake Chad	1,910,000	2,497,918	2,394,200
Niger	2,200,000	2,261,763	2,117,700
Amur	1,900,000	1,929,981	1,884,000
Ganges-Brahmaputra-Meghna	1,600,400	1,667,438	1,675,700
Volga	Not international in 1978	1,410,994	1,553,900
Zambezi	1,419,960	1,332,574	1,388,200
Aral Sea	Not reported separately in 1978	1,317,433	1,319,900

TABLE 2.5   Number of International River Basins, by Continent

Continent	United Nations (1978)	Wolf et al. (1999)
Africa	57	60
North and Central America	33	39
South America[a]	36	38
Asia	40	53
Europe	48	71
Totals	214	261

Sources: See column headings.

[a]The Jurado River, shared by Colombia and Panama, is included here for both assessments in South America.

TABLE 2.6   International Rivers Shared by Five or More States

River Basin	Number of States	States Sharing the Basin
Danube	17	Romania, Hungary, Yugoslavia (Serbia and Montenegro), Austria, Germany, Bulgaria, Slovakia, Bosnia–Herzegovina, Croatia, Ukraine, Czech Republic, Slovenia, Moldova, Switzerland, Italy, Poland, Albania
Congo	11	Democratic Republic of Congo, Central African Republic, Angola, Republic of the Congo, Zambia, United Republic of Tanzania, Cameroon, Burundi, Rwanda, Gabon, Malawi
Niger	11	Nigeria, Mali, Niger, Algeria, Guinea, Cameroon, Burkina Faso, Benin, Ivory Coast, Chad, Sierra Leone
Nile	10	Sudan, Ethiopia, Egypt, Uganda, United Republic of Tanzania, Kenya, Democratic Republic of Congo, Rwanda, Burundi, Eritrea
Rhine	9	Germany, Switzerland, France, Netherlands, Belgium, Luxembourg, Austria, Liechtenstein, Italy
Zambezi	9	Zambia, Angola, Zimbabwe, Mozambique, Malawi, United Republic of Tanzania, Botswana, Namibia, Democratic Republic of Congo
Amazon	8	Brazil, Peru, Bolivia, Colombia, Ecuador, Venezuela, Guyana, Suriname
Lake Chad (internal drainage)	8	Chad, Niger, Central African Republic, Nigeria, Algeria, Sudan, Cameroon, Libya
Tarim	7	China, Kyrgyzstan, Pakistan, Tajikistan, Kazakhstan, Afghanistan, India[b]
Volta	6	Burkina Faso, Ghana, Togo, Mali, Benin, Ivory Coast
Aral Sea (internal drainage)	6	Kazakhstan, Uzbekistan, Kyrgyzstan, Tajikistan, Turkmenistan, China
Ganges/Brahmaputra/ Meghna	6	India, China, Nepal, Bangladesh, Bhutan, Myanmar
Jordan	6[a]	Jordan, Israel, Syria, West Bank, Lebanon, Egypt
Mekong	6	Laos, Thailand, China, Cambodia, Vietnam, Myanmar
Tigris-Euphrates/ Shatt al Arab	6	Iraq, Turkey, Iran, Syria, Jordan, Saudi Arabia
Kura-Araks	6	Azerbaijan, Georgia, Iran, Armenia, Turkey, Russia
Neman	5	Belarus, Lithuania, Poland, Russia, Latvia
Vistula/Wista	5	Poland, Ukraine, Belarus, Slovakia, Czech Republic
La Plata	5	Brazil, Argentina, Paraguay, Bolivia, Uruguay

Source: Wolf et al. 1999.

[a]Overall political status of the West Bank was undetermined in mid-1999. The new registry also lists the area of the Golan Heights, which will ultimately be part of either Syria or Israel or a combination of the two.

[b]Some area in Chinese control claimed by India.

Libya is a member of the Lake Chad system. While some of these newly "discovered" riparians play very limited roles in actually generating water in these basins, as parties to a watershed they are likely to argue that they have the right to participate in negotiations over watershed management.

The Geopolitics of International River Basins

The growing literature on the connections between international conflicts and water resources primarily deals with international river basins (see, for example, Gleick 1993, McCaffrey 1993, Biswas 1994, Dabelko and Dabelko 1996, Wolf 1998). Similarly, the bulk of the legal tools that have been developed to address water disputes also deal with international river basins. As a result, the new data described above will be vitally important for understanding and resolving future political and military conflicts over water.

The extent to which fresh water is shared internationally is, however, only one factor affecting water-related disputes. Interstate conflicts are a function of many things, including religious animosities, ideological disputes, border arguments, and economic competition. While resource and environmental factors appear to be playing an increasing role in such disputes, it is difficult to disentangle the many intertwined causes of conflict. The ultimate goal of any assessment of the risks of conflict over water resources must include efforts to reduce those risks. I also note that there are whole classes of security concerns related to fresh water that are independent of traditional state borders and politics, including subnational disputes, ethnic disagreements, and the use of water systems as tools or targets of regional or subnational conflicts. These concerns are discussed in greater depth in Gleick (1998).

There is a long history of water-related disputes, from conflicts over access to adequate water supplies to intentional attacks on water systems during wars. In the first edition of *The World's Water*, published in 1998, two detailed chronologies of water-related disputes were published as part of the chapter on water and conflict. The first detailed water-related conflicts in the ancient Middle East, from 3000 B.C. to 300 B.C. The second described water-related conflicts since A.D. 1500. The present volume includes an updated version of the latter chronology (see Water Briefs), describing water-related conflicts since A.D. 1500.

Water and water-supply systems have been the roots and instruments of war. Access to shared water supplies has been cut off for political and military reasons. Sources of water supply have been among the goals of military expansionism. Inequities in water use have been the source of regional and international frictions and tensions. And there are growing concerns that terrorist acts are increasingly being directed at water systems. In late 1998, a guerrilla leader in Tajikistan reportedly threatened to blow up a dam on the Kairakkhum channel if certain political demands were not met (WRR 1998). In July 1999, a bomb was discovered at a reservoir in South Africa (Chikanga and Momberg 1999; Webb 1999). These kind of conflicts will continue—and in some places grow more intense—as growing populations demand more water for agricultural, industrial, and economic development, and perhaps more importantly, as water systems for managing or delivering water become increasingly valuable resources. While various regional and international legal mechanisms exist for reducing water-related tensions, most of these mechanisms deal solely with interstate disputes (see, for example, the 1997 United Nations Convention on the Law of the Non-navigational Uses of International Watercourses [United Nations 1997b]. The full text of this convention can be found at www.worldwater.org). And these rarely receive the international support or attention necessary to resolve many conflicts over water.

Despite progress in this area, there are concerns that existing international water law may be unable to handle the strains of ongoing and future problems. New disputes are also arising in other regions, such as southern Africa, where several major ongoing controversies remain unresolved. It is also possible that international water law is an inappropriate mechanism for addressing some problems, particularly subnational and local disputes. Many political entities claiming water rights or allocations will not be served by the International Court of Justice, the 1997 UN Convention, or even international treaties. Improvements in international and national water law will help, as might new approaches for managing water competition.

In addition to improving legal approaches in this area, efforts by UN and international aid agencies to ensure access to clean drinking water and adequate sanitation can reduce competition for limited water supplies, the economic and social impacts of widespread waterborne diseases, and injustice associated with inequities in access and use of water. In regions with shared water supplies, third-party participation in resolving water disputes, either through UN agencies or regional commissions, can also effectively end conflicts.

SUMMARY

New data and information are presented here on the world's stocks and flows of fresh water and on the extent to which that water is shared by national governments. Continued improvements in this information can help address uncertainties and disputes over freshwater resources. Such disputes are caused by many factors, including religious animosities, ideological disputes, arguments over borders, and economic competition. After more than a decade of academic semantic arguments, political rethinking, and increased participation in security discussions by environmental scientists and analysts, it is widely accepted that resource and environmental factors—particularly those associated with fresh water—play a tangled but definite role in local, regional, and even international disputes. Not all water-resources disputes will lead to violent conflict; indeed most will lead to negotiations, discussions, and nonviolent resolutions. But where water is a scarce resource vital for economic and agricultural production, shared water has evolved into an issue of high politics, and the probability of water-related disputes is increasing.

Recent experience suggests that conflicts may be more likely to occur on the local and regional level than between nations, and in developing countries where common property resources like water may be both more critical to survival and less easily replaced or supplemented. As a result, better information is still required on disputed waters within existing national borders—not just information on shared international watersheds. Moreover, water-related threats to security depend not just upon the geophysical characteristics of watersheds and the hydrologic cycle, but also upon the economic, cultural, and sociopolitical factors at work in a given country or region. Policymakers must be alert to the likelihood of disagreements over water resources and to the kinds of legal, economic, political, and technical responses that can minimize the risk of conflict.

REFERENCES

Biswas, A.K. (ed.). 1994. *International Waters of the Middle East: From Euphrates-Tigris to Nile.* Oxford University Press, Bombay and New Delhi, India.

Chikanga, K., and E. Momberg. 1999. "Outrage over Pta 'reservoir bomb.'" *Pretoria Citizen* (July 21), p. 3.

Dabelko, G.D., and D.D. Dabelko. 1996. "Environmental security: Issues of conflict and redefinitions." *Environment and Security,* Vol. 1, pp. 23–49.

Gleick, P.H. 1993. "Water and conflict." *International Security,* Vol. 1, No. 1, pp. 79–112 (Summer 1993).

Gleick, P.H. 1998. *The World's Water 1998–1999: The Biennial Report on Freshwater Resources.* Island Press, Washington, D.C.

International Hydrological Programme (IHP). 1998. "Water: A looming crisis?" H. Zebidi, ed. *Proceedings of the International Conference on World Water Resources at the Beginning of the Twenty-first Century.* June 3–6, UNESCO, Paris, France.

McCaffrey, S. 1993. "International water law." In P.H. Gleick (ed.), *Water in Crisis: A Guide to the World's Fresh Water Resources.* Oxford University Press, New York, pp. 92–104.

Revenga, C., S. Murray, J. Abramovitz, and A. Hammond. 1998. *Watersheds of the World: Ecological Value and Vulnerability.* World Resources Institute/Worldwatch Institute. Washington, D.C.

Shiklomanov, I.A. 1993. "World fresh water resources." In P.H. Gleick (ed.), *Water in Crisis: A Guide to the World's Fresh Water Resources.* Oxford University Press, New York, pp. 13–24.

Shiklomanov, I.A. 1998. Archive of world water resources and world water use. Global Water Data Files: State Hydrological Institute. Data archive on CD-ROM from the State Hydrological Institute, St. Petersburg, Russia.

United Nations. 1958. *Integrated River Basin Development.* UN Publications, sales no. 58.II.B.3. United Nations Publications, New York.

United Nations. 1978. *Registry of International Rivers.* Prepared by the Centre for Natural Resources, Energy and Transport of the Department of Economic and Social Affairs. Pergamon Press, Oxford, United Kingdom.

United Nations. 1997a. *Comprehensive Assessment of the Freshwater Resources of the World.* Commission for Sustainable Development, Stockholm Environment Institute, Stockholm, Sweden.

United Nations. 1997b. *Convention on the Law of the Non-navigational Uses of International Watercourses.* UN General Assembly A/51/869 (April 11). United Nations Publications, New York.

Van Hylckama, T.E.A. 1970. "Water balance and earth unbalance." International

Association of Scientific Hydrologists. Publication No. 93. *Symposium on World Water Balance,* Vol. 2, pp. 434–444.

Webb, B. 1999. "Outrage after bomb find at city reservoir." *Pretoria News* (July 21), p. 1.

Wolf, A.T. 1998. "Conflict and cooperation along international waterways." *Water Policy,* Vol. 1, No. 2, pp. 251–265.

Wolf, A.T., J.A. Natharius, J.J. Danielson, B.S. Ward, J. Pender. 1999. "International river basins of the world." *International Journal of Water Resources Development,* Vol. 15, No. 4 (December).

World Rivers Review (WRR). 1998. "Dangerous dams: Tajikistan" *World Rivers Review,* Vol. 13, No. 6, p. 13 (December).

THREE

Pictures of the Future: A Review of Global Water Resources Projections

If we don't think about where we want to go, we'll end up where we are heading.

CHINESE PROVERB

Making predictions is very difficult, especially about the future.

CASEY STENGEL

Water planners are among the few natural resource managers to think more than a few years into the future. The time required to design and build major water infrastructure, and the subsequently long lifetimes of dams, reservoirs, aqueducts, and pipelines, require planners to take a relatively long view. But what will future water demands be? How can they be predicted, given all the uncertainties involved in looking into the future? At the global level, various projections and estimates of future freshwater demands have been made over the past half century, some extending out as much as 60 or 70 years. These projections have invariably turned out to be wrong. In the past few years, a renewed interest in global water issues has stimulated new thinking about how to make such projections. New efforts are taking advantage of advances in computer capabilities, the availability of better water data, and new concepts of scenario development. This chapter offers an analysis and review of the history of several major scenario projections and discusses the differences in their basic assumptions, methodologies, and approaches. The differences among the projections limit the validity of direct quantitative comparisons,

Portions of this chapter were prepared as part of the World Water Vision of the 21st Century (www.watervision.org), a global consultation project that aims to develop knowledge, raise awareness of water issues, produce a consensus on a vision for the year 2025 and contribute to a Framework for Action linked to the Global Water Partnership (www.gwpforum.org). That work was carried out for the Vision Scenario Development Panel under a Stockholm Environment Institute contract funded by the Swedish International Development Cooperation Agency (SIDA).

but a general comparison highlights basic concepts in our understanding of future global water use.

The future is largely unknowable. But humans have always thought about possible futures, explored plausible paths, and tried to identify risks and benefits associated with different choices. In recent years this has led to a growing interest in scenarios, forecasting, and "future" studies (see, for example, Schwartz 1991). This sort of planning has more than academic implications. In the water sector, expectations about future water demands and supplies drive huge financial expenditures for water-supply projects. These projects, in turn, cause significant human and ecological impacts. At the same time, failing to make necessary investments can lead to the failure to meet fundamental human water needs. The challenge facing water planners is to balance the risks and benefits of these kinds of efforts.

While most water projections are usually done by local water agencies, municipalities, or companies, a number of more comprehensive, global-scale assessments have also been done in an effort to get a broader picture of critical water concerns. Most of these studies typically consist of separate regional or sectoral projections conflated to provide a global view. Each regional or sectoral estimate carries its own sets of assumptions and methods.

Projections of global water use have become increasingly sophisticated in approach and detailed in their temporal and spatial scale. Most of the earliest projections used variants on the same methodology—future water use was based on population projections; simple assumptions of industrial, commercial, and residential water-use intensity (e.g., water per unit population or income); and basic estimates of future crop production as a function of irrigated area and crop yield. Early scenarios were typically single, "business-as-usual" projections with no variants. Most scenarios ignored water requirements for instream ecological needs, for navigation, for hydropower production, for recreation, and so on. More recently, studies have been published describing a wider range of results under a wider range of assumptions. Projections have begun to include reassessments of actual water needs and water-use efficiencies, dietary requirements, cropping patterns and types, and ecosystem functions. Box 3.1 discusses different definitions of "water use."

Large-scale water-use projections have also become increasingly sophisticated due to the growing capability of easily accessible computers to handle significant numbers of calculations and the growing availability of water-use data. Assessments that used to be done for continental areas or on a national basis are now being done for watersheds on smaller and smaller temporal and spatial scales.

Data Constraints

The greatest constraints on future improvements in water forecasts now come not from computer capability but from limitations on the quality, availability, and regional resolution of water data and from difficulties in doing certain kinds of assessments. Some of the most important data problems are listed here.

- *Serious gaps in regional-scale hydrological data still exist and are unlikely to be filled soon.* While precipitation, temperature, and runoff are relatively well measured in developed countries, many regions of the world suffer both from gaps

BOX 3.1

Definitions

There continues to be tremendous confusion in the water literature about the terms "use," "need," "withdrawal," "demand," "consumption," "consumptive use" and so on. Great care should be used when interpreting or comparing different studies. For the purposes of this chapter I have tried to standardize terms as follows: "Withdrawal" refers to water removed from a source and used for human needs. Some of this water may be returned to the original source with changes in the quantity and quality of the water. The term "consumptive use" or "consumption" refers to water withdrawn from a source and made unusable for reuse in the same basin, such as through irrecoverable losses like evaporation, seepage to a saline sink, or contamination. Consumptive use is sometimes referred to as "irretrievable losses." The term "water use," while common, is often misleading or at best uninformative, referring at times to consumptive use and at times to withdrawals. "Need" for water is also subjective, but typically refers to the minimum amount of water required to satisfy a particular purpose or requirement. It also sometimes refers to the desire for water on the part of a water user. "Demand" for water is used to describe the amount of water requested or required by a user, but the level of demand for water may have no relationship to the minimum amount of water required to satisfy a particular requirement.

in present-day instrumental coverage and from lack of any long-term records. And even in richer countries, pressures to cut funds for observation and monitoring stations threaten the continuity of time-series data.

- *Certain types of water-use data are not collected or reliable.* Far less data are collected on water use than on water supply and availability. Domestic water use is often not measured directly, and details on how the water is used are rarely collected. A recent survey conducted for the American Water Works Association on U.S. domestic water use is a rare exception (http://www.waterwiser.org/frameset.cfm?b=2), and even this study was limited in scope. Industrial and commercial water use is inventoried infrequently or not at all. Agricultural water-use data are even more uneven and unreliable. Groundwater withdrawals are rarely measured or regulated. Even when water-use data are collected, information on changing water-use patterns over time is often not available, making analysis of trends difficult.

- *Some countries or regions restrict access to water data.* In this era of easy Internet access, some countries still refuse to share water-related data with neighbors or even their own scientists. In regions where water is shared internationally, nations are tempted to restrict information where there is a perceived political advantage in doing so. Singapore, for example, tightly restricts any information on its water agreements with Malaysia or on its external dependence for vital water supplies. This kind of problem led the framers of the 1997 Convention on the Law of the Non-navigational Uses of International Watercourses to include a provision calling for open sharing of data. Article 9 of the Convention states:

States shall on a regular basis exchange readily available data and information on the condition of the watercourse, in particular that of a hydrological, meteorological, hydrogeological and ecological nature and related to the water quality as well as related forecasts. . . . [States] shall employ their best efforts to collect and, where appropriate, to process data and information in a manner which facilitates its utilization by the other watercourse States to which it is communicated. (http://www.worldwater.org)

- *Some water uses or needs are unquantified or unquantifiable.* In all likelihood, some uses and needs are unlikely ever to be accurately determined or included in scenario projections. For example, ecological needs, recreational uses, water for hydropower production or navigation, and reservoir losses to seepage or evaporation are often difficult to calculate with any accuracy. Nevertheless, these water uses and activities will eventually need to be quantified and incorporated into future estimates if true water planning is to be done.

As a result of these limitations, analysts should not assume that increasing model or scenario sophistication will lead to more accurate forecasts. In the end, even "perfect" models supplied with imperfect data will be of limited value. Any scenarios must still be treated as "stories," as possible futures to be explored, with the understanding that choices we make today will determine which path we end up following, and which future we move toward.

Forty Years of Water Scenarios and Projections

During the latter part of the twentieth century, water-resources planning focused on using or making projections of variables such as populations, per capita water demand, agricultural production, levels of economic productivity, and so on. These projections are used to estimate future water demands and then to evaluate the kinds of systems or structures necessary to meet those demands. As a result, traditional water planning tends to project future water demands as variants or extensions of current trends, independent of any analysis of specific water needs. Often these projections are done independently of estimates of actual regional water availability. The planning process then consists of developing suggestions of alternative ways of bridging the apparent gaps between this idealized projection of demand and anticipated supply.

On a regional scale, the official California Water Plan is an example of this kind of approach. Every several years starting in 1957 the California Department of Water Resources (CDWR) has issued a "Water Plan." The most recent version, released in late 1998, is similar in form and approach to every one of the seven earlier versions (CDWR 1998). Using fairly constant water-intensity projections (in this case water use per person) coupled with projected increases in population, CDWR concludes that California water problems and policies in 2020 will be little changed from today. Farmers are assumed to grow the same kinds of crops on about the same amount of land. The growing urban population will continue existing patterns of water use, with relatively minor changes in some residential water-use technology and efficiency. Water used by aquatic ecosystems will remain the same or even decrease. And the projections of total demands exceed available supplies by several billion cubic meters, a shortfall projected in every California Water Plan since the first.

On a larger scale, almost every early global water projection used a similar approach and reached similar conclusions, leading many observers to worry about major shortfalls and shortages in the future. In some areas of the world, such shortages and shortfalls are already manifest, and new problem areas are likely to emerge in coming years. But it is also important to note that every one of the early global water projections estimated far greater demands for water than have actually materialized, many of them by a substantial margin. This suggests that the traditional methods used by water-scenario developers are missing some critically important real-world dynamics. Figure 3.1 and Table 3.15 show more than 25 different water projections for various points during the twenty-first century along with an estimate of actual water use up to 2000. As the figure shows, every one of the earlier projections greatly overestimated future water demands by assuming that use would continue to grow at, or even above, historical growth rates. Actual global water withdrawals in the 1990s were only around half of what they were expected to be 30 years earlier. The inaccuracy of these past projections highlights the importance of developing better methods for making projections of future

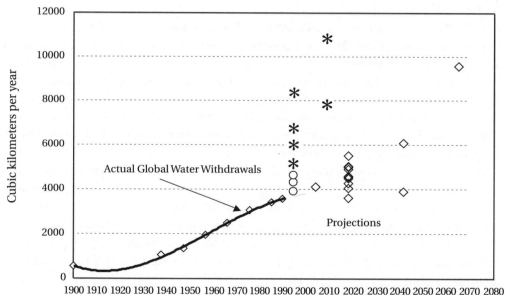

FIGURE 3.1 Water Scenarios: Projected and Actual Global Water Withdrawals

More than 25 different projections of future water withdrawals are plotted here. The data come from Table 3.15. Also plotted are actual water withdrawals through 1995. Note that the earliest projections consistently and significantly predicted far greater water demands than actually materialized. While some recent projections see a leveling off of future demands, most still show large increases.

Key: Diamond: Post-1995 projections
      Circle: 1980 to 1995 projections
      Asterisk: Pre-1980 projections

needs. Some of the most comprehensive global water assessments, including some recent progress in improving forecasting techniques, are described in the following text. Whether these more recent assessments will prove any more accurate than their older cousins will only be revealed over time.

Nikitopoulos (1962, 1967)

In the 1960s, a Greek hydrologist estimated total global water withdrawals by major sector for the year 2000 (Nikitopoulos 1967). In his work he assumed average annual domestic, industrial, and agricultural water needs on a per capita basis and then adopted population estimates to get projected future withdrawals. He did an additional estimate of ultimate water needs based on average annual per capita use for domestic, industrial, and agricultural purposes assuming a standard of living for all humans equal to the existing standard in the United States. In this study, domestic consumption was estimated to be 500 liters per capita per day (lpcd), with another 500 lpcd required for industrial water use. For agricultural water use, the study assumed that 700 m^2 of irrigated land would be needed to grow food for each person, with an average of 1,000 mm/year of irrigation water required in addition to rainfall; this amounts to an additional 700 m^3/person/year. Total annual ultimate water use per person in 2000 was thus estimated to be about 1,065 m^3 per person or a total of 6,730 km^3 (Nikitopoulos 1962).

L'vovich (1974)

In 1974, L'vovich published an assessment of global water-resources supply and use that remains one of the most comprehensive ever done (L'vovich 1974). Detailed assumptions were made for a variety of human uses to the year 2000, including domestic and industrial water use, irrigated and nonirrigated agricultural water demands, and hydropower, navigation, and fishery water requirements. L'vovich developed two different scenarios for 2000: a "business-as-usual" scenario and a "rational use" scenario. The work of L'vovich has formed the basis of much of the global water modeling done since then.

In his business-as-usual scenario, L'vovich assumed that average domestic per capita withdrawals would increase to 400 lpcd with the widespread adoption of central sewerage. He assumed each of the projected 6.3 billion humans in the year 2000 would use this amount, leading to a total domestic annual water withdrawal of 920 km^3. Consumptive use of water in this sector was assumed to be less than 20 percent of withdrawal, but the remaining water is assumed to be sewage, requiring substantial additional volumes of runoff for "dilution."

For power production, L'vovich assumed that water use per kilowatt-hour of electricity produced would decrease, as power-plant efficiencies increased. For closed-cycle cooling plants, he assumed about 3 liters of water are evaporated per kWh of electricity generated, with far higher levels "withdrawn" for use and returned to the source. In open-cycle or once-through plants, total consumptive use is lower but total withdrawals are higher. Assuming that future consumptive use in this sector followed existing patterns, L'vovich estimated that water consumption for the energy sector would increase by a factor of 20 to 200 km^3/year, while withdrawals increase to 3,100 km^3/year. He then notes, however, that this figure is unacceptably large because of the temperature discharge constraints, and reassesses power needs by assuming increases in efficiency. In this case he assumes that water consumption per unit energy decreases to half of its 1960s value. Water withdrawals and consumption would thus increase by a factor of 7 between the early 1970s and 2000, while world energy use would increase 17-fold. In this case, total water withdrawals for power-plant cooling would be about 1,550 km^3/year and consumptive use about 100 km^3/yr (L'vovich 1974).

Gross industrial water "use" in the year 2000 under current trends was assumed to increase by a factor of 15 between the early 1970s and the year 2000 to 6,000 km^3, while actual industrial withdrawals were projected at 3,000 km^3/year, implying a reuse factor of 2. L'vovich is one of the few analysts to consider this issue of reuse. Consumptive industrial water use in the year 2000 was projected to be 10 percent of gross use, or 600 km^3/year. He assumes extremely large additional volumes of water are needed to dilute industrial pollutants. Because of the huge volumes of water required in this "business as usual" scenario, L'vovich developed a "rational use" scenario for industrial and public water use, in which no sewage water is generated (all water is recycled and reused) and all industrial and power water supply needs are met with closed systems. In the rational use scenario, all industrial, power, and residential needs are assumed equal to consumptive use plus some minor additional withdrawals. This greatly reduces total "water supply" requirements, as shown below, and is one of the few early examples of an alternative scenario.

In the "business as usual" scenario, projected agricultural water use (irrigated agriculture) in the year 2000 was determined by assuming a per capita annual food consumption of 800 kg of grain (twice the 1970s amount) feeding 6.3 billion people, for a total annual production requirement of food crops (grain equivalent) of approximately 5 billion tons. Irrigated agriculture is assumed to furnish 40 percent of this grain, or 2 billion tons. Using crop yields of 4 tons per hectare, the total irrigated area needed would be 500 million hectares, of which 425 million hectares would be irrigated with clean water, and 75 million hectares with sewage water. (Around 270 million hectares were actually irrigated in 1997—see Table 10 in the Data Section.) L'vovich assumes that 8,000 cubic meters of water per hectare will be required for irrigation. Thus total withdrawals of water for irrigation would be 4,500 km^3/year, of which 4,000 would be consumed (L'vovich 1974). Because part of this water is wastewater enriched in organic fertilizer, L'vovich adds a correction for the possible increases in crop yields on lands using this water and recalculates total water needs downward to 4,250 km^3 withdrawn and 3,850 km^3 consumed. These data are for food crops only. If "technical" crops, such as for natural fibers, are included, L'vovich estimated total withdrawals of irrigation water in the year 2000 would be 4,400 km^3/yr and total consumptive use would be 4,000 km^3/yr.

L'vovich calculated total water consumed for rainfed agriculture by making assumptions about water requirements to grow grains, determining the area of land planted to food crops, and making assumptions about future increases in grain yield per hectare. Yields on rainfed lands were assumed to triple and reach 1.8 tons per hectare while total grain production was assumed to experience a four- to five-fold increase to 3 billion tons. For future projections, he extrapolates a general trend of reducing water consumption per unit agricultural output (i.e., higher efficiency) but a general increase in total water used for plant growth. Assuming that 1.7 billion hectares of land are planted for food crops, total consumption of water on unirrigated

TABLE 3.1   L'vovich Year 2000 Conventional
Assumptions Projection (km^3/year)

	Withdrawals (km^3/yr)	Consumption (km^3/yr)
Residential/drinking water	920	180
Livestock	150	100
Power industry	3,100	270
Industrial	3,000	600
Irrigated agriculture[a]	4,400	4,000
Nonirrigated agriculture[b]	700	700
Hydropower and navigation	500	500
Fishery and sports fishing	175	85
TOTAL[c]	12,770	6,350

Source: L'vovich 1974. L'vovich included many different possible scenarios. This one, from Tables 29, 30, and 32 in his book, reflects his business-as-usual variant.

[a]Includes food and fiber crops, and 450 km^3 of wastewater.

[b]An additional 1,240 km^3 of rainfall are used for nonirrigated agriculture. The 700 km^3 listed here is the reduction in runoff from expansion of rainfed agriculture and hence is counted as additional human water use over current conditions.

[c]These numbers do not always correspond to L'vovich's text. Readers should go to the text for details and clarification.

farmland will be about 2,500 km^3/yr instead of the present 1,260 km^3/yr. This additional expenditure of water will come in part from arid lands and in part from land with sufficient water. According to L'vovich, an overall decrease in natural runoff of 700 km^3/yr would be required to meet this need, so it is included in his totals.

Other human activities, including hydropower, navigation, and fisheries, will also require water withdrawals and use. Future evaporative losses will increase with an increase in reservoir area. L'vovich projected that a total of 500 km^3/yr of evaporative losses from reservoir surfaces would occur by 2000, with additional withdrawals and consumption for fisheries, livestock, and navigation. Tables 3.1 and 3.2 summarize L'vovich's conventional and "rational use" projections.

Kalinin and Shiklomanov (1974) and De Mare (1976)

Kalinin and Shiklomanov's work was initially published in Russian in 1974 and De Mare described it in 1976. They rely heavily on trends from numerous statistics and special reports concerning present (1970s) and future water use. In addition to domestic, industrial, and agricultural water use, they calculate reservoir losses from increased evaporation. Kalinin and Shiklomanov state "for reliable forecasts it is necessary to have information about the tendencies and possible changes of specific water consumption," which the authors note are difficult to consider or predict. Per capita values for consumptive uses and withdrawals are adopted for various regions and end-use sectors and multiplied by population projections. For industrial water use, they assume a North American standard of 1,200 m^3/p/year. Total 2000 water withdrawals were estimated to be 5,970 km^3. De Mare adopted most of their estimates.

TABLE 3.2   L'vovich Year 2000 "Rational Use" Projection (km³/year)

	Withdrawals (km³/yr)	Consumption (km³/yr)	Discharge of sewage (km³/yr)
Water supply (all types)	1,500	1,050[a]	0
Irrigated agriculture	3,950	4,000[b]	400[c]
Nonirrigated agriculture[d]	700	700	0
Hydropower and navigation	500	500	0
Fishery and sports fishing	175	85	90
TOTAL[e]	6,825	6,335	490

Source: L'vovich 1974. L'vovich included many different possible scenarios. This one, from Table 32 in his book, includes a more efficient variant.

[a]Excluding 450 km³ of sewage used for irrigation.

[b]Including 450 km³ of sewage used for irrigation.

[c]Polluted by agricultural chemicals.

[d]An additional 1,200 km³ of rainfall is used for nonirrigated agriculture. The 700 km³ listed here is the reduction in runoff from expansion of rainfed agriculture.

[e]These numbers do not always correspond to L'vovich's text. This table is repeated verbatim from the original source. Readers should go to the text for details and clarification.

In 1976, De Mare (1976) produced an assessment of global water use in the year 2000 based on several existing assessments of world water resources. The original purpose of the paper was to serve as input for the 1977 UN Water Conference in Mar del Plata, Argentina, and work was carried out under contract with the UN and in close contact with the secretariat of the International Federation of Institutes for Advanced Study (IFIAS).

De Mare assumed it is unlikely that regions with high "specific" (per capita) consumption would reduce their domestic use—a standard assumption of traditional water planners. In his assessment, most of the developing regions were assumed to reach 200–300 lpcd for domestic needs, with a higher per capita water use in the industrialized regions. De Mare assumed industrial water use in the year 2000 would also vary depending on region, with a range from 100 to 2,000 m³ per person per year. The high value was given for industrialized North America, which assumes a significant increase from actual per capita industrialized water use in the region. De Mare assumed global industrial water withdrawals would be 1,775 km³ per year by the year 2000.

Most agricultural figures from Kalinin and Shiklomanov were adopted directly by De Mare. Data on water losses from reservoirs, typically left out of these kinds of projections, were included using the assumptions from Kalinin and Shiklomanov. Table 3.3 summarizes De Mare's year 2000 per capita water withdrawal assumptions by region. Table 3.4 summarizes total water withdrawal by region, also for the year 2000.

Falkenmark and Lindh (1974)

Falkenmark and Lindh published several different estimates of water withdrawals and consumption in the mid-1970s. These are summarized in Table 3.5 and

TABLE 3.3   De Mare Year 2000 per Capita Water Withdrawal (cubic meters per person per year) Projections by Region and Sector

Region	Domestic m³/p/yr	Industrial m³/p/yr	Agricultural m³/p/yr	Reservoir Losses m³/p/yr
Europe	150	400	185	10
USSR	130	500	1,310	70
Asia	75	150	5,585	25
Africa	50	100	400	85
North America	260	2,000	1,050	110
South America	20	200	190	35
Oceania	110	700	750	150

Source: De Mare 1976.

TABLE 3.4   De Mare Year 2000 Total Water Withdrawal Projections

Region	Total Water Withdrawal (km³/yr)
Europe	405
USSR	640
Asia	3,140
Africa	520
North America	1,025
South America	290
Oceania	60
Total	6,080

Source: De Mare 1976.

Note: Includes water losses from reservoirs.

Table 3.6. In two 1974 papers, their estimates relied on the varying rural and urban needs of the time (Falkenmark and Lindh 1974a,b). Using United Nations urban–rural population projections (United Nations 1974), they assumed domestic water needs would be 400 lpcd for urban areas and 200 lpcd for rural areas. For other water uses, they relied on the same approach as L'vovich, with minor variations. For example, for industrial uses, Falkenmark and Lindh base their estimate on the use of water at that time by Swedish industry—500 m³/p/yr. Applying this figure to the entire world's population, the world's available freshwater resources would all be needed for wastewater transport, that is, for dilution and disposal. Therefore, their alternative was to assume that 90 percent of industrial wastewater could be recycled so that only 10 percent of the wastewater plus irretrievable losses from production (estimated at an additional 20 percent) would have to be replaced by fresh water. Consumptive use of water by industry thus amounts to 30 percent of the total, or 150 m³/p/yr.

TABLE 3.5   Falkenmark and Lindh Year
2000 Water Withdrawal and Consumption
Projections (km³/yr)

Region	Without Wastewater Reuse	With Wastewater Reuse
Europe	741	536
Asia	4,826	3,465
USSR	430	312
Africa	1,044	742
North America	437	317
South America	859	616
Australia/Oceania	46	33
Total (rounded)	8,380	6,020

Source: Falkenmark and Lindh 1974 a, b.

Note: Excludes explicit water losses from reservoirs.

TABLE 3.6   Falkenmark and Lindh Year 2015 Water
Withdrawal Projections

Sector	2015 Withdrawal (no industrial reuse) km³/yr	2015 Withdrawal (90% industrial reuse) km³/yr
Domestic	890	890
Industrial	4,100	1,145
Agricultural	5,850	5,850
Total	10,840	7,885

Source: Falkenmark and Lindh 1974a, b.

Notes: Excludes explicit water losses from reservoirs. Year 2015 population is assumed to be 8,155 million.

For estimating agricultural needs, Falkenmark and Lindh assumed that 12 people could be supported by the agricultural production of one hectare of cropped land, requiring 700 to 900 mm/yr of irrigation water. This corresponds to a per capita consumptive use of 585 to 750 m³/yr for agriculture.

For 2000, total water withdrawals were projected to be between 6,000 and 8,400 km³/yr (see Table 3.5). The lower value includes wastewater reuse. By 2015, withdrawals, even with 90 percent industrial reuse, were projected to reach nearly 8,000 km³/yr. Table 3.6 shows the sectoral projections from Falkenmark and Lindh. Both sets of estimates exclude water lost from reservoirs by evaporation.

World Resources Institute (1990) and Belyaev (1990)

Year 2000 projections for water withdrawal, consumptive use, and waste in return flow are summarized in one of the annual reports by the World Resources Institute (WRI 1990, pp.167–73, Table 8). Of the 3,500 km³ withdrawn for human use in 1990,

TABLE 3.7   World Resources Institute Year
2000 Water Withdrawal and Consumption
Projections

Region	Withdrawals (km³/year)	Consumptive Use (km³/year)
Europe	404	158
Asia	2,160	1,433
USSR	533	286
Africa	289	201
North America	946	434
South America	293	165
Oceania	35	22.5
Total (rounded)	4,660	2,700

Source: From Table 10.3, WRI 1990.

TABLE 3.8   Belyaev Year 2000 Water Withdrawals
and Consumption Projections

Region	Water Withdrawals (km³/year)	Water Consumption (km³/year)
Europe	381–481	143–148
Asia	2,020–2,040	1,315–1,320
Africa	220–225	133–138
North America	840–850	332–342
South America	230–240	110–120
Oceania and Australia	28.5–29	17
USSR	475–485	235–240
Total (rounded)	4,195–4,350	2,285–2,320

Source: Belyaev 1990.

around 2,100 km³ are used consumptively. The remaining 1,400 km³ are returned to rivers and lakes. In the WRI projection, global withdrawals are expected to rise 2 to 3 percent annually until the year 2000. Primary data and water-use projections for 2000 were developed by region and sector by a team of Soviet hydrologists (directed by A.V. Belyaev) and from a United Nations conference paper by Asit Biswas. Table 3.7 summarizes the WRI estimates. Little detail is given in the WRI source on the background assumptions made.

Belyaev's estimate from the USSR Academy of Sciences, Institute of Geography, Moscow (described in WRI 1990), also offers an analysis of water withdrawals and consumption for 2000. Single projections are made for the irrigation, domestic, and municipal sectors. A range is provided for the industrial sector. The values summarized in Table 3.8 are the sums of the sector estimates.

Shiklomanov (1993, 1998)

In 1987 Shiklomanov and Markova (from the State Hydrological Institute in St. Petersburg—then Leningrad) published a set of estimates of current and projected water-resources use by region and sector (see Shiklomanov 1993 for an English-language version of this work). This work was the most comprehensive since

L'vovich's work nearly 20 years earlier. Water use was broken down for the agricultural, industrial, and municipal sectors, and included water lost from reservoir evaporation. Both water withdrawal and consumptive uses were estimated for the years 1990 and 2000. Shiklomanov and Markova used population and economic factors as driving variables, with detailed assessments completed for many regions around the world, which were then aggregated to continental scale. For some regions, improvements in water-use efficiency are implicit in their simple assumption that water use continues at current rates (a declining per capita water use).

Increases in water requirements over 1980 levels are projected for all areas, with the largest increases in South America and Africa. Decreases in overall consumptive use are possible, this study suggests, due to improvements in industrial reuse of water. Agricultural water withdrawals are projected to decrease as a fraction of total water withdrawals because industrial water withdrawals increase at a faster rate. Evaporative losses from reservoirs exceed industrial and municipal consumptive uses, but remain far below agricultural consumptive uses. Table 3.9 shows projected water withdrawals and consumption by region.

Shiklomanov and the Russian group have continued to refine their assessments, releasing a new comprehensive analysis in 1998 as part of the Comprehensive Assessment of the Freshwater Resources of the World prepared for the Commission on Sustainable Development of the United Nations. This new work (see Table 3.10) considerably reduces past projections, such that their year 2025 estimates for water withdrawals and consumptive use are actually below their earlier estimates for the year 2000 (Shiklomanov 1998).

Table 3.11 presents a similar assessment from Shiklomanov, but disaggregated by water use category. This table shows that agricultural water use continues to make up around 65 percent of all water withdrawals and 85 percent of all consumptive use of water. As this volume of *The World's Water* went to press, no details were available about specific regional assumptions, though the complete report is to be published by UNESCO by late 2000.

TABLE 3.9   Two Different State Hydrological
Institute Year 2000 Water Withdrawal and
Consumption Projections

Region	2000 Water Withdrawals (km³/year)	2000 Water Consumption (km³/year)
Europe	444	109
Asia	3,140	2,020
Africa	314	211
North America	796	302
South America	216	116
Oceania and Australia	47	22
USSR	229	113
Total	5,186	2,893

Source: Shiklomanov and Markova 1987.

Note: Includes about 210 cubic kilometers in water losses from reservoirs.

TABLE 3.10 Shiklomanov Dynamics of Water Withdrawal and Consumption, by Continent (1900–2025) (km³/year)

Continent	Historical Estimates of Use								Forecasted Use		
	1900	1940	1950	1960	1970	1980	1990	1995	2000	2010	2025
Europe											
Withdrawal	37.5	71	93.8	185	294	445	491	511	534	578	619
Consumptive Use	17.6	29.8	38.4	53.9	81.8	158	183	187	191	202	217
North America											
Withdrawal	70	221	286	410	555	677	652	685	705	744	786
Consumptive Use	29.2	83.8	104	138	181	221	221	238	243	255	269
Africa											
Withdrawal	41.0	49.0	56.0	86.0	116	168	199	215	230	270	331
Consumptive Use	34.0	39.0	44.0	66.0	88.0	129	151	160	169	190	216
Asia											
Withdrawal	414	689	860	1,222	1,499	1,784	2,067	2,157	2,245	2,483	3,104
Consumptive Use	322	528	654	932	1,116	1,324	1,529	1,565	1,603	1,721	1,971
South America											
Withdrawal	15.2	27.7	59.4	68.5	85.2	111	152	166	180	213	257
Consumptive Use	11.3	20.6	41.7	44.4	57.8	71.0	91.4	97.7	104	112	122
Australia and Oceania											
Withdrawal	1.6	6.8	10.3	17.4	23.3	29.4	28.5	30.5	32.6	35.6	39.6
Consumptive Use	0.6	3.4	5.1	9.0	11.9	14.6	16.4	17.6	18.9	21	23.1
Total (rounded)											
Withdrawal	579	1,065	1,366	1,989	2,573	3,214	3,590	3,765	3,927	4,324	5,137
Consumptive Use	415	704	887	1,243	1,536	1,918	2,192	2,265	2,329	2,501	2,818

Source: Shiklomanov 1998.

Note: Includes about 270 cubic kilometers in water losses from reservoirs for 2025.

TABLE 3.11 Shiklomanov Dynamics of Water Withdrawal and Consumption, by Type of Use (1900–2025)

						Assessment						Forecast	
Sector	1900	1940	1950	1960	1970	1980	1990	1995	2000	2010	2025
Population											
(million people)			2,542	3,029	3,603	4,410	5,285	5,735	6,181	7,113	7,877
Irrigated land area											
(million hectares)	47.3	75.9	101	142	169	198	243	253	264	288	329
Agricultural use											
(km³/year)											
Withdrawal	525	897	1,122	1,544	1,821	2,179	2,408	2,488	2,560	2,737	3,097
Consumptive use	406	681	849	1,170	1,392	1,688	1,895	1,939	1,970	2,093	2,331
Industrial use											
(km³/year)											
Withdrawal	37.8	127	181	333	546	699	691	732	768	884	1,121
Consumptive use	3.36	9.49	14.4	23.8	36.9	59.0	73.7	79.4	84.6	103	133
Municipal use											
(km³/year)											
Withdrawal	16	36.8	53.1	83.5	130	207	322	357	389	468	649
Consumptive use	4.17	9.04	13.9	20.1	30.8	41.8	54.1	58.9	64.4	70.5	84.0
Reservoirs											
(km³/year)											
Consumptive use	0.3	3.7	10.1	29.2	76.2	129	169	188	210	235	270
Total (rounded)											
(km³/year)											
Withdrawal	579	1,065	1,366	1,989	2,573	3,214	3,590	3,765	3,927	4,324	5,137
Consumptive use	415	704	887	1,243	1,536	1,918	2,192	2,265	2,329	2,501	2,818

Source: Shiklomanov 1998.

Gleick (1997)

Using a disaggregated "end-use" approach instead of traditional supply/demand projections, this author developed a sustainable water "backcast" for the year 2025. Model assumptions included future water use by region and sector under a set of explicit sustainability criteria and limits. In a "Vision" scenario, total domestic water use in 2025 is estimated using two assumptions: (1) the world's entire population has access to a "basic water requirement" (from Gleick 1996) of 50 lpcd to meet basic needs; and (2) regions using more than that amount in 1990 implement water-efficiency improvements that reduce per capita domestic water use toward the level presently used in the more efficient nations of Western Europe—around 300 lpcd. The net result is that total domestic water needs in 2025 are not substantially different than current estimates—approximately 340 km^3 per year. The overall distribution of that water use is far more equitable than today's distribution.

Agricultural sector projections are based on specific human dietary needs in each region and the water requirements to produce calories of specific food types. Model assumptions include successful efforts to reduce per capita meat consumption in Europe and North America combined with overall increases in calorie consumption in the developing world. (See Chapter 4 for a discussion of the water required to grow food.) In the agricultural scenario, all regions are assumed to reach a minimum of 2,500 calories per person per day by the year 2025 and those regions currently consuming more than 3,000 calories per person per day experience dietary changes that reduce per capita daily consumption toward the 2,500-calorie level. Meeting these needs will require not just production goals, but policies for open food markets and transfers.

More dramatic, however, are the projected reductions in water needed to grow these diets. These reductions are the result of changes in the water-intensive components of diets, particularly meats. A North American diet that currently requires over 5,000 lpcd to grow today can be reduced to less than 3,500 lpcd by modestly reducing the meat component. This is still higher than any other diet but a considerable water savings nevertheless. Similar reductions are developed for each region. Additional assumptions are included about changes in irrigation efficiency, cropping intensities, and irrigated area.

Even with these assumptions, the Gleick Vision scenario projects that overall irrigation requirements would go up substantially between now and 2025. This seems an inevitable result of the anticipated increase in population. Nevertheless, the increases that occur could be far smaller than the increases anticipated by some of the conventional development scenarios described earlier. This approach produces a projection of agricultural irrigation water consumption of 2,930 km^3/year in 2025.

Gleick also argues that future industrial and commercial water demands could look significantly different than today's because of shifts in energy technologies, increases in water-use efficiency, and a change in industrial makeup. At the same time, an increased use of recycled water could further reduce total industrial withdrawals. As developing countries industrialize, they have the potential to "leapfrog" directly to more efficient technologies. The opportunity to bypass certain styles of development would permit many nations to move directly to industries and energy systems that are less consumptive of water. Both of these trends can also be seen in the scenario work done by Raskin et al. (1997).

TABLE 3.12   Gleick Year 2025
"Vision" Water Withdrawal Projections

Water Sector	Withdrawals (km³/yr)
Agriculture	2,930
Industrial	1,000
Domestic	340
Reservoir evaporation	225
Total	4,495

Source: Gleick 1997.

The 1997 Gleick study assumes that major reductions in industrial water demand in the industrialized nations can occur, driven by increases in the efficiency of water use and shifts in the structure of industry. At the same time, equity considerations argue for an increase in minimum industrial water use in developing countries. Urban water use in developing countries was assumed to increase to at least 100 cubic meters per person per year—a level described by Shuval (1994) as an appropriate minimum level for a moderately efficient industrialized nation. By 2025, in Gleick's Vision scenario, total industrial water withdrawals remain virtually the same as 1990 levels at around 1,000 km³/year but the per capita industrial water-use distribution is far more equitable than in the 1990s. Per capita industrial water use in almost all developed regions would drop—most dramatically in Europe and North America—and increase in Asia, Africa, and Latin America on both a per capita and an absolute level. Despite the improvements described in this projection, industrial water-use efficiency in most developed countries would still not reach the level of Japan's or California's in the early 1990s, as measured by both per capita industrial water use and industrial economic productivity per unit of water used. These measures suggest that even greater improvements could be achieved without imperiling a region's economic well-being.

Total global water withdrawals for 2025, presented in Table 3.12, are projected in this scenario to be approximately 4,500 km³ in 2025, one of the lowest total water-use levels of any global projection.

Raskin et al. (1997, 1998)

Paul Raskin and a group of researchers at the Stockholm Environment Institute/Tellus Institute in Boston developed a set of scenarios of future water use in an effort to explore future conditions (Raskin et al. 1995, 1997). Using a computer model tool developed for projecting resource demands under different socioeconomic conditions, they evaluated possible water withdrawals separately for the agricultural, industrial, and domestic sectors and for a variety of regions around the world. The principal drivers of these scenarios are demographic and macroeconomic projections, coupled with estimates of water "intensity," defined as water use per person or per unit of economic production. For some regions and sectors, scenarios were developed where water intensities are projected to decline, reflecting improving water-use efficiency. These intensities are then combined with future population and gross domestic product (GDP) estimates. The initial estimates were done for 2025. In the follow-up report (Raskin et al. 1998), projections are done for both 2025 and 2050 under a "Reference" (business as usual) scenario and a "Policy Reform" scenario designed to meet specific sustainability targets.

For the domestic sector, water intensity in North America is assumed to decrease toward the average in Organization for Economic Cooperation and Development (OECD) countries, while domestic intensities in developing countries increase toward the OECD average. Overall, this leads to substantial increases in total domestic withdrawals. Similar assumptions are made for the industrial sector, where water intensities in developed countries such as the United States are beginning to decline as water-intensive industries are replaced by low water-using industries and as industrial water-use efficiency improves. These trends are assumed to continue, but are swamped by major increases in total industrial water use in the developed countries, which is projected to rise dramatically because of growth in economic development and populations.

Agricultural water-use projections are done differently in the work by Raskin and colleagues than in most conventional scenarios, which simply make assumptions about land and water requirements per person or use broad assumptions about water needs per ton of agricultural product. The scenarios by Raskin et al. include more detail on irrigation water intensities, crop yields, cropping intensities, and trends toward improved irrigation efficiency (Leach 1995, Raskin et al. 1998). Combining their various regional assumptions with population projections, Raskin et al. (1997) show global irrigated land area growing at 0.3 percent annually between 1990 and 2025, an increase in irrigation efficiency of 8 percent over this period, and rising cropping intensities. Overall, the 1997 scenarios report that total freshwater withdrawals under a "mid-range" scenario will rise to 5,000 km³ by 2025 from their estimated base of 3,700 km³ in 1990. Their 2025 "low-range" and "high-range" scenarios for 2025 are 4,500 km³ and 5,500 km³, respectively. No water losses from reservoirs are included. The 1998 scenarios have lower total withdrawals, but start from a lower 1990 base water use, making straightforward comparisons with other studies difficult.

Alcamo et al. (1997)

The Center for Environmental Systems Research of the University of Kassel, Germany, developed a global water model, Water—Global Assessment and Prognosis (WaterGAP), and evaluated water use and availability for nearly the entire land surface of the world. Version 1.0 of this model, described in Alcamo et al. (1997), works on a watershed basis and takes into account socioeconomic factors that affect domestic, industrial, and agricultural water use, as well as physical and climatic factors that affect surface runoff and groundwater recharge (a new version will be available in 2000). Calculations are done on a grid cell scale of 0.5 degrees longitude and latitude and aggregated to the watershed and country scale. Three different scenarios are developed—low, medium, and high water-use cases—and this is one of the only efforts that include projections past 2050. Domestic water use in a country is estimated by multiplying population by an assumed per capita water use. Industrial water use is estimated by multiplying industrial GDP by water intensity (water use per unit GDP). Per capita use and water-intensity assumptions for the medium and low scenarios assume different levels of improvements in water-use efficiency as

TABLE 3.13   Alcamo et al. Global Projections for Water Use by Sector, 1995, 2025, and 2075 Medium Scenario (km³/yr)

	Total	Agricultural	Domestic	Industrial
1995 Estimate	3,046	2,022	296	728
2025 Medium estimate	4,580	1,724	621	2,235
2075 Medium estimate	9,496	1,826	1,290	6,380

Source: Alcamo et al. 1997.

Note: Excludes water losses from reservoirs

a function of income. Agricultural water use in WaterGAP is split into water requirements for livestock and water requirements for irrigation. Livestock water use is assumed to vary only with livestock population; water use per head is assumed to remain at the 1995 levels. Irrigation water estimates are developed by multiplying water use per hectare times estimates of irrigated area. Water use per hectare is assumed to be a function of climate, cropping intensity, and water-use efficiency. These factors are varied for the low, medium, and high scenarios. Table 3.13 shows the 1995, 2025, and 2075 global water use estimates for the medium scenario, along with a breakdown of water use by sector. Additional regional data are presented in the original source.

Seckler et al. (1998)

The International Water Management Institute (IWMI) released a study in mid-1998 assessing world water demand and supply to the year 2025 under different scenarios. IWMI created a simulation model based on a conceptual and methodological structure that mixes various strategies from earlier assessments. It includes a submodel of the irrigation sector that the authors describe as more thorough than any previously used (Seckler et al. 1998). Two alternative scenarios are developed, with different assumptions only about the productivity of agricultural water use: the first scenario is a "business as usual" (BaU) scenario; the second assumes a high degree of effectiveness in the use of irrigation water. While an updated study should be released in the near future, this review covers the 1998 version available at the time this chapter was prepared.

Projections are made for three sectors: agricultural irrigation, domestic, and industrial water use (see Table 3.14). Irrigation is a function of irrigated area, withdrawals of water per hectare of irrigated area, reference evapotranspiration rates for

TABLE 3.14   Seckler et al. Global Projections for Water Use by Sector: 1990 and 2025 BaU and High Irrigation Efficiency Scenarios (km³/yr)

	Total	Agricultural	Domestic and Industrial
1990 Estimate	2,907	2,084	823
2025 Business as usual	4,569	3,376	1,193
2025 High irrigation efficiency	3,625	2,432	1,193

Source: Seckler et al. 1998.

Note: These totals exclude water losses from reservoirs. Their 1990 starting point is lower than that in most studies.

different countries and seasons, and irrigation effectiveness. Two separate irrigation scenarios are developed. In both, per capita irrigated area is assumed to be the same in 2025 as it was in 1990. Thus differences in water use between the two scenarios depend exclusively on assumptions about the change in basin irrigation efficiencies. In the BaU scenario, irrigation effectiveness in 2025 is assumed to be the same as in 1990, so future irrigation withdrawals are determined purely by multiplying 1990 irrigation withdrawals by population increases. The second scenario assumes that most countries achieve an irrigation effectiveness of 70 percent, with some differences for particular countries.

Seckler et al. note the domestic water assumptions of Gleick's (1996) "basic water requirements" paper and double domestic water use in countries reported to be using less than 10 cubic meters per person annually. For countries using more than 10 cubic meters per person annually, Seckler et al. project 2025 demand on the basis of a relationship between per capita GDP and per capita withdrawals provided by Mark Rosegrant of the International Food Policy Research Institute, modified for some regional differences. Domestic and industrial uses are capped in countries with a high GDP at the 1990 level. Overall 2025 withdrawals increase 45 percent over 1990 values, a smaller increase than population growth because per capita increases in low water-using countries are offset by decreases in per capita use in high water-using countries. It is important to note that the base 1990 water-use estimate of Seckler et al. is far below that of other analysts, leading to a significant reduction in their estimate of future water use compared to other projections. They estimate 1990 water use at 2,900 km^3; Raskin et al. (1997) estimated 1990 water use at 3,700 km^3; Shiklomanov (1998) estimated it at 3,590 km^3.

Analysis and Conclusions

Many conventional development water scenarios have been prepared over the past quarter century. Looking at the major studies that have been done reveals two noteworthy trends. First, the earlier projections routinely, and significantly, overestimated future water demands because of their dependence on relatively straightforward extrapolation of existing trends. Second, the methods and tools used for forecasting and scenario analysis have been getting more and more sophisticated, permitting a wider range of exploratory scenarios and a better understanding of the driving factors behind changes in demands for water.

The projections just reviewed are presented graphically in Table 3.15 and Figure 3.1. This figure also shows actual (estimated) water withdrawals over time up to the year 2000. As these data show, the earlier projections greatly overestimated future water demands by assuming that use would continue to grow at, or even above, historical growth rates. Actual global withdrawals for the mid-1990s were in fact only about half of what they were expected to be thirty years ago. The reasons for this vary, ranging from the failure to keep up with basic water needs in many parts of the world to major improvements in the efficiency with which water is used in all sectors. Our difficulty in making accurate predictions highlights the importance of developing better methods for making projections of future needs.

Such methods are beginning to appear, as advances in computer modeling and speed develop and as better water-use data are collected and made available. Nev-

TABLE 3.15   Summary of Various Global Water Forecasts

Author	Publication Year	Forecast Year	Withdrawal (km³/yr)
Nikitopoulos	1967	2000	6,730
L'vovich "Rational Use"	1974	2000	6,325[a]
L'vovich "Conventional"	1974	2000	12,270[a]
Kalinin and Shiklomanov	1974	2000	5,970
Falkenmark and Lindh	1974a, b	2000	6,030
Falkenmark and Lindh	1974a, b	2000	8,380
Falkenmark and Lindh	1974a, b	2015	10,840
Falkenmark and Lindh	1974a, b	2015	7,885
De Mare	1976	2000	5,605[a]
Belyaev	1990	2000	4,350
World Resources Institute	1990	2000	4,660
Shiklomanov and Markova	1987	2000	4,976[a]
Shiklomanov	1998	2000	3,717[a]
Shiklomanov	1998	2010	4,089[a]
Shiklomanov	1998	2025	4,867[a]
Raskin et al. "Low"	1997	2025	4,500
Raskin et al. "Mid"	1997	2025	5,000
Raskin et al. "High"	1997	2025	5,500
Gleick "Sustainable Vision"	1997	2025	4,270[a]
Alcamo et al. "Medium 2025"	1997	2025	4,580
Alcamo et al. "Medium 2075"	1997	2075	9,496
Raskin et al. "Reference 2025"	1998	2025	5,044
Raskin et al. "Reference 2050"	1998	2050	6,081
Raskin et al. "Policy Reform 2025"	1998	2025	4,054
Raskin et al. "Policy Reform 2050"	1998	2050	3,899
Seckler et al. "Business as Usual"	1998	2025	4,569
Seckler et al. "High Irrigation Efficiency"	1998	2025	3,625

Notes:

1990 Withdrawals (estimated by Shiklomanov 1993), 4,130 km³.

1990 Withdrawals (estimated by Shiklomanov 1998), 3,590 km³.

[a]These studies included estimates for water lost from reservoir evaporation. In order to make more consistent comparisons here with those studies that failed to estimate reservoir evaporative losses, those estimates are subtracted from total withdrawals. The numbers here thus represent withdrawals *without* reservoir evaporation. See the text for full citations and for specific assumptions underlying each projection.

ertheless, past experience with water projections teaches us that they must always be considered as possibilities, not as predictions. Water planners should use them as tools for evaluating the risks and benefits of alternative water policies, rather than as straitjackets that limit our ability to respond to uncertainties and future surprises. As the proverb states: Seeing the future is good, but preparing for it is better.

REFERENCES

Alcamo, J.P. Döll, F. Kaspar, S. Siebert. 1997. "Global change and global scenarios of water use and availability: An application of WaterGAP 1.0." Wissenschaftliches Zentrum für Umweltsystemforschung, Universität Gesamthochschule Kassel, Germany.

Belyaev, A. V. 1990. Cited in *World Resources 1990–1991,* pp. 165–178. Washington, D.C., Oxford University Press, New York.

Bigwas, A.K. 1978. *United Nations Water Conference: Summary and Main Documents,* Vol. 2. Pergamon Press, Oxford, U.K.

California Department of Water Resources (CDWR). 1998. *The California Water Plan Update.* Bulletin 160-98, Sacramento, California.

De Mare, L. 1976. "Resources—Needs—Problems: An assessment of the world water situation by 2000." Institute of Technology/University of Lund, Sweden.

Falkenmark, M., and G. Lindh. 1974a. "How can we cope with the water resources situation by the year 2050?" *Ambio,* Vol. 3, Nos. 3–4, pp. 114–122.

Falkenmark, M., and G. Lindh. 1974b. "Impact of water resources on population." Swedish contribution to the UN World Population Conference, Bucharest.

Gleick, P.H. 1996. "Basic water requirements for human activities: Meeting basic needs." *Water International,* Vol. 21, pp. 83–92 (June).

Gleick, P.H. 1997. *Water 2050: Moving Toward a Sustainable Vision for the Earth's Fresh Water.* Working Paper of the Pacific Institute for Studies in Development, Environment, and Security, Oakland, California. Prepared for the Comprehensive Freshwater Assessment for the United Nations General Assembly and the Stockholm Environment Institute, Stockholm, Sweden (February).

Kalinin, G.P., and I.A. Shiklomanov. 1974. "USSR: World water balance and water resources of the earth," USSR National Committee for IHD, Leningrad (in Russian)

Leach, G. 1995. *Global Land and Food in the 21st Century: Trends and Issues for Sustainability.* Polestar Series Report No. 5. Stockholm Environment Institute. Boston, Massachusetts.

L'vovich, M.I. 1974. *World Water Resources and Their Future.* 1979 English translation, Raymond Nace (ed.). American Geophysical Union, Washington, D.C.

Nikitopoulos, B. 1962. "The influence of water on the distribution of the future earth's population." Athens Technological Organization/Center for Ekistics, RR-ACE: 125 (COF) (October 7).

Nikitopoulos, B. 1967. "The world water problem–water sources and water needs." Athens Technological Organization/Center for Ekistics, RR-ACE: 106 and 113 (COF).

Raskin, P., G. Gallopin, P. Gutman, A. Hammond, and R. Swart. 1998. *Bending the Curve: Toward Global Sustainability.* Polestar Series Report No. 8. Stockholm Environment Institute. Boston, Massachusetts.

Raskin, P., P. Gleick, P. Kirshen, G. Pontius, K. Strzepek. 1997. *Water Futures: Assessment of Long-Range Patterns and Problems.* Background Document for Chapter 3 of the *Comprehensive Assessment of the Freshwater Resources of the World.* Stockholm Environment Institute, Boston, Massachusetts.

Raskin, P., E. Hansen, R. Margolis. 1995. *Water and Sustainability: A Global Outlook.* Polestar Series Report No. 4. Stockholm Environment Institute, Boston, Massachusetts.

Schwartz, P. 1991. *The Art of the Long View.* Currency/Doubleday Press, New York

Seckler, D., U. Amarasinghe, D. Molden, R. de Silva, and R. Barker. 1998. "World water demand and supply, 1990 to 2025: Scenarios and issues." Research Report 19, International Water Management Institute, Colombo, Sri Lanka.

Shiklomanov, I.A. 1993. "World fresh water resources." In P.H. Gleick (ed.), *Water in Crisis: A Guide to the World's Fresh Water Resources.* Oxford University Press, New York, pp. 13–24.

Shiklomanov, I.A. 1998. "Assessment of water resources and water availability in the world." Report for the Comprehensive Assessment of the Freshwater Resources of the World, United Nations. Data archive on CD-ROM from the State Hydrological Institute, St. Petersburg, Russia.

Shiklomanov, I.A., and O.A. Markova. 1987. *Specific Water Availability and River Runoff Transfers in the World.* Gidrometeoizdat, Leningrad (in Russian).

Shuval, H. 1994. "Proposed principles and methodology for the equitable allocation of water resources shared by the Israelis, Palestinians, Jordanians, Lebanese, and Syrians." In J. Isaac and H. Shuval (eds.), *Water and Peace in the Middle East.* Elsevier, Amsterdam, pp. 481–495.

United Nations. 1974. *Concise Report on the World Population Situation in 1970–75 and Its Long Range Implications.* Dept. of Economic and Social Affairs. ST/ESA/SER.A/56, New York.

World Resources Institute. 1990. *World Resources 1990–1991.* Oxford University Press, New York.

Water for Food: How Much Will Be Needed?

One of the most vital questions facing society is whether or not we will be able to produce enough food for future populations. Reviewing the literature and science behind this problem, evaluating the resources that are available or may be needed, and predicting the behavior and actions of growers, governments, and markets addressing food needs, leads to one definitive answer: we do not know. Those who reach a firm conclusion that we either *will* or *will not* be able to feed future populations are unreasonably optimistic about their ability to predict the future.

Directly related to this question is a parallel question facing the world's water community: how much water will be needed to produce this food, and where will it come from? The answer to this is equally uncertain. We do not know how much water will be needed for future food production and whether it will be available given absolute constraints and competing human and ecosystem needs.

If we cannot know the answer to these questions with certainty, it is vital that our institutions and organizations at least work to understand the pieces of the puzzle and develop the flexibility to act as needed. This chapter explores several of these pieces in an effort to understand the implications of future food needs for our freshwater resources and agricultural water policies. We know the task of meeting agricultural water requirements will not be an easy one. The stakes are high, with direct implications for the quality of life and life itself for billions of people now alive and soon to be born.

Feeding the World Today

Great progress in feeding the earth's burgeoning population has been made in the past several decades. Between 1970 and 2000, the number of people on earth has grown by 2.3 billion, and average food production and distribution have more than kept pace. Most of the improvements in feeding people have occurred in Asia (particularly India) where the number of undernourished people dropped from 475 million in 1969–71 to 268 million in 1990–92 (Pandya-Lorch 1999). Much of this improvement has been the result of expansions in irrigated area coupled with the use of high-yielding varieties of crops. Yet while enough food is produced today to meet the basic needs of all of the world's 6 billion people, that food is not evenly distributed or consumed. The UN Food and Agriculture Organization (FAO) estimates that in the developing world as a whole, nearly 830 million people remained chronically undernourished in the mid-1990s. Sixty-two percent of these people were in East and South Asia—45 percent in China and India alone. Another 26 percent were in Sub-Saharan Africa, where the fraction of the population that is undernourished rose over the past 20 years from 38 to 43 percent and 215 million people fail to get an adequate number of calories (UNFAO 1996a, 1998). Thirty percent of the preschool children in this region are underweight; 60 percent are underweight in South Asia (Pandya-Lorch 1999).

As distressing as this picture is, even more disturbing trends are surfacing. The rapid growth in food production evident over the past several decades has begun to slow. Cereal production grew at a rate of 2.6 percent per year between 1967 and 1982. In the late 1980s and early 1990s, this rate dropped to around 1.3 percent per year. Per capita grain production is now actually decreasing, as population growth outstrips growth in grain production (Brown et al. 1995, Rosegrant et al. 1997). Per capita grain production in Africa is down 12 percent from 1981 (Kendall and Pimentel 1994). The rate of land degradation in some regions is accelerating. And competition for limited water supplies is growing, with more water moving from the agricultural sector to cities.

These data have raised some alarm in the global food community that food production may be unable to keep pace with future needs. Organizations like the International Food Policy Research Institute (IFPRI), a think tank devoted to evaluating these issues, and the FAO project continuing difficulties in meeting local or regional food needs (see, for example, Rosegrant et al. 1995, Pinstrup-Anderson et al. 1997). But enormous uncertainties complicate such projections.

On the water side of the equation, food is produced with natural precipitation, groundwater pumped from aquifers, and artificial irrigation using waters brought from a distance. In 1998, 18 percent of all cropland was irrigated, and these lands produced 40 percent of all food grown. Total irrigation water applied in 1999 was estimated to be around 2,500 km^3, 65 percent of the water withdrawn by humans for all purposes. Even more significant, agricultural irrigation accounts for 85 percent of all water *consumed* and thereby made unavailable for other uses (Shiklomanov 1998). Without this irrigation water, the food produced by natural precipitation would be insufficient to feed the world's current population. As it is, Postel (1998, 1999) suggests that nearly 500 million people live in countries with insufficient water to produce their own food requirements and that as much as a quarter of the entire international trade in food goes to meet these needs.

How much water is needed to grow food? The estimates of Shiklomanov just described, and others, offer only one part of the picture. But there are many missing parts. Official estimates of agricultural water use, for example, do not include the water used by rainfed agriculture, which provides 60 percent of all the world's food supply. No estimates adequately separate water withdrawn from water actually consumed. Inadequate information is available on the water content of food traded around the world. And differences in regional climatic conditions, diets, agricultural practices, and other factors all affect how much water is needed to produce food.

TABLE 4.1 Total Water Required to Produce
Regional Diets, Late 1980s

Region	Water to Produce Average Diet (liters/p/d)	Water to Produce Average Diet (m³/p/yr)
Africa, South of Sahara	1,760	640
Centrally Planned Asia	2,530	920
Eastern Europe	3,910	1,430
Former USSR	4,300	1,570
Latin America	2,810	1,030
Middle East/North Africa	2,940	1,070
OECD-Pacific/Oceania	3,310	1,210
South and East Asia	2,110	770
Western Europe	4,690	1,710
North America	5,020	1,830

Notes: Includes both rainfall and irrigation water. Assumes variations in regional irrigation efficiencies and diets. Data on diets come from UNFAO 1999. Water requirements per calorie developed from Gleick 1997. See Table 4.2 for typical regional diets.

In a first attempt to include many of these factors, Table 4.1 and Figure 4.1 present an estimate of the amount of total water required to produce the vastly different regional diets currently being consumed. These estimates were developed for this chapter assuming variations in regional diets, the amount of water required to grow different kinds of food, regional irrigation efficiencies, and rough estimates of the fraction of food lost between harvesting and the dinner plate. A current average North American diet requires over 1,800 m³/p/yr from both natural rainfall and irrigation; the diet of an average African living in the sub-Saharan region is produced with less than 650 m³/p/yr. Part of the difference comes from the much larger number of calories consumed by a North American; part of it comes from the large fraction of that diet that comes from water-intensive meat production. But regional climates, technology, and farming practices also have an effect on these numbers.

Feeding the World in the Future: Pieces of the Puzzle

The question of whether we will be able to produce enough food for the world's people is really many questions, some of which are addressed in the following sections. The answers are sometimes technical, sometimes economic, sometimes environmental, and often subjective. But thinking about these issues permits us to begin to craft solutions and to move toward meeting basic needs.

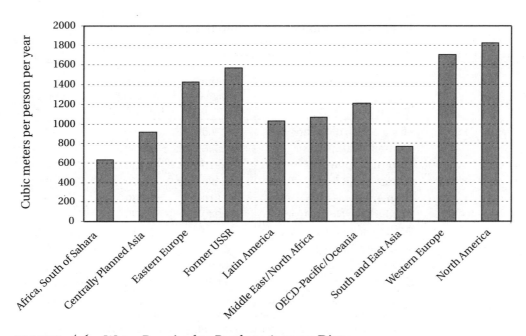

FIGURE 4.1   Water Required to Produce Average Diets

An estimate of the total amount of water required to produce the average diets consumed in various regions of the world is shown here. This figure includes both irrigation water and rainfall and is computed using variations in regional climate, irrigation efficiencies, and other factors.

How Many People Will There Be to Feed?

One of the most basic pieces of the puzzle about how much food will have to be produced in the future is the number of people that will have to be fed. Official and unofficial estimates of future populations abound and vary greatly. The best guess of the current population of the world is that it reached 6 billion people in late 1999, an increase of 4.4 billion in the twentieth century alone. By 2050, the UN's medium estimate projects that the world's population will exceed 8.9 billion, with a range of population of between 7.3 and 10.7 billion; a difference of more than 3 billion people (UN 1998). Table 4 (in the Data Section) and Figure 4.2 show the increase in global and regional populations from 1750 to 2050, using the UN future medium projections for 2050. This vast spread, even ignoring all the other uncertainties involved, greatly complicates both planning and estimating whether or not there will be a problem meeting future food needs. Fertility rates, as typically measured by the number of children per woman, have dropped more quickly in some regions than anticipated, particularly where education, family planning, and contraceptives have been made available. But in other regions, nearly unrestrained population growth continues. It thus appears likely that the world will have 3 billion more mouths to feed—and potentially as many as 4.7 billion more—by 2050. And 95 percent of the population increase will be in developing countries, where the greatest food and water problems already exist.

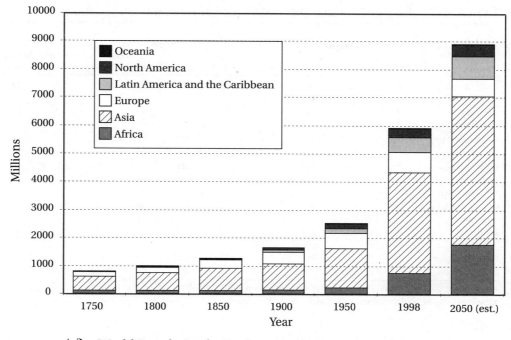

FIGURE 4.2 World Population by Region: 1750–2050

Estimated population, in millions, by region, from 1750 to the present, with the medium United Nations projections for 2050. *Source:* Data Section, Table 4.

How Much Food Will People Need and Want to Eat?

Almost as important as the number of people is how much food these people will need and desire. Needs and wants are not the same thing. Richer countries and individuals can, and do, eat considerably more than required to satisfy basic food needs. Here, however, a minimum amount of food can be—and has been—established. The UNFAO sets minimum requirements for a healthy and productive life at 2,200 to 2,300 calories per person per day. If the world population in 2050 is between 7.3 and 10.7 billion people, meeting a minimum caloric requirement of 2,200 calories per day will require the production of 16 to 24 trillion consumable calories per day worldwide. Adopting the UN mid-range population estimate of 8.9 billion people in 2050 means the world will have to produce nearly 20 trillion consumable calories per day. Current production is around 14 trillion calories.

There is already a significant difference between how much food is needed to survive and how much food richer, healthier populations demand. The UNFAO estimates that average caloric intake today varies among regions from 2,190 in poorer countries to 3,345 calories in the wealthier, industrialized nations (Table 4.2). Current average consumption of food products in the mid-1990s was around 2,670 calories per person per day. Food, however, is not evenly distributed or consumed. Even today, around 830 million people in 42 countries receive fewer than the minimum number of calories required for a healthy diet (Ghassemi et al. 1995, UNFAO 1996b, 1998, 1999). Of these countries 29 were in Africa and 6 were in Asia.

TABLE 4.2   Average Regional Diets, 1989 (Calories per person per day)

Region	Maize	Millet, Sorghum, Soybean	Wheat	Rice	Pulses	Roots	Vegetables	Fruits	Vegetable Oils	Animal Fat
Africa, South of Sahara	351	237	195	281	81	332	21	90	169	22
Centrally Planned Asia	118	43	380	1143	51	144	43	40	59	39
Eastern Europe	143	0	1095	32	39	100	74	74	269	274
Former USSR	3	16	1043	72	17	180	56	54	248	219
Latin America	274	13	438	256	69	124	30	133	220	63
Middle East/ North Africa	84	28	1103	203	69	68	70	137	339	47
OECD-Pacific/ Oceania	42	12	492	315	25	326	49	147	209	98
South and East Asia	87	41	315	1075	51	62	39	57	190	30
Western Europe	37	0	647	43	25	151	70	123	383	243
North America	27	2	537	112	28	81	105	146	364	129

Notes: These data were compiled from FAO regional data for the late 1980s. "All others" includes sugar, non-alcohol stimulants, spices, offals, treenuts, other meats, and other aquatic.

Table 4.2 shows the average estimated calorie intake for the major regions of the world. Tables 4.3 and 4.4 show the number of people estimated to be undernourished in the early and mid-1990s using a variety of measures. Table 4.4, in particular (from Kates 1996), expresses the complicated nature of food shortages and hunger.

What Kind of Food Will They Eat?

Preferences for food vary from region to region and culture to culture. Food analysts note that urban populations have different food preferences and that the rapid growth of urban centers coupled with rising incomes in both rural and urban areas will affect food demands. Urbanization especially leads to a shift to more diversified diets away from grain staples to processed foods, meats, milk, and fruits and vegetables (Pandya-Lorch 1999).

The calories needed or desired by people can be provided in many ways, with different combinations of livestock, crop, and fish products. Table 4.2 shows a breakdown of how today's diet is met in the major regions of the world. Even a cursory glance at these data reveals tremendous variations in preferences and food priorities.

A crucial factor in food preferences is the growing consumption of meat. About 40 percent of world grain production already goes to feed livestock. One estimate is that the grain fed to livestock in the United States is equivalent to the amount needed to feed 400 million people on a vegetarian diet (Kendall and Pimentel 1994). Producing that meat requires that significant amounts of certain crops be grown to feed animals rather than humans, reducing the overall number of people that the planet can support or the level of support they can receive. Table 4.2 and Figure 4.3 show the fraction of meat and vegetable calories consumed regionally. This ranges

Sheep	Beef	Pig	Poultry	Eggs	Milk	Fish	Alcohol	Sweet-ener	All Others	Total Calories/day	Percentage of Calories from Meat
14	36	25	15	6	67	25	51	145	25	2191	10
68	45	104	10	17	36	25	52	82	43	2541	15
20	79	202	47	50	250	15	165	381	37	3345	28
15	205	91	42	59	244	73	133	458	25	3253	30
11	89	50	65	20	156	30	78	401	34	2555	19
48	41	2	45	24	125	15	12	308	51	2819	13
57	90	96	64	28	118	71	80	329	43	2691	24
3	23	61	38	18	60	51	43	202	38	2485	12
30	116	307	47	49	342	54	189	413	80	3350	36
21	232	202	136	46	267	48	154	431	68	3133	35

TABLE 4.3   Number of People Undernourished, Millions

Region	1990–1992	1994–1996
Sub-Saharan Africa	196	210
Near East and North Africa	34	42
East and Southeast Asia	289	258
South Asia	237	254
Latin America and the Caribbean	64	63
Total	820	827

Source: UNFAO 1998: http://www.fao.org/News/FACTFILE/FF9808-E.HTM

from over 35 percent in Europe and North America to under 15 percent in many developing countries.

Meat consumption patterns also change over time. These changes will have a dramatic effect on overall grain requirements and, ultimately, overall water requirements. Figure 4.4 shows the trends in U.S. and global per capita meat consumption between 1960 and 1998. While per capita meat consumption in the United States has leveled off in recent years, it still remains far above the global average. What this figure doesn't show, however, is the change in type of meat consumption. In the United States, poultry consumption has more than doubled in the past 40 years and now exceeds the consumption of both beef and pork. Beef consumption has dropped more than 25 percent from its peak in the mid-1970s (see Figure 4.5). Of great importance to both future food needs and the water requirements to meet those needs is the trend toward greater per capita meat consumption in developing countries where incomes are rising rapidly (e.g., China) (Postel 1999).

TABLE 4.4   Recent Estimates of World Hunger

Dimension of Hunger	Population Affected (millions)	(percent)	Year	Source
Starvation				
Famine (population at risk)	15–35	0.3–0.7	1992	WHP
Related deaths per year	0.15–0.25	≤.01	1990s	RPG
Undernutrition (chronic and seasonal)				
Household	786	20	1988–90	FAO
Children	184	34	1990	ACC/SCN
Micronutrient Deficiencies				
Iron deficiency (women aged 15–49)	370	42	1980s	ACC/SCN
Iodine deficiency (goiter)	655	12	1980s	ACC/SCN
Vitamin A deficiency (children under 5 yrs of age)	14	3	1980s	ACC/SCN
Nutrient-Depleting Illness				
Diarrhea, measles, malaria (deaths of children under 5)	4.7	0.7	1990s	UNICEF
Parasites (infected population)[a]				
Roundworm	785–1,300	15–25	1980s	WB
Hookworm	700–900	13–17	1980s	WB
Whipworm	500–750	10–14	1980s	WB

Source: Original table from Kates 1996, with permission.

[a]Includes those people expected to have multiple infections.

Abbreviations: ACC/SCN (Advisory Committee on Coordination—Subcommittee on Nutrition of the United Nations); FAO (Food and Agriculture Organization of the United Nations); RPG (Refugee Policy Group); WHP (World Hunger Program, Brown University); WB (World Bank).

How Much Land Will Be Available—and Used—to Produce Food?

How much cropland is there today and what is the potential to expand that area? Globally, about 1,510 million hectares are under cultivation. Another 3,000 million hectares are categorized as pasture and rangeland (Scherr 1999, UNFAO 1999). In developing countries, approximately 885 million hectares are currently in crop production (UNFAO 1995). Worldwide, there are over 2,600 million hectares of land on which crops might achieve reasonable yields. If more land can easily be brought into production, it will be much easier to increase the total amount of food available. Overall, FAO estimates that expansion of arable land could account for 21 percent of the necessary increase in food production for future populations by 2010 (UNFAO 1995), though others are far less optimistic, and there are doubts about how productive these new lands might be (Carruthers 1993, Brown and Kane 1994, Kendall and Pimentel 1994). Much of the best land has already been brought into production, and much of the remaining uncultivated land is in remote areas, under ecologically important forest cover, or has marginal soils and farming conditions.

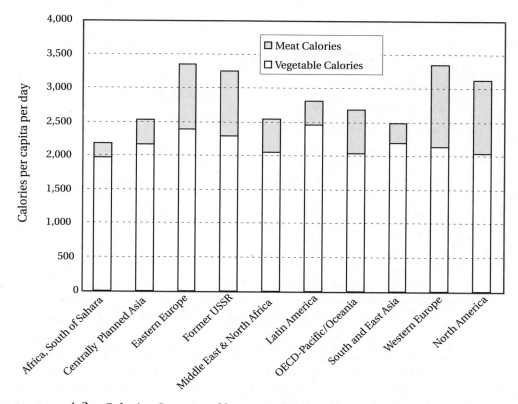

FIGURE 4.3 Calories Consumed by Region and Type

Meat and vegetable calories consumed by region. *Source:* Table 4.2.

Any significant expansion of cropland would thus come at the expense of lands that are often essential for other purposes.

The availability of cropland, and whether that land is actually used to grow food, are different questions. There is growing competition for land, particularly from urban development and industrial and commercial use. Loss of prime agricultural lands to nonagricultural purposes is happening in Indonesia, the United States, China, and elsewhere. Some efforts are underway to keep good agricultural lands in production through development easements, zoning rules, and tax breaks, but pressures are growing to build housing and industry on croplands around the world.

The growth of area under crops is not presently keeping up with population growth. Total cropland per capita dropped from 0.31 hectares per person in 1983 to 0.26 hectares per person just a decade later, to under 0.25 hectares per person in 2000. The global average area of grain harvested per person was 0.22 hectares per person in 1950 and 0.16 hectares in 1981, and shrank to 0.12 hectares in 2000 (USDA 1996, Gardner 1997).

What Quality Will That Land Be?

Soil is a renewable resource, but only on a very long time scale. Soils take hundreds or thousands of years to create through natural processes of erosion and organic decay and accumulation. Soil quality and the potential to produce crops vary

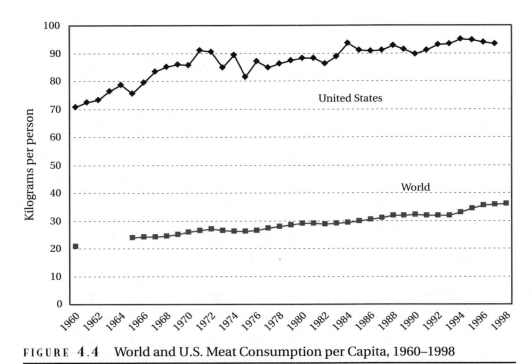

FIGURE 4.4   World and U.S. Meat Consumption per Capita, 1960–1998

The per-capita production of meat, globally, from 1960 to 1998, compared to per-capita consumption figures for the United States. *Sources:* UN Food and Agricultural Organization 1999, U.S. Department of Agriculture 1998.

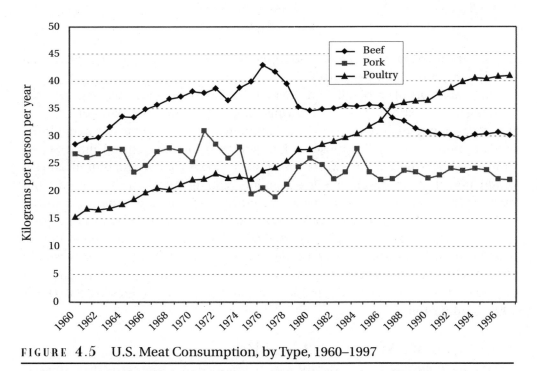

FIGURE 4.5   U.S. Meat Consumption, by Type, 1960–1997

Trends in U.S. consumption of meat by type, ready-to-cook weight. Note the great increase in poultry consumption during this period and the decreases in consumption of pork and beef. *Source:* http://www.tcfa.org/cnsmptn.html

enormously from location to location and from soil type to soil type. Top-quality lands are typically brought into production earlier because of their higher economic potential to produce food and revenue. As more and more land has been brought under production, the average quality of land has decreased, reducing potential productivity per hectare.

Agricultural activities also affect the productivity of soils over time. Growing crops depletes soil fertility by consuming nutrients, which can then reduce crop yields. Irrigation can both add and remove salts depending on practices and local conditions. Poor management practices lead to erosion, loss of soil cover, and soil compaction. Without proper management and the constant input of nutrients and energy in the form of fertilizers, pesticides, and irrigation, crop production drops over time. While these kinds of impacts have occurred since ancient times, there is little agreement about the extent to which land degradation is now occurring or how that degradation will affect future productivity. Some researchers argue that degradation rates have been overestimated and are relatively unimportant to overall crop production (Crosson 1995). Others argue that loss of soil fertility poses a serious future threat and that it has already contributed to a slowdown in the growth of agricultural productivity (Brown and Kane 1994, Pingali and Heisey 1996).

There is no doubt that soil quality is diminishing in many areas around the world. Soil quality on one quarter of the world's soils has experienced some degradation, and the pace of degradation has accelerated over the past 50 years (Scherr 1999). Annual crops tend to degrade soils more than perennial crops, and common property lands generally suffer more degradation than privately managed lands. Because the poor are particularly dependent on annual crops, poor soil quality tends to affect poorer people more. One estimate is that about 75 percent of agricultural lands in Central America have suffered major degradation, 20 percent in Africa, and 11 percent in Asia (Scherr 1999) with higher levels in particular countries. Several different estimates of soil degradation are presented in the Data Section (Table 12).

One of the most critical soil-quality problems—particularly related to water—is the increase in concentration of total dissolved solids, commonly referred to as salinization. Natural processes or human activities can salinize lands. In extreme cases, irrigation water can contain as much as one to 3.5 tons of salt per 1,000 m^3. Since crops can need 6,000 to 10,000 m^3 of water per hectare, land can receive tens of tons of salt per hectare (Crosson and Rosenberg 1989). As this water evaporates, it leaves behind the salts. Excessive waterlogging of soils can also salinize lands by bringing salts to the surface and concentrating them. Tables 13 and 14 in the Data Secton describe the extent of salinization worldwide.

Human-induced salinization has a long history. In the later part of the fourth millennium B.C. in the Tigris and Euphrates river basins, early cities flourished based on irrigated agriculture growing primarily wheat. Over centuries, dissolved salts in the irrigation water increased soil salinity. By around 3500 B.C., wheat was increasingly being replaced with more salt-tolerant barley. One thousand years later, wheat accounted for only one sixth of the total production of wheat and barley. By 1700 B.C., the cultivation of wheat had been abandoned—a decline attributed to decreasing soil fertility from salinization (Ghassemi et al. 1995). According to

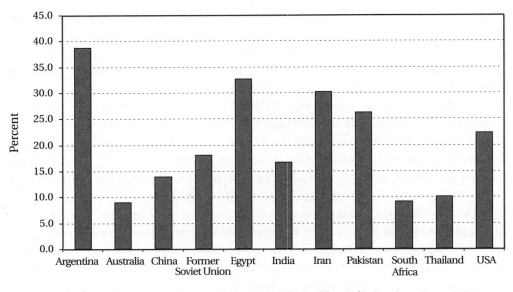

FIGURE 4.6   Percentage of Irrigated Land Affected by Salinization, Late 1980s

Source: Ghassemi et al. 1995.

FAO, 2 million hectares of land per year are currently lost to production due to salinization, while more land suffers from reduced yields (UNFAO 1995). Figure 4.6 shows the percent of irrigated land estimated to be affected by salinization for 11 of the world's most important food growing countries.

Another serious soil problem is erosion—the loss of soils from water and wind action. Soil erosion occurs on agricultural land without vegetative cover to protect it or because of certain agricultural practices related to how land is prepared and managed. One estimate is that topsoil on cultivated land is being lost 16 to 300 times faster than it is being replaced (Barrow 1991). A shift away from traditional agricultural practices in Africa and elsewhere is leading to greater pressures on soils there and to an increase in soil erosion rates. Studies in the United States suggest that the loss of just 2.5 centimeters of topsoil reduces corn and wheat yields by 6 percent (Brown and Young 1990).

What Will Crop Yields Be?

Crop yields are usually measured in a weight or volume of food produced per unit cropland. This is one of the most sensitive variables in any future food projections, and widely varying assumptions lead to widely varying conclusions. In a 1995 study of food production to the year 2010, 66 percent of the growth in food production in developing countries is projected to come from future increases in crop yields (UNFAO 1995). Improvements of this magnitude have been accomplished in the past. Between 1960 and 1990, yields of rice, wheat, and maize doubled, with greater increases in Asia and Latin America, and smaller increases in Africa. In Asia, wheat yields increased during this period from 0.7 to 2.6 metric tons per hectare (t/ha). In China, yields of rice—the most important single crop in the region—rose to 5.7 t/ha from 2.3 t/ha (WRI 1996). There is also optimism about the potential for further improvements because of large differences from region to region. Current yields of rice, for example, vary enormously, from a country average of below 4 t/ha to nearly 7 t/ha. Some yields as high as 10 t/ha have been achieved (UNFAO 1995). These differences lead some analysts to project great further improvements as the average moves up toward the current best levels. Others note, however, that the current "best" yields simply cannot be attained everywhere, because of climatic, soil, and related agronomic factors (Postel 1999, personal communication).

There is also no necessary connection between past trends and future increases in yield, and the rates of increase for yields are now slowing. Table 4.5 and Figure 4.7 show the slowing in the growth of grain yields for wheat, rice, and maize from 1951 to 1995. Yields for wheat actually decreased between 1990 and 1995, while maize yield increases dropped almost to zero. A large part of this decrease resulted from dramatic changes in production in the former Soviet Union and some bad harvests in the United States during the early 1990s. But even correcting for the strong influence of these kinds of short-term events, global yield growth is slowing (Ingco et al. 1996). And these regional fluctuations demonstrate how vulnerable food security can be.

Total yield of crops is only one measure of productivity. Another is yield per person. In order to keep up with growing populations, the increase in yields must grow as fast as population. Recent evidence, however, shows that per capita increases in certain key crop yields are slowing. World grain production in 1950 was less than 250 kg per person. Over the next thirty-five years grain production expanded far faster than population, and by 1984 global production was 346 kg per person. Increases in production then slowed: by 1990 production was 336 kg per person and it dropped even further by 1996 to only 313 kg per person (Brown 1997).

In Asia, where fertilizer, pesticide, and irrigation use are high, farmers are approaching maximum yields expected from current crop types. Some argue that further dramatic increases are unlikely, given the wide penetration of the highest-yielding varieties already and the stagnant yields in crops like rice in experimental research stations (Brown and Kane 1994). Advances in biotechnology and crop genetics may help reverse this trend, but again, there is controversy over this option. Many plant breeders believe that significant yield increases can still be achieved. Wheat breeders at the International Maize and Wheat Improvement Center, for example, have expressed confidence that "the genetic yield potential is not near any definite limit" (van Ginkel 1996). But additional improvements, even in laboratory situations, are proving hard to obtain. Research by the International Rice Research Institute (IRRI) and others has found significant slowing in the rate of yield increases of wheat and rice under experimental conditions (McCalla 1994).

TABLE 4.5   Percentage Changes in Average Annual Grain Yields, 1951–1995

Grain	1951–1960	1960–1970	1970–1980	1980–1990	1990–1995
Wheat	1.84	3.06	1.99	2.89	−0.64
Rice	1.27	2.40	1.63	2.34	0.93
Maize	2.74	2.48	2.84	1.01	0.37

Source: UNFAO data for 1951–1990; USDA data for 1990–1995. From Ingco et al. 1996.

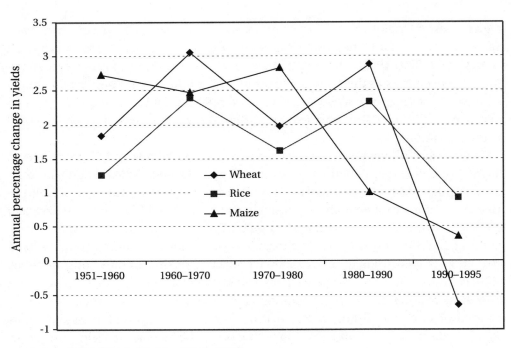

FIGURE 4.7 Grain Yields, 1951–1995

Annual average percentage changes in grain yields for wheat, rice, and maize, over each five-year period. *Source:* Table 4.5.

How Many Crops Can Be Produced on the Land?

This variable, often called cropping intensity, is a measure of how many times in a year a crop can be grown. Some crops are annuals—being planted and harvested once a year. Some crops can be grown and harvested more than once a year, multiplying the productivity of each hectare of land. Thus, if some fraction of agricultural area is multiple-cropped, the estimate for total acreage worldwide in production can be misleading. There has been a trend toward higher cropping intensity lately as new, faster-growing varieties have become available. For example, a rice variety produced by the IRRI, called IR-8, ripens 20 to 30 days faster than traditional varieties (WRI 1998). FAO estimates that 13 percent of the future growth in food production will come from multiple cropping or fewer fallow periods (UNFAO 1995). Higher cropping intensity, however, puts additional pressure on soils and water and can accelerate soil-quality degradation. Multiple cropping typically requires irrigation and the ability to store water in wet seasons for use in the drier periods.

What Fraction of Crop Production Is Actually Eaten by Humans?

Most projections or predictions of future food production stop at how much food can be grown in the field. But this is only a piece of the puzzle. Not all crops grown can be successfully harvested. Severe weather can wipe out a crop or reduce its value. There are inefficiencies in harvesting that leave part of every crop in the field. Insects, diseases, and weeds destroy around a third of all crop production. For example, as much as 24 percent of grains harvested in Kenya are subsequently lost

TABLE 4.6 Post-Harvest Losses
from Rice Production

Category	Percentage of Rice Lost
Harvest	1 to 3
Handling	2 to 7
Threshing	2 to 6
Drying	1 to 5
Storage	2 to 6
Transport	2 to 10
Total Range	10 to 37

Source: http://www.fao.org/News/FACTFILE/
FF9712-E.HTM

Notes: In some regions of Africa and Latin
America as much as 50 percent of cereal and
legume harvests are lost. Losses of 10 to 15 percent are considered "common."

to rodents, insects, fungi, and other pests (Berck and Bigman 1993). After harvesting, as much as 20 percent of crop production in developing countries may be lost
during distribution and marketing (WRI 1996). Altogether, estimates of total
postharvest rice losses vary from 10 percent to as much as 37 percent (Table 4.6),
while losses from grain and legume harvesting as high as 50 percent have been
reported (UNFAO 1997a).

Relatively little work has been done on estimating possible improvements in harvesting technology or on evaluating other methods of reducing crop losses. The
application of biotechnology may have an important effect on reducing these
losses. While the science of genetic manipulation of food crops is at an early stage,
and some have grave doubts about the relative risks and benefits, there is a possibility that new crop varieties with high yields of product or resistance to pests and
disease will greatly help in meeting future needs.

Not all food harvested by a farmer is actually eaten. For some food types, a significant fraction cannot ever be eaten, such as seeds, skin, and other inedible
portions. For many crops, this fraction can be large. For example, half of the
weight of some fruits is regularly discarded. Even large amounts of food that
could be eaten may end up on the floor of a food processing plant or as waste in
some other process. Some fraction of food purchased is lost during preparation;
some additional fraction is lost during serving. This "waste" of food is extremely
hard to estimate and varies enormously from society to society. A 1997 study of
some of these losses estimated that 27 percent of all edible food available for
human consumption in the United States was lost at the retail, consumer,
and food service levels (http://www.econ.ag.gov/epubs/pdf/foodrevw/Jan97/
index.htm).

Finally, some purchased food goes to feed pets. While in most societies this quantity is small, and is often composed of food that humans would not eat, some fraction of high-quality food goes to meet this need, directly competing with human
requirements.

How Much Water Will Be Needed to Grow Food?

The preceding discussion has so far only lightly touched on our central question of water. Many of the factors raised above directly or indirectly affect how much fresh water will be needed by agriculture, where that water will be needed, and what quality it will have to be. This question, too, can be broken into several intricately related issues.

How Much Water Is Necessary to Grow Different Crops?

Different crops require different amounts of water. Crops transpire amounts of water that depend on their physiological requirements, climate, soil conditions, and location. The total amount of water required to produce food depends on the kinds of crops that will be grown, their water requirements, and agricultural practices. Growing a ton of corn may require between 1,000 and 2,000 tons of water; a ton of rice may require 2,000 to 5,000 thousand tons of water; a ton of cotton may require as much as 15,000 tons of water (Pimental et al. 1997). Table 4.7 lists approximate water requirements to produce a kilogram of various important crops. Actual values vary with a wide range of factors.

There is potential to increase crop yields as a function of water inputs. For example, water-use efficiency of rice production can be increased by controlling losses in rice fields, seeding rice directly into fields, and exploring hybrid crop types. Some advanced rice types can reduce water needs by 20 to 30 percent. When combined with new irrigation or seeding methods, the overall amount of water required to

TABLE 4.7   Approximate Crop Water Requirements to Produce Food Harvested

Crop/Food	Water Requirement (kilograms of water per kilogram of food produced)
Potato	500 to 1,500
Wheat	900 to 2,000
Alfalfa	900 to 2,000
Sorghum	1,100 to 1,800
Corn/Maize	1,000 to 1,800
Rice	1,900 to 5,000
Soybeans	1,100 to 2,000
Chicken	3,500 to 5,700
Beef	15,000 to 70,000

Sources: Pimentel et al. 1997; Tuong and Bhuiyan 1994; UNFAO 1999, author's calculations.

Notes: These are approximate values, and they vary significantly by region, climate, irrigation methods, and other factors.

grow rice can be reduced by as much as 50 percent (Tuong and Bhuiyan 1994). Rice is still produced with rainfall on around 40 million hectares. Yields on these lands remain low and irregular compared with irrigated lands, but even rainfed rice could be far more productive. In water-short regions, improvements in such rainfed production would be extremely valuable (Tuong and Bhuiyan 1994).

One of the fastest-growing food-production systems in the world is aquaculture—the raising of fish for food. While aquaculture currently only provides about 20 percent of total fish production (UNFAO 1997b), that share has risen from 13 percent in 1990 and continues to rise rapidly. Global food production from aquaculture rose on average more than 11 percent annually between 1990 and 1995 (Ahmed 1997). Moreover, fish captures from most ocean stocks are at, or above, sustainable limits, putting more pressure on aquaculture to meet growing demands. The per capita fish catch from marine and inland waters has not increased since the 1980s, hovering around 16 kilograms per person, down from the peak of over 17.4 kilograms per person in 1988 (McGinn 1999).

Substantial amounts of water are required for raising and processing fish artificially. Competition for water between aquaculturalists raising shrimp and growers raising rice in Asia has already been reported. Water diversions for shrimp ponds in Thailand have lowered groundwater levels (WRI 1998). Between 1987 and 1993, Thailand lost more than 17 percent of its ecologically valuable mangrove forests to shrimp ponds and these ponds contribute large quantities of waste to rivers and coastal waters. Not all aquaculture must compete with irrigation. Some farmers are experimenting with innovative uses of reservoirs and canals to grow fish, and some are even combining rice and fish production (Meinzen-Dick 1999).

Livestock also require water, sometimes substantial amounts. Tables 4.8 and 4.9 show two different estimates for the water used by livestock in developed and developing countries. One of the major uncertainties in predicting future water needs for food production, as discussed earlier, is the question of meat consumption. Table 4.7 shows that producing a kilogram (or a calorie) of meat requires far more water than producing the same amount of calories or weight of vegetable or

TABLE 4.8 Water Use by Livestock, Typical Developed Country Estimates

Livestock	Liters per Head per Day
Milk Cows	154
Steers	51
Bulls	97
Calves	25
Other Cattle	64
Pigs	9
Sheep	3
Horses	68
Hens/Chickens	0.3
Other Poultry	0.5

Source: Environment Canada 1986.

TABLE 4.9   Water Use by
Livestock, Typical Developing
Country Estimates

Livestock	Liters per Head per Day
Sows	80 to 90
Boars	16 to 18
Hogs	30 to 33
Water Buffalo	60 to 66
Mules, Horses	40 to 45
Poultry	0.2
Sheep, Goats	5
Oxen	16 to 18

Source: World Bank 1987.

grain crops. Table 4.1 indicates that the water required to grow an average North American diet exceeds 5,000 lpcd, while the water required to produce an average diet in the sub-Saharan region of Africa is under 2,000 lpcd. The largest difference between these two numbers is the much larger amount of meat eaten in North America. As (and if) meat demands grow, water demands will also grow, decreasing the overall water productivity of agriculture, that is, the amount of food produced per unit water.

How Will These Lands Be Watered?

Because evapotranspiration requirements vary with location and climate, future estimates of water needs depend on assumptions about where food production will take place. Crops grown in a warmer climate typically require more water per unit yield than the same crops grown in a temperate climate. Crops receive water from both natural precipitation and human-built systems.

Worldwide in 1997 an estimated 267 million hectares of land were irrigated, out of a total of around 1,500 million hectares of cropland—around 18 percent. This is nearly a doubling from 138 million hectares in 1960 (UNFAO 1999) (see Figure 4.8 and Tables 10 and 11 in the Data Section). This irrigated land produced an estimated 40 percent of total crop production. Because irrigated land is typically so much more productive than unirrigated land, there is a strong desire to increase irrigated acreage as a major factor in increasing food production. There are great disparities in the distribution of irrigated lands. Around 60 percent of the world's irrigated land is in Asia, while Africa and South America together have less than 10 percent.

In the 1970s, the area of irrigated land expanded faster than 2 percent per year. This rate slowed to around 1.8 percent annually in the 1980s and has now fallen to around 1.4 percent per year. FAO estimates that the rate of expansion will continue to drop to less than 1 percent per year in the next decade (UNFAO 1995, 1999). It is proving increasingly difficult to expand irrigated acreage because of the rapidly rising costs of large-scale irrigation systems, the increasingly difficult sites that have to be developed, the depletion of aquifers, the marginality of the lands remaining to be brought into production, and growing conflicts over water priorities and needs.

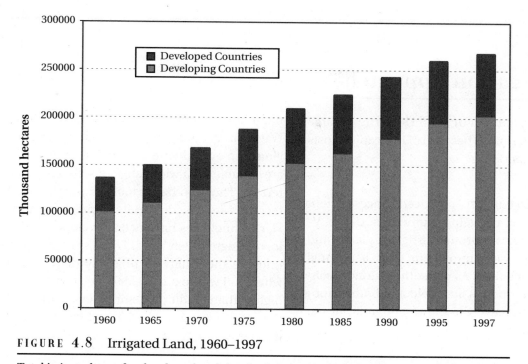

FIGURE 4.8   Irrigated Land, 1960–1997

Total irrigated area for developed and developing countries, from 1960 to 1997, the last year for which reliable data are available. *Source:* UNFAO 1999.

How Will Crops Be Irrigated? Can Irrigation Efficiency Improve?

Growing attention is being paid to the consumptive use of agricultural water, compared to both total withdrawals of water for agriculture and the consumptive use of other water-using sectors of society. Without a doubt, one of the greatest concerns of water analysts is the very high apparent consumption of water in agriculture, which makes it unusable by others. This can be contrasted with the very high withdrawal but very low consumptive use of water by, for example, electric power plants for cooling. Water in such cases typically becomes available for reuse.

There are many forms of irrigation and types of irrigation technology, ranging from poorly controlled flood irrigation to computer-controlled drip irrigation. Box 4.1 describes typical irrigation methods. Different crops have different irrigation requirements. Different irrigation technologies have different irrigation efficiencies and vastly different economic costs. The efficiency with which irrigation water is used varies by region, crop, agricultural practice, and technology. In many places less than half of all water applied to a field is used productively by the crop. The rest is lost to unproductive evaporation or to groundwater, where it may or may not be recovered for other uses.

In recent years there has been growing interest in improving our understanding of the concepts of efficiency, water loss, and agricultural water use. Seckler (1996) and others have pointed out that there is an important difference between on-farm efficiencies and overall system efficiencies in agriculture. While one farm may only effectively use 40 to 50 percent of the water it withdraws, the remaining water might be used downstream by the next farmer or it might seep into groundwater pumped

BOX 4.1
Irrigation Approaches

Surface irrigation or field flooding of large areas is widely used because of low capital costs and long experience.

Furrow irrigation, practiced since ancient times, involves digging numerous U- or V-shaped furrows through irrigated land and introducing water into them from a channel at the top of a field. As with other surface techniques, water ponds on a field and both evaporates unproductively to the air and infiltrates to groundwater.

Border irrigation involves flooding land in long parallel strips separated by earth banks built lengthwise in the direction of the slope of the land. Water flows from the highest point in the field to the lowest.

Basin irrigation is similar to border irrigation but includes earth banks constructed crosswise to those used for border irrigation, dividing a field into a series of basins that can be separately irrigated.

Sprinkler systems vary in type, capabilities, and cost. Typical systems include water pumps, main lines, smaller lines, and sprinkler heads. Water sprays over crops or land surfaces providing an even distribution of water. Sprinklers are suitable for a wide range of soils, crops, and terrain. They can be portable or permanent. In recent years, a wide variety of precision sprinkler technologies have been developed and are increasingly widely used. These systems greatly improve the ability to precisely deliver water when and where it is needed, at a cost typically higher than traditional sprinklers. Some fraction of the water applied is still lost to unproductive evaporation.

Drip irrigation involves the use of low-volume water emitters placed precisely where crops need water. Components of drip systems include pumps, filters, main lines, lateral lines, and emitters or drip heads. Chemicals can often be added to the water and precisely applied at the same time as the water. When soil moisture is carefully monitored, applications of water can be very carefully controlled. Drip systems are suitable for permanent crops, such as vineyards and orchards, and increasingly are being applied to row crops. In California, farmers are even experimenting with using drip systems on crops such as cotton.

by a neighbor, making the apparent system efficiency much higher. This distinction between field efficiency and basin-scale efficiency is important, and under certain circumstances it is very useful. In recent years it has been applied in Asia and Africa. Among other things, this distinction can help identify where improvements in water-use efficiency may be most appropriate and valuable (Keller and Keller 1995, Seckler 1996, Molden 1997).

While more water is applied to fields, almost universally, than is necessary, not all excess water can be saved. Some serves the valuable purpose of leaching salts from the soils. Depending on the basin, a significant fraction may already be committed to other users downstream. Saving that water and reallocating it to other needs may thus deprive a current user. These issues have led to an ongoing discussion of whether efficiency improvements can produce any water that can be reallocated to other users for other purposes. For example, some California water planners dismiss the potential of water-use efficiency improvements because, they argue, such improvements will not produce much "real" water. Their use of the term "real" water, however, is misleading. All water from conservation activities is "real." The appropriate goal is to evaluate the potential for water that can be reallocated to other users and for water conservation savings that can reduce future increases in demand and provide other benefits.

Two additional critical observations are required. First of all, a distinction must be made between regions with fixed demand and regions with growing demand. Where demand is constantly growing, what is needed is either "new" water (i.e., water that can be reallocated to other uses or users) or real reductions in demand. Efficiency improvements in inland areas may not generate "new" water, but they produce real one-for-one reductions in anticipated demand for water. Efficiency improvements that reduce the need for applied water also produce other benefits as well, including improvements in human health, more reliable instream flows, ecosystem and habitat restoration, reductions in the cost of treating drinking water, less environmental contamination by agricultural chemicals, and reductions in the economic costs of multiple unnecessary withdrawals of water.

Secondly, agricultural water-use efficiency improvements can always generate some "new" water. One problem with past analyses is that evapotranspiration (ET) and other losses are too often lumped into a simple fixed depletion value when, in fact, they are made up of several distinct processes. Molden and others distinguish these types of depletions based on their beneficial use (Molden 1997). Process depletion is defined as that amount of water diverted and depleted to produce an intended good, such as water transpired by crops and incorporated into the plant tissue. Non–process depletion is when water is depleted, but not by the process that it was intended for, such as evaporation from soil and free water surfaces and evaporation of spray drift. In other words, evaporation and transpiration are distinguishable, and separating ET into its component parts would shift the focus to beneficial uses of irrigation water by allowing the user to manipulate nonbeneficial losses.

Both unproductive evaporation and plant transpiration can be reduced with different technologies, water policies, and agricultural practices. Reductions in unproductive evaporation can be achieved by reducing surface water exposure, evaporation from soils, and misapplication of irrigation water. Reductions in transpiration can result from changes in crop types, the introduction of more water-efficient varieties, land fallowing, and land retirement. Far more attention to these issues is needed and can have a major effect on agricultural water requirements.

Improved water-use efficiency in agriculture can result in more food production without increasing overall water demands. Such gains require education, investment, and better crop and water management. Table 4.10 shows sample gains that

TABLE 4.10 Examples of Productivity Gains
from Shifting to Drip from Conventional Surface
Irrigation, India, Mid-1990s

Crop	Change in Crop Yield (percent)	Change in Water Use (percent)	Change in Overall Productivity[a] (percent)
Banana	52	−45	173
Cabbage	2	−60	150
Cotton	27	−53	169
Cotton	25	−60	255
Grapes	23	−48	134
Potato	46	0	46
Sugarcane	6	−60	163
Sugarcane	20	−30	70
Sugarcane	29	−47	91
Sugarcane	33	−65	205
Sweet potato	39	−60	243
Tomato	5	−27	49
Tomato	50	−39	145

Source: Data from various Indian field sources, from Postel 1999.

[a] Measured as crop yield per unit water applied.

have been achieved in India by converting to drip systems. These data show improvements in both yields and reduction in water use, leading to large—sometimes huge—gains in overall productivity. Similar kinds of gains have been seen wherever more efficient irrigation technology has been applied. Table 4.11 shows the differences in labor requirements for alternative irrigation technologies, suggesting that more efficient technologies can also reduce labor costs.

In the developed world there has been a slow shift toward more efficient irrigation technology as information has become available, as technology has improved, and as economic and policy incentives for change have intensified. Yet enormous potential still remains. In Australia, for example, furrow and flood irrigation was still used on 74 percent of agricultural lands in 1990, while drip was used on only 6 percent (Table 4.12 and Figure 4.9). In California, drip irrigation was used on an estimated 13 percent of all irrigated lands in 1991, but 30 percent of orchards and 45 percent of vineyards were still using flood irrigation, which has relatively high unproductive evaporative losses (Snyder et al. 1996; Gleick 1999).

While the trend toward more efficient irrigation technology is likely to continue, a gap is likely to remain between the theoretical maximum efficiency and actual field implementation. All too often, even sophisticated water planners ignore these factors and the potential for reducing these losses.

Improving modern irrigation is only one way to get water to plants. Most of the world's poorer people do not have access to expensive irrigation technology and will not benefit from centralized large-scale irrigation projects. Traditional water harvesting techniques often play an important role for these populations. UNDP

TABLE 4.11 Labor Requirements for Different Irrigation Technologies

Irrigation Method	Type	Labor Requirements (hours per irrigation application per hectare)
Surface	Furrow	1–3
	Border	0.5–1.5
	Basin	0.1–1
Sprinkler	Hand-moved laterals	1–2.5
	Tractor-moved laterals	0.5–1
	Self-moved systems	0.05–0.3
	Permanent systems	0.05–0.2
Drip	Orchard	0.1–0.3
	Row crops	0.1–0.3

Source: Ghassemi et al. 1995.

TABLE 4.12 Australian Irrigation Method by State, 1990

State	Percentage of Irrigated Area		
	Furrow/ Flood	Sprinkler	Drip/ Other
New South Wales	84	13	3
Victoria	90	8	2
Queensland	46	43	11
South Australia	33	51	16
Western Australia	48	26	26
Tasmania	11	77	12
Northern Territory	18	28	54
Australia Total	74	20	6

Source: AATSE 1999.

Note: Some data rounded.

(1994), NRC (1996), Postel (1999), and others describe these in detail, along with information on their potential for boosting food supplies in precisely those regions where water may be a limiting factor in production.

Where seasonal precipitation or runoff permits, such as in river valleys with cultivated wetlands, valley bottoms, river deltas, and lands that are periodically flooded, no expensive irrigation technologies may be necessary. Instead, traditional methods that take advantage of the periodic nature of water availability can provide reasonable yields for local needs. Parts of the Niger and Senegal river valleys in Africa have used these methods for centuries, but the risk of crop failures can be relatively high.

An improvement over pure reliance on the variable climatic conditions in these regions can be seen in parts of India, Bangladesh, and sub-Saharan Africa where some human-powered irrigation methods are combined with seasonal

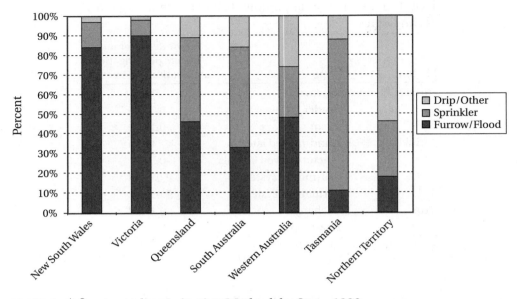

FIGURE 4.9   Australian Irrigation Method, by State, 1990

Fraction of irrigation in each State in Australia provided by drip, sprinklers, and furrow//flood. *Source:* AATSE 1999.

recession farming or river delta cultivation. Human-powered irrigation techniques include treadle and hand pumps and the *shadouf* (beam, fulcrum, and bucket) found in northern Africa, particularly along the Nile. Such approaches are suitable for very small farms (under a hectare) with shallow groundwater wells or local irrigation canals (Postel 1999). A further improvement replaces human effort with animal power or low-cost mechanical pumps. Where animals or mechanical pumps can be used, the area irrigated can often be expanded somewhat, though these approaches are still best suited to small-scale plots. Both human-powered and animal or mechanical powered pumps still make use of natural water availability with no artificial storage, such as shallow groundwater or irrigation ditches.

The above approaches have traditionally relied upon furrow or flood irrigation, where water is applied to row crops and distributed by gravity. A more modern improvement on these traditional methods includes exploring the possibility of using low-cost microirrigation systems (such as bucket, small sprinklers, or drip irrigation) that make more efficient use of the limited water supply. These approaches are particularly suited to hilly or terraced lands where flood irrigation does not work well and regions with perennial but limited water availability. Postel (1999) notes that locations where these approaches may be most appropriate include much of northwest, central, and southern India, Nepal, Central Asia, parts of China, and Latin America.

A further step up the technological ladder melds small-scale storage systems with local irrigation systems. In Sri Lanka, irrigation using integrated tanks and local distribution lines, in use for thousands of years, have been displaced by "modern" (albeit often inappropriate) systems (Mendis 1999). Other approaches use check dams to increase the capture of flood flows in waddis in the Middle East, terracing, and local percolation ponds. Where water availability is strongly seasonal or in drought-prone areas, these "conjunctive use" systems integrate groundwater and surface water use in an effective manner.

What Kind of Water Will Be Used to Grow Food?

Not all water is the same. Agriculture can use direct rainfall, direct streamflow, water stored in lakes and reservoirs, high-quality fossil groundwater, rapid-recharge groundwater, brackish surface or groundwater, reclaimed and recycled wastewater, and under some circumstances, even untreated wastewater or seawater. The kinds of water that will be desired, available, and used to grow the crop mix of the future remain highly uncertain. In Israel, for example, there is talk of devoting all primary water supplies to urban uses, with agriculture being forced to use only reclaimed and recycled water. In California, standards have been passed describing the kinds of crops that can be grown with different kinds of wastewater. For example, undisinfected secondary recycled water can be used on orchards and vineyards where the water doesn't come into contact with the fruit, on fodder and seed crops, and on food crops that undergo processing that destroys pathogens (see Table 7.3 in Chapter 7). Where water comes into contact with food crops that are to be eaten directly, a higher standard of water is required—in California this water must be treated to tertiary standards. The ability to use reclaimed water to meet these kinds of needs greatly increases overall water availability.

There may have to be significant shifts away from current sources of water. For example, many regions currently depend on unsustainable groundwater for crop production, where the extraction of groundwater is occurring faster than natural recharge. In the North China plain, groundwater levels drop as much as 4 meters per year. In parts of southern India, groundwater levels drop 2 to 3 meters or more per year. This cannot be sustained, and eventually crop production based on these sources will have to be cut back or stopped. Will alternative sources of water be available? If so, with what level of reliability? At what water quality? With what implications for long-term soil fertility or human health?

How Will Climatic Change Affect These Factors?

No discussion of water needs for agriculture can be complete without raising the issue of climatic change. We now have the potential to change the climate of the earth; indeed, most climate scientists now believe that the effect of human emissions of greenhouse gases can already be discerned (IPCC 1996). What the ultimate impacts of climate change will be on water availability, water quality, agricultural water demand, crop yields, and overall food production are still the domain of speculation and research. Nevertheless, answering the initial questions about how much food will be needed to feed the world and how much water it will take to grow it requires that we also know the answers to how climate change will affect the food/water equation. Among the many factors that must be considered are the role of increasing CO_2 on crop water needs and crop yields, the direct effects of temperature and precipitation changes on yields, and how severe events may change.

Climate change will alter rainfall, temperature, runoff, and soil moisture, as well as the nature and severity of extreme events. Increased evaporation and transpiration from soils and plants will cause moisture stress and changes in flowering, pollination, and grain-development success (Rosenzweig and Parry 1994, Rosenzweig and Hillel 1995). Variability of climate already affects food production, and any significant change in climate will affect local agriculture and global food availability and distribution. Extensive studies have been done on this issue, but considerable uncertainties limit current projections, particularly uncertainties about the regional changes in precipitation patterns and overall water availability. These climatic uncertainties only compound other uncertainties inherent in socioeconomic characteristics of agricultural systems and world food markets.

CONCLUSIONS

One of the most important questions facing the world is how to produce enough food to meet current and future human needs. Fundamentally related to this is whether sufficient water will be available to produce this food, and whether that water can be used in ways that don't conflict with other vital human and ecosystem water needs. This chapter has presented a series of related issues that must be resolved, or at least considered, before we can answer these questions. No definitive conclusions can be reached because of the myriad factors that affect the outcome, but it seems clear that many current agricultural water trends and policies are leading us in the wrong direction.

It would be a mistake to leave the impression that the production of food to meet the demands of growing populations is simply a series of technical factors related to land area, water availability, agricultural yields, irrigation technology, and crop genetics. Feeding the world's people in fact requires a complex set of institutions, effective and equitable international trade, the exchange of technology and ideas, food distribution systems, and economic structures. Too little attention has been paid in the past to how best to integrate human and ecosystem needs for water into this already complex equation.

But ultimately, there are two slightly different questions that must also be asked and answered. How much food will be needed, not to meet overall demands, but to meet the basic needs of all humans? and how much water will it take to meet that need? These two questions emphasize the difference between human desires and basic needs.

Finally, despite the fact that no clear answer can be given to most of the questions raised here, it should be apparent that there are many things that can be done to reduce the risks that insufficient water will be available to meet the demands of the agricultural sector in the future. We must begin serious efforts to integrate information on diets, basic food needs, meat consumption, agricultural water-use efficiency and technology, cropping patterns, land use, genetic engineering, water pricing, and even population policies. Addressing these problems in isolation only increases the risk that basic human needs will continue to be unmet for hundreds of millions of people.

REFERENCES

Ahmed, M. 1997. "Policy issues deriving from the scope, determinants of growth, and changing structure of supply of fish and fishery products in developing countries." International Consultation on Fisheries Policy Research in Developing Countries: Issues, Priorities, and Needs. June 2–5. Hirtshals, Denmark.

Australian Academy of Technological Sciences and Engineering (AATSE). 1999. *Water and the Australian Economy.* Australian Academy of Technological Sciences and Engineering, Parkville, Australia.

Barrow, C.J. 1991. *Land Degradation.* Cambridge University Press, Cambridge, United Kingdom.

Berck, P., and D. Bigman. 1993. *Food Security and Food Inventories in Developing Countries.* Cab International, Wallingford, United Kingdom.

Brown. L. 1997. "Facing the Prospect of Food Scarcity." In L. Brown (ed.), *State of the World 1997.* W.W. Norton and Company, New York, pp. 23–41.

Brown, L.R., and H. Kane. 1994. *Full House: Reassessing the Earth's Population Carrying Capacity.* Worldwatch Institute, Washington, D.C.

Brown, L.R., and J.E. Young. 1990. "Feeding the world in the nineties." *State of the World 1990,* Worldwatch Institute. W.W. Norton, New York, pp. 59–78.

Brown, L.R., N. Lenssen, and H. Kane. 1995. *Vital Signs 1995: The Trends That Are Shaping Our Future.* W.W. Norton, New York.

Carruthers, I. 1993. "Going, going, gone! Tropical agriculture as we know it." *Tropical Agriculture Association Newsletter,* Vol. 13, No. 3, pp. 1–5.

Crosson, P.R. 1995. "Future supplies of land and water for world agriculture." In N. Islam (ed.), *Population and Food in the Early Twenty-First Century: Meeting Future Food Demand of an Increasing Population.* International Food Policy Research Institute, Washington, D.C.

Crosson, P.R., and N.J. Rosenberg. 1989. "Strategies for agriculture." *Scientific American,* September, pp. 128–135.

Environment Canada. 1986. *Canada Water Yearbook 1985.* Water Use Edition, Ottawa, Canada.

Gardner, G. 1997. "Preserving Global Cropland." In L. Brown (ed.), *State of the World 1997.* W.W. Norton, New York, pp. 42–59.

Ghassemi, F., J.A. Jakeman, and H.A. Nix. 1995. *Salinisation of Land and Water Resources: Human Causes, Extent, Management and Case Studies.* Centre for Resource and Environmental Studies, University of New South Wales Press, Ltd, Sydney, Australia.

Gleick, P.H. 1997. "Water 2050: Moving toward a sustainable vision for the earth's fresh water." A Working Paper of the Pacific Institute for Studies in Development, Environment, and Security, prepared for the Comprehensive Freshwater Assessment for the United Nations General Assembly and the Stockholm Environment Institute, Stockholm, Sweden (February).

Gleick, P.H. 1999. "Crop shifting in California: Increasing farmer revenue, decreasing farm water use." In L. Owens-Viani, A.K. Wong, and P.H. Gleick (eds.), *Sustainable Use of Water: California Success Stories.* Pacific Institute for Studies in Development, Environment, and Security, Oakland, California, pp. 149–164.

Ingco, M.D., D.O. Mitchell, and A.F. McCalla. 1996. *Global Food Supply Prospects.* World Bank Technical Paper No. 353. World Bank, Washington, D.C.

Intergovernmental Panel on Climate Change (IPCC). 1996. *Climate Change 1995: The Science of Climate Change.* Contribution of Working Group I to the Second

Assessment Report of the Intergovernmental Panel on Climate Change. Cambridge University Press, New York.

Kates, R.K. 1996. "Ending hunger: Current status and future prospects." *Consequences*, Vol. 2 No. 2 (Summer). Available online at: http://www.gcrio.org/CONSEQUENCES/vol2no2/article1.html.

Keller, A., and D. Keller. 1995. "Effective efficiency: A water use efficiency concept for allocating freshwater resources." Center for Economic Policy Studies, Winrock International, Arlington, Virginia.

Kendall, H.W., and D. Pimetal. 1994. "Constraints on the expansion of the global food supply." *Ambio.* Vol. 23, No. 3, May.

McCalla, A.F. 1994. "Agriculture and food needs to 2025: Why we should be concerned." Crawford Memorial Lecture, October 27. Consultative Group in International Agricultural Research. World Bank, Washington, D.C.

McGinn, A.P. 1999. "Fisheries falter." In L.R. Brown, M. Renner, and B. Halwell (eds.), *Vital Signs 1999.* W.W. Norton, New York, pp. 34–35.

Meinzen-Dick, R. 1999. Personal communication.

Mendis, D.L.O. 1999. *Eppawala: Destruction of Cultural Heritage in the Name of Development.* DLO Mendis, Ratmalana, Sri Lanka (May).

Molden, M. 1997. "Accounting for water use and productivity." Research report, International Irrigation Management Institute (IIMI), Sri Lanka.

National Research Council (NRC). 1996. *A New Era for Irrigation.* National Research Council, National Academy Press, Washington, D.C.

Pandya-Lorch, R. 1999. "Prospects for food security." Study week on Science for Survival and Sustainable Development, March 12–16, Vatican City, the Vatican.

Pimental, D., J. Houser, E. Preiss, O. White, H. Fang, L. Mesnick, T. Barsky, S. Tariche, J. Schreck, and S. Albert. 1997. "Water resources: Agriculture, the environment, and society." *Biosciences* Vol. 47, No. 2, pp. 97–106.

Pingali, P.L., and P. Heisey. 1996. "Cereal crop productivity in developing countries: Past trends and future prospects." Paper presented at the Global Agricultural Science Policy for the Twenty-first Century, August 26–28, Melbourne, Australia.

Pinstrup-Anderson, P., R. Pandya-Lorch, M.W. Rosegrant. 1997. The world food situation: Recent developments, emerging issues, and long-term prospects." 2020 Vision Food Policy Report. International Food Policy Research Institute (IFPRI), Washington D.C.

Postel, S.L. 1998. "Water for food production: Will there be enough in 2025?" *BioSciences*, Vol. 28, pp. 629–637.

Postel, S.L. 1999. *Pillar of Sand: Can the Irrigation Miracle Last?* Worldwatch Books. W.W. Norton, New York.

Rosegrant, M.W., M. Agcaoili-Sombilla, N.D. Perez. 1995. "Global food projections to 2020: Implications for investment, food, agriculture, and the environment." Discussion Paper 51. International Food Policy Research Institute (IFPRI), Washington D.C.

Rosegrant, M.W., M.A. Sombilla, R.V. Gerpacio, and C. Ringler. 1997. "Global food markets and U.S. exports in the 21st century." International Food Policy Research Institute (IFPRI), Washington D.C.

Rosenzweig, C., and D. Hillel. 1995. "Potential impacts of climate change on agriculture and food supply." *Consequences*, Vol. 1, No. 2, (Summer). Available online at: http://www.gcrio.org/CONSEQUENCES/summer95/agriculture.html.

Rosenzweig, C., and M.L. Parry. 1994. "Potential impacts of climate change on world food supply." *Nature*, Vol. 367, pp. 133–138.

Scherr, S.J. 1999. "Soil degradation: A threat to developing country food security by 2020?" International Food Policy Research Institute (IFPRI), 2020 Brief 58, Washington, D.C.

Seckler, D. 1996. "The new era of water resources management: From 'dry' to 'wet' water savings." Research Report 1, International Irrigation Management Institute (IIMI), Sri Lanka.

Shiklomanov, I.A. 1998. "Assessment of water resources and water availability in the world." Report for the Comprehensive Assessment of the Freshwater Resources of the World, United Nations. Data archive on CD-ROM from the State Hydrological Institute, St. Petersburg, Russia.

Snyder, R.L., M.A. Plas, and J.I. Grieshop. 1996. "Irrigation methods used in California: Grower survey." *Journal of Irrigation and Drainage Engineering*, Vol. 122, No.7/8, pp. 259–262.

Tuong, T.P., and S. Bhuiyan. 1994. "Innovations toward improving water-use efficiency of rice." World Bank 1994 Water Resources Seminar, December 13–15, Landsdowne, Virginia.

United Nations. 1998. 1998 Revision of the World Population Estimates and Projections. Population Division, Department of Economic and Social Affairs. http://www.popin.org/pop1998/.

United Nations Development Programme (UNDP). 1994. *Sustainable Human Development and Agriculture*. UNDP Guidebook Series. United Nations, New York.

United Nations Food and Agriculture Organization (UNFAO). 1995. *World agriculture: Toward 2010, an FAO Study*. N. Alexandratos (ed.). J. Wiley and Sons, Chichester, United Kingdom, and the United Nations Food and Agriculture Organization, Rome, Italy.

United Nations Food and Agriculture Organization (UNFAO). 1996a. "Mapping undernutrition—an ongoing process." World Food Summit poster, FAO, Rome, Italy.

United Nations Food and Agriculture Organization (UNFAO). 1996b. "Food, agriculture, and food security: Developments since the World Food Conference and prospects." World Food Summit Technical Background Document 1. FAO, Rome, Italy.

United Nations Food and Agriculture Organization (UNFAO). 1997a. "Estimated post-harvest losses of rice in Southeast Asia." http://www.fao.org/News/FACTFILE/FF9712-E.HTM

United Nations Food and Agriculture Organization (UNFAO). 1997b. http://www.fao.org/News/FACTFILE/9712-E.HTM

United Nations Food and Agriculture Organization (UNFAO). 1998. http://www.fao.org/News/FACTFILE/FF9808-E.HTM

United Nations Food and Agriculture Organization (UNFAO). 1999. FAOSTAT database. http://faostat.fao.org.

United States Department of Agriculture (USDA). 1996. "Production, Supply and Distribution." Electronic database Economic Research Service, Washington, D.C. (May).

United States Department of Agriculture (USDA). 1998. *Livestock and Poultry: World Markets and Trade.* Foreign Agricultural Service. U.S. Department of Agriculture, Washington, D.C. (October).

van Ginkel, M. 1996. personal communication, cited in Ingco et al. 1996.

World Bank. 1987. "Community piped water supply systems in developing countries." Technical Paper No. 60. World Bank, Washington, D.C.

World Resources Institute (WRI). 1996. "Food and agriculture: Future global cereal production: Feast or famine?" *World Resources Report 1996–1997.* Oxford University Press, New York.

World Resources Institute (WRI). 1998. *World Resources: A Guide to the Global Environment.* Oxford University Press, New York.

Desalination: Straw into Gold or Gold into Water?

Saltwater, when it turns into vapor, becomes sweet and the vapor does not form saltwater again when it condenses.

ARISTOTLE, FOURTH CENTURY B.C.

If we could ever competitively, at a cheap rate, get freshwater from saltwater, that would be in the long-range interests of humanity [and] would dwarf any other scientific accomplishments.

JOHN F. KENNEDY, APRIL 12, 1961

All natural waters contain some salts—indeed, the vast majority of water on the earth is too salty for humans to use for irrigation, drinking, or most commercial and industrial purposes. Ninety-seven percent of the earth's water is found in the oceans, with a salt content of more than 30,000 milligrams per liter (mg/l). Water with a dissolved solids (salt) content below about 1,000 mg/l is typically required for domestic or commercial water supply. Drinking water should be below 500 mg/l. Because of the potentially unlimited quantities of seawater available, great effort has been spent to try to develop economically viable technologies for converting salt water to fresh water.

Desalting is an essential part of the natural hydrologic cycle. Precipitation falls to earth and moves through soils, dissolving minerals and becoming increasingly salty. Simultaneously, water is evaporating from ocean, lake, and land surfaces leaving salts behind. The resulting water vapor forms clouds that produce rain, continuing the cycle. The oceans are salty because the natural process of evaporation, precipitation, and runoff is constantly moving salt from the land to the sea, where it remains. "Desalination" refers to the wide range of technical processes designed to remove salts from water. Desalination technology is in use throughout the world for many purposes, including providing potable fresh water for domestic and municipal purposes, treated water for industrial processes, and emergency water for refugees or military operations.

TABLE 5.1   Salt Concentrations of Different Waters

Water Source or Type	Approximate Salt Concentration (grams per liter)[a]
Brackish waters	0.5 to 3
North Sea (near estuaries)	21
Gulf of Mexico and coastal waters	23 to 33
Atlantic Ocean	35
Pacific Ocean	38
Persian Gulf	45
Dead Sea	~300

Sources: OTV 1999, Gleick 1993.

[a]Slight spatial variations in salt content are found in all major bodies of water. The values in the table are considered typical.

The application of desalination technologies over the past forty years has permitted development to continue in many arid and water-short areas of the world where seawater or brackish waters are available. In particular, desalination is an important water source in parts of the arid Middle East, Arabian Gulf, North Africa, and islands where the lack of fresh water cannot be overcome with traditional water supply or transfers from elsewhere. Increasingly, other regions are exploring the use of desalination as sources of emergency or even basic water supplies as the prices slowly drop toward the cost of more traditional alternatives. Table 5.1 categories the salt content of various waters.

History of Desalination and Current Status

As the quote from Aristotle at the beginning of this chapter reveals, the concept of removing salt from water is an ancient one. For centuries, however, the goal was getting the salt from the water, not the water from the salt, since pure salt was of enormous value. As populations and demands for fresh water expanded, however, entrepreneurs began to look for ways of producing fresh water in remote locations and, especially, on naval ships at sea. In 1790, U.S. secretary of state Thomas Jefferson received a request to sell the government a distillation method to convert salt water to fresh water. A British patent was granted for such a device in 1852 (Simon 1998). The first place to make a major commitment to desalination was the island of Curaçao in the Netherlands Antilles. Desalination plants have operated there since 1928 (Birkett 1999). One of the first plants to desalinate large quantities of seawater was built in 1938 in what is now Saudi Arabia. Considerable research was conducted on desalination during World War II to meet military needs in water-short regions. That work continued after the war in various countries, including concentrated efforts by the United States. The U.S. Congress passed the Saline Water Act (PL 448) in 1952, which created and funded the Office of Saline Water (OSW). The first place to build a modern desalination plant may have been Kuwait, which operated a large distillation facility in the early 1960s.

In the 1960s, Senator, and then President, John F. Kennedy made a variety of speeches on desalination and clearly believed that a successful, commercial desalination system would be of enormous benefit for the whole world. Such a system "can do more to raise men and women from lives of poverty than any other scientific advance" (Kennedy 1961). During the 1960s and '70s considerable government investments in desalination continued to be made in the United States.

In the early 1970s, the federal Saline Water Conversion Act (PL 92-60) led to successor organizations like the Office of Water Research and Technology (OWRT). In 1977, the United States spent almost $144 million for desalination research (Simon 1998). Altogether, the U.S. government funded research and development equivalent in today's dollars to over US$1.5 billion. These government funds helped lay much of the basic research and development framework for the different technologies now used for desalting seawater and brackish waters.

In 1982, however, the Reagan administration cut federal funding for nonmilitary scientific research of almost every kind, including desalination work, and the Office of Water Research and Technology was closed. The next fourteen years saw practically no U.S. support for desalination. In 1996, Senator Paul Simon revived interest in federal support for a modest desalination research program, authoring the Water Desalination Act (PL 104-298). This bill was signed into law and authorized $5 million a year for six years for desalination research and studies, together with another $25 million over a four-year period (fiscal years 1999 to 2002) for demonstration projects. The bill required 50 percent cost sharing from the private sector and the support of multiple technologies. Despite the authorization, however, the president has never requested—and the Congress has never appropriated—the full amount authorized by the legislation. For the 1999 fiscal year, the U.S. government appropriated only $2.5 million; for fiscal year 2000, only $1.3 million were appropriated—about a tenth of what some estimate could effectively be spent on needed research projects (ADA 1999, Price 1999).

The slow rate of government support around the world for desalination research has certainly slowed the pace of commercialization, and desalination remains an expensive source of water. Increasingly, however, private commercial investments are advancing the technology and expanding operating experience, and the price is continuing to fall. While demand management and efficiency options are still far more cost-effective, desalination may ultimately prove to be the best new supply option for high-valued uses of water in coastal regions where efforts to maximize efficiency are being made and where absolute supply constraints are severe. In some regions of the world, this is already the case (United Nations 1987, McRae 1998).

A variety of desalting technologies have been developed over the years. Early plants were mostly based on large-scale thermal evaporation or distillation of seawater, mimicking the natural hydrologic cycle. Some early distillation plants were used to desalt brackish water, but high costs prevented widespread adoption of this approach. In the 1970s, more plants began to be installed using membrane technology, which mimics the natural biological process of osmosis. When electrodialysis was introduced, it could desalt brackish water much more economically, and many applications were found for it. Similarly, reverse osmosis (RO), which separates water and salt ions using selective membranes, was initially used primarily for desalting brackish waters, although the process is increasingly used to purify seawater.

Desalting equipment is now used in about 120 countries. By January 1, 1998, a total of 12,451 desalting units (100 m³/d or larger) had been installed or contracted.[1] These plants have a total capacity of around 22,700,000 m³/d. Around 13,300,000 m³/d of this capacity is used to desalinate seawater and 9,400,000 m³/d to desalinate water of other qualities (http://www.ida.bm/html/inventory.htm, Wangnick 1998). The principal techniques for desalinating water involve distillation and reverse osmosis. Forty-four percent of the total installed or contracted capacity is based on multistage flash distillation (MSF) and 40 percent is based on RO. Figure 5.1 shows total desalination capacity by process. The trend over the last decade shows a steady shift toward the construction of reverse osmosis facilities, and it is likely that within a very few years, RO capacity will exceed MSF capacity.

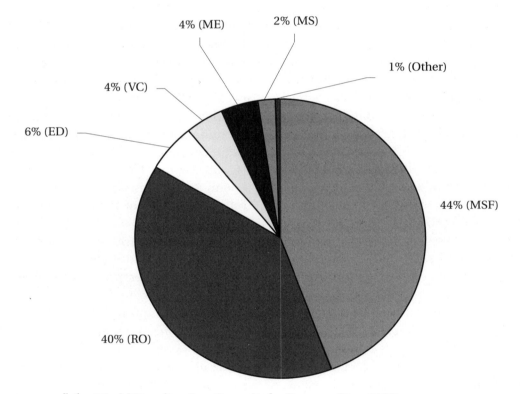

FIGURE 5.1   World Desalination Capacity by Process, June 1999

A variety of methods are used to desalinate water (see text for detail). This figure shows the proportional breakdown of total desalination capacity by process. Table 20 in the Data Section has details on actual installed and contracted volumes.

Key: Multistage flash distillation (MSF)
Reverse osmosis (RO)
Electrodialysis (ED)
Vapor compression (VC)
Multi-effect distillation (ME)
Membrane softening (MS)

Of the total desalinated water produced, 60 percent came from seawater. A quarter of all source water was brackish water and the remainder came from other sources. Figure 5.2 shows the breakdown of water sources as of mid-1999.

More than half of all desalination capacity is in the Middle East/Arabian Gulf/North Africa regions. Nearly one quarter of all capacity is in Saudi Arabia, followed by 16 percent in the United States, 10 percent in the United Arab Emirates, and 7 percent in Kuwait (Wangnick 1998). Figure 5.3 shows those countries with more than 1 percent of global desalination capacity, as of mid-1999. The majority of distillation plants are installed in Saudi Arabia, Kuwait, and the United Arab Emirates; most reverse osmosis plants and vapor compression plants are in the United States. It is important to note that many island communities, not on this list, rely on desalination for a large fraction of their total water needs (see Table 19 in the Data Section).

While desalination provides a substantial part of the water supply in certain oil-rich Middle Eastern nations, globally, installed desalination plants have the capacity to provide just two thousandths of total world freshwater use. Total global water withdrawals are estimated to be 3,900 km³/yr (see Chapter 3), while the total production capacity of desalination plants is approximately 8 km³/yr.

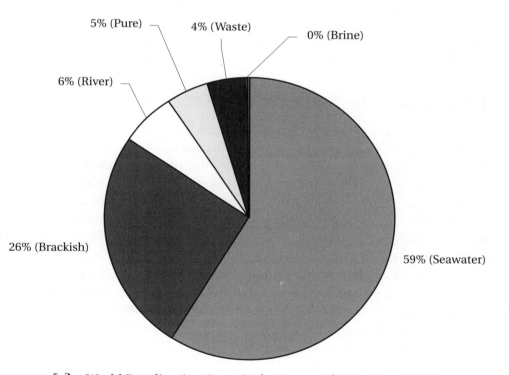

FIGURE 5.2   World Desalination Capacity by Source of Water, June 1999

Water from many different sources, and of different qualities, is processed by desalination plants. Nearly 60 percent of all desalination capacity is used to desalt seawater. Nearly 30 percent of desalination capacity desalts brackish water. River water, municipal water, and wastewaters are also occasionally desalted, but in relatively low volumes. "Pure" refers to the production of high-quality waters for human consumption or certain special industrial processes. *Source:* Data Section, Table 21.

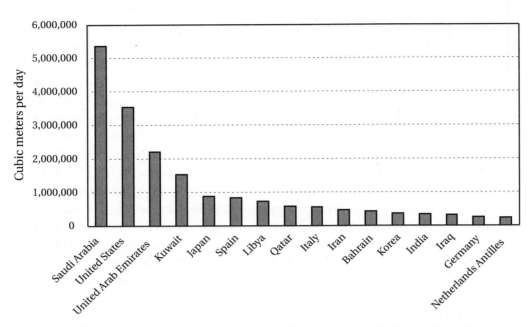

FIGURE 5.3 Countries with More than One Percent of Global Desalination Capacity, June 1999

A few countries have the majority of global desalination capacity. Shown here are all countries with greater than 1 percent of global desalination capacity as of mid-1999. Note that countries near the Arabian Gulf dominate—Saudi Arabia, United Arab Emirates, and Kuwait. See Table 19 in the Data Section for a complete list of desalination capacity by country.

Desalination Technologies

There is no single best method of desalination. A wide variety of desalination technologies effectively take salts out of salty water (or fresh water out of salty water), producing a water stream with a low concentration of salt (the product stream) and another with a high concentration of remaining salts (the brine or concentrate). Many different approaches can be used to separate water and salt, but they all require significant amounts of energy. Various distillation and membrane technologies are widely used for seawater desalting, while reverse osmosis and electrodialysis are often used to desalt brackish water. Significant numbers of plants using vapor compression and other methods have also been built (USAID 1980, Wangnick 1998, 1999). Ultimately, the selection of a desalination process depends on site-specific conditions, economics, the quality of water to be desalinated, the purpose for which the water is to be used, and local engineering experience and skills. The major desalination processes are described below.

Thermal Processes

The distillation process mimics the natural water cycle by producing water vapor that is then condensed to form fresh water. In the simplest approach, water is heated to the boiling point to produce the maximum amount of water vapor. Adjusting the atmospheric pressure of the water being boiled affects the boiling temperature. By decreasing pressure, the boiling point can be reduced, which provides two major benefits: it permits multiple boilings and it reduces problems asso-

TABLE 5.2   Boiling Temperature of
Water at Different Pressures

Pressure (bar)	Temperature (degrees C)
2	120
1[a]	100
0.25	65
0.1	45

[a]Normal atmospheric pressure

ciated with scaling. Just over half of the world's desalted water is produced with processes that use heat to distill fresh water from seawater or brackish water.

Multiple boiling refers to a process where water is boiled in a series of vessels, operating at successively lower temperatures and pressures. The process of reducing the pressure promotes continued boiling at lower and lower vaporization temperatures (see Table 5.2). If the pressure were reduced enough, the point at which water would be boiling and freezing at the same time would be reached. The concept of distilling water with a vessel operating at a reduced pressure has been used for well over a century.

Scaling occurs when substances like carbonates and sulfates found in seawater come out of solution and cause thermal and mechanical problems. One of the most significant concerns is gypsum, a hydrate of calcium sulfate ($CaSO_4$) that forms from solution when water approaches about 95°C. Gypsum is the main component of concrete and can coat pipes, tubes, and other surfaces. It is very difficult to remove. Scale reduces the effectiveness of desalination operations by restricting flows, reducing heat transfer, and coating membrane surfaces. Ultimately scaling increases costs. By keeping the temperature and boiling point low, the formation of scale can be slowed.

Multiple Effect Distillation

Multiple-effect distillation (MED) is a thermal method that has been used successfully for well over 100 years, substantially predating MSF (Birkett 1999). MED takes place in a series of vessels and reduces the ambient pressure in subsequent effects. This approach reuses the heat of vaporization by placing evaporators and condensers in series and is based on the principle that vapor produced by evaporation can be condensed in a way that uses the heat of vaporization to heat salt water at a lower temperature and pressure in each succeeding chamber. This permits seawater to undergo multiple boilings without supplying additional heat after the first effect. In MED plants, the salt water enters the first effect and is heated to the boiling point. Salt water may be sprayed onto heated evaporator tubes or may flow over vertical surfaces in a thin film to promote rapid boiling and evaporation.

Only a portion of the salt water applied to the tubes in the first effect evaporates. The rest moves to the second effect, where it is applied to another tube bundle heated by the steam created in the first effect. This steam condenses to fresh water, while giving up heat to evaporate a portion of the remaining salt water in the next effect. The condensate from the tubes is recycled. There are 8 to 16 effects in a typical large plant.

Although some of the earliest distillation plants used MED, MSF units with lower costs and less tendency to scale have increasingly displaced this process. In the past few years, interest in the MED process has been renewed and new concepts and technologies are being explored. Recent plants have been built to operate with a top temperature (in the first stage) of about 70°C (158°F), which reduces the potential for scaling within the plant but increases the need for additional heat transfer area in the form of tubes. MED plants are typically built in units of 1,000 to 10,000 m^3/day for smaller towns and industrial uses. MED plants continue to be built, but their overall numbers remain relatively small compared to MSF plants (http://www.ida.bm/html/abc.htm).

Multistage Flash Distillation

The process that accounts for the greatest installed desalting capacity is multistage flash distillation (MSF). Like all evaporative processes, MSF can produce high-quality fresh water with a salt concentration of only 10 parts per million or even less, from salt concentrations as high as 60,000 to 70,000 ppm total dissolved solids. In MSF, evaporation (flashing) occurs from the bulk liquid, not on a heat-exchange surface, as is the case with MED. This minimizes scale and is the principal reason MSF displaced MED in the 1960s (Birkett 1999). As a result, MSF has been the primary technology used for desalinating seawater since its commercialization in the 1960s.

In MSF distillation, seawater is heated in a series of stages. Typical MSF systems consist of many evaporation chambers arranged in series, each with successively lower pressures and temperatures that cause sudden (flash) evaporation of hot brine, followed by condensation on tubes in the upper portion of each chamber. The steam generated by flashing is condensed in heat exchangers that are cooled by the incoming feed water going to the brine heater. This simultaneously warms up the feed water, reducing the total amount of thermal energy needed. Generally, only a small percentage of this water is converted to steam (water vapor), depending on the pressure maintained in each stage. MSF plants may contain between 4 and 40 stages, but most typically are in the range of 18 to 25.

The large energy requirement for MSF is a significant drawback in certain regions. At present, typical distillation technologies require around 200 kJ to desalinate a liter of seawater (Gleick 1994), though improvements in techniques and increased efficiency of equipment are already reducing this. To properly evaluate comparative energy needs of different plants, however, requires separating out thermal, mechanical, and electrical energy needs. MSF plants are typically built today in large units from 10,000 m^3/day to over 35,000 m^3/day, with several units grouped together.

Vapor Compression Distillation

Vapor compression (VC) distillation is generally used for small- and medium-scale seawater desalting units. These units also take advantage of the principle of reducing the boiling point temperature by reducing ambient pressure, but the heat for evaporating the water comes from the compression of vapor rather than the direct exchange of heat from steam produced in a boiler. The two primary methods used to condense vapor to produce enough heat to evaporate incoming seawater are mechanical compression or a steam jet. The mechanical compressor can be electrically driven, making this process the only one to produce water by distillation solely with electricity (http://www.ida.bm/html/abc.htm).

The compressor creates a vacuum in the vessel and then compresses the vapor taken from the vessel and condenses it inside of a tube bundle also in the same vessel. Seawater is sprayed on the outside of the heated tube bundle where it boils and partially evaporates, producing a stream of fresh water. Steam jet-type VC units, also called thermocompressors, create lower ambient pressure in the main vessel. This mixture is condensed on the tube walls to provide the thermal energy (heat of condensation) to evaporate the salt water being applied on the other side of the tube walls. VC units are usually built in the 250 to 2,000 m³/day range. They are often used for tourist resorts, small industries, and remote sites.

Membrane Processes

Other than thermal desalination, the major approach for removing salt ions from water uses membranes that selectively permit or prohibit the passage of certain ions. Membranes play an important role in the separation of salts in natural processes of dialysis and osmosis and this natural principle has been adapted in two commercially important desalting processes: electrodialysis and reverse osmosis. These two approaches now account for 45 percent of all desalination capacity, and they have typically been used to desalinate brackish water. Electrodialysis uses an electrical current to move salt ions selectively through a membrane, leaving fresh water behind. In reverse osmosis, pressure is used to force fresh water to move through a membrane, leaving the salts behind. Both of these concepts have been understood for a century, but their commercialization was slowed until the technology for creating and maintaining membranes was developed. In recent years, great advances in RO technology have been achieved and more new membrane capacity is added annually than distillation capacity.

Electrodialysis and Electrodialysis Reversal

Electrodialysis is an electrochemical separation process, in which electrical driving forces are used to separate ions in ion-exchange membranes. The process was commercially introduced in the mid-1950s. The development of electrodialysis provided a cost-effective way to desalt brackish water and spurred considerable interest in the use of membranes. Among the advantages of electrodialysis are the ability to produce more product and less brine, energy needs and costs proportional to the salts removed, the ability to treat water with a higher level of suspended solids than reverse osmosis, and less need for chemicals for pretreatment. Electrodialysis has been used to produce water for cooling tower makeup and fish farming, treat industrial wastes, satisfy municipal needs, and concentrate polluted groundwater, as well as for other purposes. In an interesting application, this approach has also been used to remove salts and sodium from wastewater, which is then used to irrigate bananas in Tenerife (von Gottberg 1999).

Electrodialysis depends on the following principles: salts dissolved in water are naturally ionized; these ions are attracted to electrodes with an opposite electric charge; and membranes can be constructed to permit selective passage of either anions (negatively charged ions) or cations (positively charged ions). Brackish water is pumped at low pressure between stacks of flat, parallel, ion-permeable membranes, and electric current flows across these parallel channels, pulling ions through the membranes. Membranes are arranged with anion-selective membranes alternating with cation-selective membranes. Water flows along the face of these alternating pairs of membranes and one channel carries feed (and product) water while the next carries brine. As the electrodes are charged, the anions in the feed water are attracted and diverted toward the positive electrode. This dilutes the salt content of the water in the product water channel. The anions pass through the anion-selective membrane, but cannot pass through the cation-selective membrane. Cations move in the opposite direction through the cation-selective membrane to the concentrate channel on the other side where they are trapped. Concentrated and diluted solutions are thus created in the spaces between the alternating membranes. These spaces are called cells. A cell pair consists of two cells, one from which the ions migrated (the dilute cell for the product water) and the other in which the ions concentrate (the concentrate cell for the brine stream). A basic electrodialysis unit or "membrane stack" consists of several hundred cell pairs bound together with electrodes on the outside. Feed water passes simultaneously in parallel paths through all of the cells to produce continuous flows of fresh water and brine (IDA 1999).

In the early 1970s, a modification of straight electrodialysis was introduced—electrodialysis reversal (EDR). An EDR unit operates on the same principle as a standard electrodialysis plant except that both the product and the brine channels are identical in construction. Two to four times an hour, the polarity of the electrodes is reversed, and the brine channel and product water channel flows are switched. Immediately following the reversal of polarity and flow, the ions are attracted in the opposite direction across the membrane stack and product water is used to clean out the stack and lines. After flushing for a few minutes, the unit resumes producing water. The reversal process breaks up and flushes out scale and other deposits in the cells. Flushing also allows the unit to operate with fewer pretreatment chemicals and minimizes fouling of the membranes.

EDR systems can operate on feed water with higher turbidity and are less prone to biofouling than RO systems. Experience suggests that EDR can achieve higher water recovery than RO systems. The energy cost of electrodialysis rises with the concentration of the salts in the water. The major energy requirement is the direct current used to separate the ionic substances in the membrane stack.

Reverse Osmosis

RO uses semipermeable membranes that pass water but retain salts and solids when pressure is applied to a solution with concentrations of salt. Pure water molecules from the solution are forced through the membrane. Because RO is a separation process rather than a filtration process, the volume of water entering the system and that which ends up as product differ. The amount of pure water that can be obtained varies between 30 and 85 percent of the feed water, depending on the initial water quality, the desired water quality of the product, and the technology involved. In comparison to distillation and electrodialysis, RO is relatively new, with successful commercialization occurring in the early 1970s. Installations of RO plants now exceed that of distillation. As in electrodialysis, the energy requirement for RO depends directly on the concentration of salts in the feed water, and no heating or phase change is necessary for this separation. The major energy required for desalting is for pressurizing the feed water. RO facilities are most economical for desalinating brackish water.

In commercial RO plants, saline feed water is pressurized and brought into contact with a semi-permeable membrane. As a portion of the water passes through the membrane, the remaining feed water increases in salt content. A portion of this feed water is discharged without passing through the membrane. Without this controlled discharge, the pressurized feed water would continue to increase in salt concentration, leading to precipitation of supersaturated salts and increased osmotic pressure across the membranes, reducing system effectiveness (IDA 1999).

An RO system is made up of the following basic components: pretreatment, high-pressure pump, membrane assembly, and post-treatment. Pretreatment of feed water is often necessary to remove contaminants and prevent fouling or growth of microorganisms in membranes. This is important because the feed water must pass through very narrow passages during the process. Usually pretreatment consists of filtration and the addition of chemicals to inhibit precipitation. A high-pressure pump generates the pressure needed to enable the water to pass through the membrane. This pressure ranges from 17 to 27 bar (250 to 400 psi) for brackish water and from 54 to 80 bar (800 to 1180 psi) for seawater (Fisia Italimpianti 1999, IDA 1999).

The membrane assembly consists of a pressure vessel and a membrane that permits the feed water to be pressurized against the semipermeable membranes. The membranes are fragile and vary in their ability to pass fresh water and reject salts. RO membranes are made in a variety of configurations. The two most commercially successful are spiral-wound and hollow-fine fiber. Post-treatment consists of preparing the water for distribution, including removing gases such as hydrogen sulfide and adjusting the pH.

Two developments have helped to reduce the operating cost of RO plants during the past decade: the development of lower-cost, higher-flux, higher salt-rejecting membranes that can operate efficiently at lower pressures, and the use of pressure recovery devices. Low-pressure membranes are being increasingly used to desalt brackish water. The pressure-recovery devices are connected to the concentrate stream and help reduce energy costs.

The largest RO plant in the world at the beginning of 1998 was located in Yuma, Arizona. This plant was designed and constructed specifically to fulfill water-quality obligations under an international treaty between the United States and Mexico on the Colorado River, and has a capacity of about 270,000 m^3/d. Ironically, this plant has never operated for any significant period because of economic, environmental, and political constraints, and because by the time it was finally built, it was technologically outmoded.

Other Processes

A number of other processes are also available to desalt waters. These processes have not achieved the level of commercial success that distillation, electrodialysis, and RO have, but they are effective under special circumstances. They include small-scale ion-exchange resins, freezing, and membrane distillation, but altogether they account for less than 3 percent of total desalination capacity (Wangnick 1999).

Ion-Exchange Methods

Ion-exchange methods use resins to remove undesirable ions in water. For example, cation-exchange resins are used in homes and municipal water-treatment plants to remove calcium and magnesium ions in "hard" water. The greater the concentration of dissolved solids, the more often the expensive resins have to be replaced, making the entire process economically unattractive compared with RO and electrodialysis. At lower concentrations, however, and for small-scale systems, these methods have proven effective. Thus ion exchange is often used for the final "polishing" of waters that have had most of their salt content removed by RO or electrodialysis (Birkett 1999).

Freezing

Extensive work was done in the 1950s and 1960s on separation technology using freezing of water but this approach has had no commercial success. Freeze separation takes advantage of the insolubility of salts in ice. When ice crystals form, dissolved salts are naturally excluded and the resulting pure ice crystals can be strained from the brine. Seawater can be desalinated by cooling the water to form crystals under controlled conditions. Before the entire mass of water has been frozen, the mixture is usually washed and rinsed to remove the salts adhering to the ice crystals. The ice is then melted to produce fresh water. The most efficient freeze methods use vapor-compression freeze-separation systems.

Freezing has some theoretical advantages over distillation, including a lower minimum energy requirement, minimal potential for corrosion, and little scaling or precipitation. Disadvantages include ice and water mixtures that are mechanically complex to handle, move, and process. A small number of demonstration plants have been built over the past forty years, but the process has not proven commercially feasible. The few demonstration plants built have now largely been abandoned. Better commercial success has been achieved in the application of freezing to the treatment of industrial wastes.

Membrane Distillation

Membrane distillation (MD) was introduced commercially on a small scale in the 1980s. The process combines the use of both thermal distillation and membranes, but it is primarily still a thermal, evaporative process. Its energy efficiency and product recovery rates have not been demonstrated to exceed those of MSF. In the process, saline water is warmed to enhance vapor production, and this vapor is exposed to a membrane that can pass vapor but not water. After the vapor passes through the membrane, it is condensed on cooler surfaces to produce fresh water.

Thus far, the process has been used only in a few areas. Compared to the more commercially successful processes, membrane distillation requires more space and more pumping energy per unit of fresh water produced. The main advantages of membrane distillation lie in its simplicity and the need for only small temperature differentials to operate. Membrane distillation probably has its best application in desalting saline water where inexpensive low-grade thermal energy is available, such as from industries or solar collectors.

Solar and Wind-Driven Systems

Solar energy has been used directly for over a century to distill brackish water and seawater. Water vapor produced by evaporation is condensed on a cool surface and the condensate collected as product water. The simplest example of this type of process is the greenhouse solar still, in which saline water is heated and evaporated by incoming solar radiation in a basin on the floor, and the water vapor condenses on a sloping glass roof that covers the basin. When commercial plate glass began to be produced toward the end of the nineteenth century, solar stills began to be developed. One of the first successful solar systems was built in 1872 in Las Salinas, Chile, an area with very limited fresh water. This still covered 4,500 square meters, operated for 40 years, and produced about 20 m^3/d of fresh water (Delyannis and Delyannis 1984). Variations of this type of solar still have been made in an effort to increase efficiency, but they all share some major difficulties, including solar collection area requirements, high capital costs, and vulnerability to weather-related damage.

Solar stills typically require a collection area of about one quarter of a square meter per liter of fresh water produced per day. Thus, 100 hectares of solar stills would produce about 4,000 m^3/d of water, with variations depending on climate and system design. This amount of area is rarely available in urban or industrial areas, but it may be available in desert or arid regions with access to saltwater sources.

These systems are relatively expensive to construct and maintain. While the principal energy input is free, the capital cost of renewable energy systems remains high. In addition, operation and maintenance expenses include preventing scale formation caused by the basins drying out and repairing broken glass or vapor leaks in the stills.

Desalting units that use more advanced solar systems to provide heat or electrical energy have also been built. Some modern desalination facilities are now being run with electricity produced by wind turbines or other solar electric technologies, such as photovoltaics. An inventory of known wind- and solar-powered desalting plants (Wangnick 1998) listed units in about 70 locations scattered over 28 countries. Most of these installations had capacities of less than 50 m^3/d. The largest solar desalination plant in operation by the end of 1998 was a plant in Libya (2,000 m^3/d), which uses wind energy systems for power. Another plant in Libya in the same location produces 1,000 m^3/d and uses photovoltaics for energy. Both of these plants began operation in 1992 and are used to desalt brackish water (Wangnick 1998).

The final cost of water from these plants depends, in large part, on the cost of producing energy with these alternative energy devices. A pilot plant combining photovoltaic electricity production with electrodialysis is in operation in Gallup, New Mexico, producing around 3 m^3/d of fresh water at a cost of around $3 per cubic meter (Price 1999). At present, this cost is prohibitive for typical water agencies, but these systems are increasingly economical for remote areas where the cost of bringing in conventional energy sources is very high. If the price of fossil fuels increases or renewable energy costs drop, such systems will look more attractive. Despite years of research, no particular advantages have appeared for coupling solar (or nuclear, for that matter) energy systems with desalination plants. Ultimately, these energy systems must prove themselves on the market before any such coupling can become attractive.

Other Aspects of Desalination

Technical factors are not the only ones that will determine the success or failure of desalination. Several others are discussed in the following section, including the relative economics of removing salt from water, problems associated with disposing of the concentrated salt brines produced, and the energy costs.

Economics

The most important factor determining the ultimate success and extent of desalination is economics. Yet no easy discussion of desalination costs is possible. Extreme caution, even skepticism, should be used in evaluating different desalination cost estimates and claims. Most cost estimates presented today are classic "apples and oranges" comparisons of projects with so many fundamental differences that direct comparisons are invalid or meaningless. Feed water as well as product water qualities vary. The cost of water depends enormously on assumptions about capital and labor costs, debt rates, and financing periods. Comparison years are rarely normalized. Sometimes costs to users are compared to costs of production. Added costs arise if the water must be moved a great distance to the point of use. Energy prices— a fundamental factor in desalination economics—are different in different regions and predictions about how those prices will change in the future are fraught with uncertainty. And hidden subsidies are often embedded in the numbers.

Experience to date suggests that desalinated water can be delivered to users at prices between $1 and $4 per cubic meter, with the great difference due to the factors listed above. Even the low end of this range remains above the price of water typically paid by urban water users, and far above the price paid by farmers. For example, growers in the western United States usually pay $0.01 and $0.05 per cubic meter for water. Even urban users rarely pay more than $0.30 to $0.80 per cubic meter.

The capital and operating costs for desalination have decreased over the years, in part because of dropping energy prices—a trend that is unlikely to continue indefinitely. The cost of desalinating water has also been reduced by technological improvements, economies of scale associated with larger plants, and improved project management and experience. The greatest progress in cost reduction has been associated with improvements in RO technology. Salt rejection, a measure of the ability to remove salt from feed water, can be as high as 99.7 percent today, up from 98.5 percent a decade ago, while the output of product from a unit of membranes has risen from 60 m³/d to 84 m³/d (Glueckstern 1999). Membrane manufacturers are now offering guarantees on membrane life of six to ten years, reflecting greater confidence in design and performance of the most sensitive technical component of the process. Optimistic observers believe the trend in declining costs will continue (Glueckstern 1999). Evidence that costs are, in fact, dropping, materialized in 1999 in a proposal to build a large plant in the United States in Florida, though this project differs in many important ways from typical ones. This project is described later in more detail.

Concentrate Disposal

A common element in all desalination processes is the production of a brine stream with a high concentration of salt. This waste stream may also contain chemicals that may have been added during the process. The stream varies in volume depending on the process, but it is almost always a significant quantity of water.

An important part of the design and operation of desalination facilities includes environmentally appropriate disposal of this wastewater. If the desalting plant is located near an ocean, disposing of desalination brine usually requires only careful attention to dilution, dissolved oxygen levels, and water temperature.

More significant problems occur when desalination facilities are built away from the coasts. Care must then be taken to prevent salt contamination of groundwater aquifers, surface streams, or lakes. Disposal may involve dilution, injection of the concentrate into a saline aquifer, evaporation, or transport by pipeline to a suitable disposal point. All of these methods add to the cost of the process.

Energy Reuse and Costs

Because of the high energy costs involved in desalination, the ability to reuse energy or minimize energy needs offers economic advantages. As noted above, many of the recent advances in reducing desalination costs are associated with reducing the energy required to take salt out of water. The theoretical minimum amount of energy required to remove salt from a liter of seawater is 2.8 kilojoules (kJ). The best plants now operating use many times this amount, though improvements in technology are continuing to reduce this spread.

Another approach is to integrate desalination systems into energy cogeneration plants. The term "cogeneration" refers to the use of a single energy source to meet multiple needs. Certain types of desalination processes, especially distillation, have been integrated into cogeneration systems. Many of the distillation plants installed in the Middle East and North Africa operate under this principle.

In a typical power plant, boilers produce high-pressure steam at about 540°C. This steam is used to run turbines, which drive electric generators. Distillation plants need steam whose temperature is about 120°C or less, which is readily obtainable by extracting the steam at the low-pressure end of a turbine after it has expanded and much of its energy has been used to generate electricity. This steam can then be run through a distillation plant to increase the temperature of the incoming water. The condensate from the steam is then returned to the boiler to be reheated for use in the turbine, as in a traditional power plant. Cogeneration systems significantly reduce the consumption of fuel when compared to the fuel needed for two separate plants. Since energy is a major operating cost in any desalination process, this leads to important economic benefits.

The Tampa Bay Desalination Plant

In March 1999, regional water officials in Florida approved plans to build an RO plant with a capacity of 95,000 m³/d—the largest such plant in the Western Hemisphere after the Yuma desalting plant, which is not operating.[2] Desalination advocates are extremely excited by the project and by the apparent breakthrough in price this plant will achieve. The desalination facility will be privately owned and operated and on completion would supplement drinking water supplies for 1.8 million retail water customers in Hillsborough, Pasco, and Pinellas counties and the cities of New Port Richey, St. Petersburg, and Tampa (U.S. Water News 1999, Wright 1999). Water-resource problems in the region to be served by the plant include groundwater overpumping and concern about meeting future demands because of population growth. The local water district, Tampa Bay Water, has made a commitment to reduce groundwater withdrawals by the end of 2002. The groundwater overdraft is adversely affecting natural wetlands in the area and leading to salinity intrusion (Wright 1999).

The planning process for the plant began in October 1996. Initial proposals were submitted in December 1997, and binding offers were received in October 1998. All four of the final proposals in the highly competitive bidding process offered water at a finished cost dramatically lower than costs at desalination plants under construction or in operation elsewhere (U.S. Water News 1999). The final choice for project developer was made in July 1999 and consisted of a consortium including Poseidon Resources and the Stone and Webster Company. Their proposal calls for construction on the plant to begin in January 2001 and for operation to begin between July and December 2002 (Heller 1999, Hoffman 1999). This group made a binding commitment to deliver desalinated water in the first year of operation at an unprecedented wholesale cost of $0.45/m³ ($1.71/1,000 gallons), with a 30-year average cost of $0.55/m³ ($2.08/1,000 gallons). Even the highest of the four bidders, however, offered a price between $0.56 and $0.67/m³, well below the cost of water from other recent desalination plants. For comparison, the Singapore Public Utility Board recently announced plans to build a desalination plant that will produce 136,000 m³/d at an estimated cost of between $1.98 and $2.31/m³ (U.S. Water News 1999). A better comparison may be plans for plants in Cyprus and Trinidad, which are expected to produce water for under $2/m³. The capital for the Tampa Bay project will come primarily from the Southwest Florida Water Management District (SWFWMD).

The plant is to be located on the site of the Big Bend Power Plant on Tampa Bay. The power plant will provide electricity and some supporting operation and maintenance functions. Intake water for the desalination plant will be taken from the cooling water discharge of the power plant, and the brine will be disposed of back into this discharge. A total of 167,000 m³/d (44 million gallons per day) of feed water will be required to produce around 95,000 m³/d (25 million gallons per day) of potable water; 72,000 m³/d of brine will be returned to the outflow of the power plant. The potable water will then be sent 22 kilometers by pipeline to the municipal water supply plant for distribution to customers.

While the cost of water from this project is clearly much lower than typical desalination costs, careful examination of the project should caution desalination advocates against excessive optimism on price. The project has a number of unique conditions that may be difficult to reproduce elsewhere. Energy costs in the region are very low—around $0.04 per kWh. Salinity of the source water from Tampa Bay is only about 26,000 ppm instead of 33,000 to 40,000 ppm typical for most seawater. Financing will be spread out over 30 years and the cost of money will be only 5.2 percent (Wright 1999). In addition, the plant will take advantage of shared infrastructure with the Big Bend Power Plant, reducing the costs of construction by a substantial amount.

Some local residents oppose construction of the plant because of concern about impacts of water withdrawals and brine disposal on local ecosystems (Karp 1999). Ironically, at the same time the desalination plant was approved, the regional water authorities agreed not to go ahead with a project to use less-expensive, highly treated wastewater (Heller 1999) and aggressive, less-expensive demand management options have not yet been exhausted.

SUMMARY

In energy-rich, arid, and water-scarce regions of the world, desalination is already a vitally important option. Desalinated water is being used as a source of municipal supply in many areas of the Caribbean, North Africa, Pacific Island nations, and the Arabian Gulf. But the goal of unlimited, cheap fresh water from the oceans expressed by John F. Kennedy at the beginning of this chapter continues to be an elusive dream. The costs of desalination remain high—higher than almost all other alternatives—largely because of the cost of energy and the high capital and maintenance costs of the complex infrastructure required. Recent advances in technology and decreases in costs are keeping alive the hopes that desalination will soon begin to make a more substantial contribution to meeting vital freshwater needs around the world.

There is no "best" method of desalination. Local circumstances and needs will always play a significant role in determining the most appropriate technology and approach for an area. Distillation and RO technologies are both widely used for seawater desalting, while RO and electrodialysis are often used to desalt brackish water. The final selection of a process depends on the careful study of site-specific conditions and economics.

Desalination research programs continue in countries such as Japan, Germany, Israel, the United States, and Saudi Arabia, and new advances continue to be made. Improvements in membrane technology are expanding membrane lifetimes and decreasing costs. Advanced energy recovery methods decrease total energy requirements. Biological systems are being integrated with membrane designs to improve separation efficiencies. And advanced energy systems for remote applications are being designed and tested. Ultimately, how important desalination will become depends on one's perspective. Despite all the progress made over the past several decades, and despite all the improvement in economics and technology, desalination still makes only modest contributions to overall water supply. By the late 1990s, the total amount of desalinated water produced in an entire year was about as much as the world used in fourteen hours. On the other hand, in some regions of the world, nearly 100 percent of all drinking water now comes from desalination—providing a vital and irreplaceable source of water. Is desalination the ultimate solution to our water problems? No. Is it a vital and irrefutable part of the puzzle? Yes. And it will certainly become more important for future populations.

NOTES

1. Excellent data on installed or contracted desalination capacity exist (Wangnick 1998). Some of these plants, however, either do not operate at all, or operate at less than full capacity. Figures on actual production of desalinated water are not collected, but it is certainly less than the total potential.
2. In late 1999, plans for a slightly larger plant in Trinidad and Tobago were announced.

REFERENCES

American Desalting Association (ADA). 1999. "ADA meets in Washington D.C." *ADA News,* Vol. 12, No. 7, Sacramento, California, p. 1.

Birkett, J. 1999. Personal communication.

Delyannis A.A., and Delyannis E. 1984. "Solar desalination." *Desalination*, Vol. 50, pp. 71–81.

Fisia Italimpianti. 1999. "Exposition of the advantages of MSF desalination for the years 2000." Fisia Italimpianti S.p.A. Genova, Italy.

Gleick, P.H. 1993. *Water in Crisis: A Guide to the World's Fresh Water Resources.* Oxford University Press, New York.

Gleick, P.H. 1994. "Water and energy." *Annual Review of Energy and Environment.* Vol. 19, Annual Reviews, Inc., Palo Alto, California, pp. 267–299.

Glueckstern, P. 1999. "Desalination today and tomorrow." *International Water and Irrigation*, Vol. 19, No. 2, pp. 6–12.

Heller, J. 1999. "Water board green-lights desalination plant on bay." *Tampa Bay Business News*, March 16. http://www.tampabay.org/press65.asp.

Hoffman, P. 1999. Personal communication, Stone and Webster Company.

International Desalination Association (IDA). 1999. Desalination factsheet. http://www.ida.bm/html/abc.htm.

Karp, D. 1999. "Proposed desal site criticized." *St. Petersburg Times*, (February 24). http://www.sptimes.com/News/22499/Hillsborough/Proposed_desal_site_c.html.

Kennedy, J.F. 1961. *Speeches of Senator John F. Kennedy: Presidential Campaign of 1960.* U.S. Government Printing Office, Washington, D.C.

McRae, W. 1998. Desalination for the twenty-first century." In *Desalination and Filtration.* New World Water, Sterling Publications, London, United Kingdom.

OTV. 1999. "Desalinating seawater." *Memotechnique*, Planete Technical Section, No. 31 (February), p. 1.

Price, K. 1999. Personal communication. U.S. Bureau of Reclamation, Denver, Colorado.

Simon, P. 1998. *Tapped Out: The Coming World Crisis in Water and What We Can Do about It.* Welcome Rain Publishers, New York.

United Nations. 1987. *Non-conventional Water Resource Use in Developing Countries.* United Nations Publication, E.87.II.A.20. New York.

U.S. Agency for International Development (USAID). 1980. *The USAID Desalination Manual.* CH2M HILL International for the U.S. Agency for International Development, Washington, D.C.

U.S. Water News. 1999. "Tampa Bay desalinated water will be the cheapest in the world." *U.S. Water News Online.* http://www.uswaternews.com/archives/arcsupply/9tambay3.html.

von Gottberg, A. 1999. "Advances in electrodialysis reversal systems." *Proceedings of the 1999 AWWA Membrane Technology Conference,* February 28–March 3, Long Beach, California.

Wangnick, K. 1998. *1998 IDA Worldwide Desalting Plants Inventory, No. 15.* Produced by Wangnick Consulting for the International Desalination Association. Gnarrenburg, Germany.

Wangnick, K. 1999. Personal communication, with updated data to June 25, 1999.

Wright, A.G. 1999. "Tampa to tap team to build and run record-size U.S. plant." *Engineering-News Record* (March 8). http://beta.enr.com/news/enrpwr8.asp.

The Removal of Dams: A New Dimension to an Old Debate

[T]his is a challenge to dam owners and operators to defend themselves—to demonstrate by hard facts, not by sentiment or myth—that the continued operation of a dam is in the public interest.

U.S. SECRETARY OF THE INTERIOR BRUCE BABBITT, 1998

Tens of thousands of big dams and hundreds of thousands of small dams have been built around the world over the past century. As the twenty-first century begins, the construction of new dams continues to play an important role in water supply for cities and farms, for electricity production, and for benefits like flood control and navigation. At the same time, a new trend is emerging: the intentional removal of dams to meet a wide range of water-management objectives, including providing water for wetlands and fisheries, improving water quality, and reducing risks to downstream communities. Growing numbers of dams are being removed when they end their useful life, cause severe environmental damage, or produce limited economic and social benefits.

This trend is a modest one so far. So few dams have been torn down that their removal is still considered big news—for example, the destruction of the Edwards Dam in Maine was the subject of a July 1999 editorial in the *New York Times* (New York Times 1999) and a September 1999 *New Yorker* article by John McPhee (McPhee 1999). Even the idea that some big dams *might* be removed can be front-page news (Verhovek 1999) or the subject of special scientific meetings: in December 1999, a symposium entitled the Effects of Dam Removal on Aquatic Ecosystems was held in Chicago, Illinois, sponsored by the American Fisheries Society. The meeting explored the effects of dam removal on aquatic ecosystems in North America, including physical, chemical, and biological impacts on streams, rivers, and impoundments.

Fights against large dams have defined much of the history of the environmental movement. Although proposals for new large dams in North America and Europe still materialize now and then, the perception of such projects has changed. It is now widely acknowledged that few, if any, new large dams will be built in most

TABLE 6.1   Dams and Reservoirs Completed in the United States, 1961–1995

	Number of Dams	Reservoir Storage Volume (km³)	Cumulative Number of Dams	Cumulative Storage Volume (km³)	Average Volume of Reservoir (million m³/dam)
1961–1965	9,687	130	9,687	130	13.5
1966–1970	9,401	180	19,088	311	19.2
1971–1975	6,628	67	25,716	377	10.0
1976–1980	3,957	51	29,673	428	12.9
1981–1985	2,269	22	31,942	450	9.5
1986–1990	2,139	18	34,081	467	8.2
1991–1995	1,044	6	35,125	473	5.6

Source: Calculated from data in U.S. Army Corps of Engineers 1996.

developed countries. In the United States, where huge numbers of dams have already been built, the rate of construction has dropped enormously. Table 6.1 shows the number of dams built during the past seven five-year periods since the early 1960s. Figure 6.1 plots the cumulative number of U.S. dams built during this time, showing how the rate of increase has dropped to nearly zero (U.S. Army Corps of Engineers 1996). Moreover, Figure 6.2 shows that the average size of the dams built during each five-year period has dropped significantly. In the 1960s, reservoir volumes averaged 15 to 20 million cubic meters each. By the 1990s, the average reservoir volume of dams built had dropped to under 6 million cubic meters each.

This same trend can be seen around the world. Figure 6.3 shows the average number of large reservoirs built annually, by region, for various time periods. Construction of large dams peaked in the 1960s at around 70 per year and has now dropped to only about 10 per year, with decreases in every region of the world. The data for this figure come from Table 16 in the Data Section.

Historically, questions about dams were limited to where to build them and how big they should be. What we have learned over the past decades, however, is that we should have been asking a far wider range of questions related to their economic and social costs and their environmental and ecological impacts. And these questions lead to the question whether certain dams should be torn down. Ultimately, for many dams that question can be answered in the affirmative. Some dams are so egregious in their impact on the environment, local communities, or fisheries that their removal must be seen as a clear benefit. Even the possibility that dams can be removed has raised the hope of many communities that have suffered from the oftentimes ignored consequences of dam construction, such as destruction of fisheries, population displacement, land inundation, and more.

A combination of social, institutional, environmental, and economic factors are responsible for helping to reframe the debate about dams. Among the reasons for increased attention to dam removal or decommissioning are safety concerns, severely declining fisheries, a growing demand for free-flowing river recreation, limited budgets for new public works projects, and new options for supplying water and energy. In the United States many dams are up for renewal of federal hydropower licenses, which offers an opportunity for dam removal advocates to

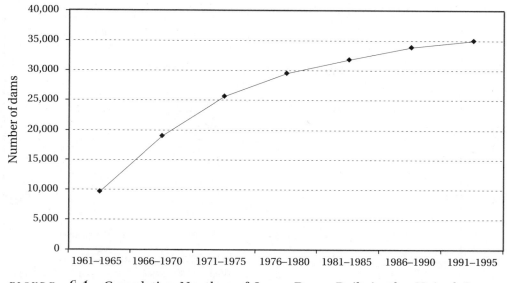

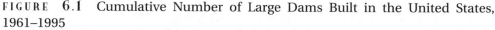

FIGURE 6.1 Cumulative Number of Large Dams Built in the United States, 1961–1995

The number of dams built in the United States over the past 40 years has continued to increase, but at a slower and slower rate. Very few large dams are now under construction. *Source:* Calculated from data in U.S. Army Corps of Engineers, 1996.

raise concerns ignored when a dam was initially constructed or to bring new issues to the attention of regulators.

The Federal Energy Regulatory Commission (FERC) licenses hydropower dams in the United States. Many hydropower dams were given original 50-year licenses back in the 1940s and hundreds of these dam licenses across the country are currently up or will soon be up for renewal. In 1986, Congress passed legislation requiring FERC, through its hydropower licensing activities, to balance power generation and environmental protection. FERC is required to review projects for impacts on recreation, fish, wildlife, and other river values, and to incorporate recommendations and suggestions of other federal and state agencies who may have interests other than hydroelectricity production. Outside advocates can also intervene and participate in the process. In most cases, new licenses are granted with no or few new conditions, such as requiring fish passages or setting minimum flows during sensitive periods. In a few instances, however, relicensing processes have begun to lead to requirements that dams significantly change their operations, physically modify their structures, or even cease operation and be removed.

Thousands of other small dams were built in the nineteenth and early twentieth centuries for logging, mining, or farming operations. Most dams are built with a life expectancy of around 50 years, and in the United States, 25 percent of all dams are now older than this. Some of these pose serious hazards to public safety. In the National Dams Inventory in the United States (http://crunch.tec.army.mil/nid/webpages/nid.html), 32 percent of the dams have a "high" or "significant" downstream hazard potential, and the majority of these dams do not have an emergency action plan in the event of failure or negligent operation. The American Society of Civil Engineers estimates that by the year 2020, the number of U.S. dams exceeding

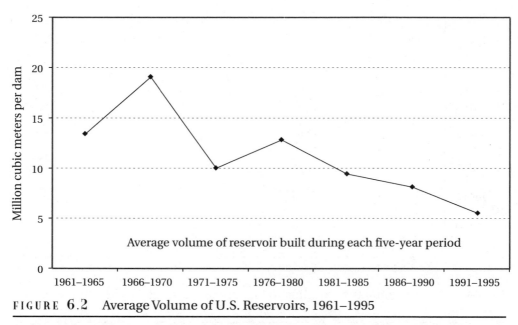

FIGURE 6.2   Average Volume of U.S. Reservoirs, 1961–1995

As the number of dams under construction in the United States has dropped, so has the average storage volume of their reservoirs. This figure shows that the average volume of reservoirs being built in the United States has decreased to around 5 million cubic meter per dam from nearly 20 million cubic meters per dam in the mid-1960s. *Source:* Calculated from data in U.S. Army Corps of Engineers, 1996.

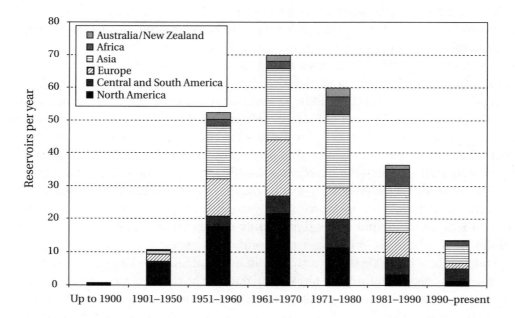
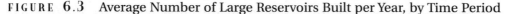

FIGURE 6.3   Average Number of Large Reservoirs Built per Year, by Time Period

The average number of large (>0.1 km³) reservoirs built annually worldwide, by region. Construction of large dams peaked in the 1960s and has dropped in every region since then. *Source:* Data Section, Table 16.

the average life expectancy will reach 85 percent (AR & TU 1999). As dams continue to age, public safety will become more important. The state of North Carolina, for example, passed a dam safety law more than 30 years ago that requires the state to "approve detailed engineering plans for the construction, repair, modification or *removal of dams* equal to or greater than 15 feet high and impounding 10 acre-feet or more, or dams that can be classified as high hazard" (State of North Carolina 1967).

Many older dams were built for purposes that can no longer be justified. Around the turn of the last century, dams were built to provide mechanical power for mills or small factories to power the industrial revolution. Often, the original factory has long been closed, leaving behind a dam with no economic rationale. In recent years, however, a driving force behind the efforts to remove dams has been ecological restoration and the revival of free-flowing rivers. Among the most severe consequences of dam construction is the alteration of the biological, chemical, and physical characteristics of a river, often leading to the elimination of native fisheries. Especially vulnerable have been anadromous fish—fish that are born in rivers, live most of their lives in the oceans, and then return to their original rivers to spawn and die.

In watersheds feeding both the Pacific and the Atlantic, anadromous salmon populations have been eliminated or threatened with extinction. On the heavily dammed Columbia and Snake Rivers in the Pacific Northwest, 95 percent of the juvenile salmon fail to survive dams and the reservoirs created by the dams. Atlantic salmon populations are blocked by more than 900 dams on New England rivers and major dams on almost all European rivers, and their populations are less than 1 percent of historic levels (Haberman 1995). Almost all of the salmon populations of California have suffered severely from dam construction and other factors, and most of the different populations are officially extinct, endangered, or threatened. The socioeconomic value of commercial and sport fishing industries and Native American tribal treaty fishing rights are gaining more attention as anadromous fisheries continue to decline. There are growing efforts to remove dams as an essential and practical strategy for restoring rivers and these fish populations. The U.S. National Park Service has removed more than 100 dams on rivers and streams affecting national parks and fisheries because of their serious adverse impacts on anadromous and freshwater fish populations. The greatest attention has focused on a few of the largest efforts in the United States and Europe, such as the hydroelectric dams on the Elwha River in Olympic National Park, on tributaries of the Loire in France, and on the Kennebec River in Maine, but there have been many other dam removal efforts around the world, and the trend is accelerating rapidly.

States are taking actions too. In the state of Wisconsin there are about 3,500 dams, built for old mills or to provide water for logging operations. Some are approaching 100 years in age and most were built of timber and rock. The Wisconsin Department of Natural Resources (WDNR) estimated that at least half of the state's deteriorating dams will need safety-related repairs costing upward of $100,000 each (Keefer 1990) and the WDNR has determined that it is more cost effective in many cases to remove, not repair, small dams with no current economic justification (Haberman 1995). In the late 1980s, the Woolen Mills Dam on the Milwaukee River in Wisconsin was removed. After being dammed for almost 70 years, the degraded Milwaukee River provided habitat for a large exotic carp population at the expense of native species. After Woolen Mills Dam was removed, water quality and fisheries habitat improved and native fish species were restored (Haberman 1995). Quality trout fisheries have been reestablished on Wisconsin's Kickapoo River, Black Earth Creek, and Tomorrow River—in part by dam removals. Dam removals have caused a surge of canoeing and kayaking on the Prairie and Apple rivers. Restoration of old impoundment sites has also created parks, such as those at Woolen Mills on the Milwaukee River, and Fulton on the Yahara River (River Alliance of Wisconsin 1999). In Minnesota the Welch Dam on the Cannon River was removed in part to meet the growing demand for river recreation near Minneapolis/St. Paul. Similarly, the state of Pennsylvania has more than 5,000 dams and in the last few years, more than two dozen old dams have been removed, with 30 others scheduled for removal (http://www.amrivers.org/membersh.html).

In addition to active dam decommissioning, there are more and more examples of natural dam removal combined with active river protection. After flood flows on the Merrimack River in New Hampshire partially destroyed the Sewalls Fall Dam near Concord in 1984, a diverse coalition of recreationists, fishermen, and community groups defeated a proposal to rebuild the dam.

Table 18 in the Data Section lists nearly 500 dams removed in the United States, from a data set collected by American Rivers and Trout Unlimited (AR & TU 1999). This is the most complete database available on dam removals, though it focuses on the United States. It can be visited at www.amrivers.org, where it will be updated regularly, or at www.worldwater.org, where information on international dam removals will also be maintained.

Economics of Dam Removal

While environmental and safety reasons are dominant factors in arguments to remove dams, economics are increasingly important. In the early days of dam construction, no economic assessments of the environmental consequences of dams were included, leading to underestimates of actual dam "costs." More recently, however, as regulatory requirements have forced the consideration of restoration or protection of aquatic ecosystems when dams are being repaired or relicensed, dam owners or developers are faced with new and unanticipated expenses related to environmental impacts, flood protection, or safety concerns. Building new fish ladders for old dams, for example, can often cost as much as the original cost of the dam. Implementing an integrated fisheries restoration project with hatcheries, spawning habitat purchase and protection, restrictions on power generation, and so on, are real economic costs imposed for environmental purposes. The cost of structural strengthening to meet new safety standards can often exceed the original costs of construction.

The economics of these kinds of projects are often extremely difficult to evaluate, because many of the costs and benefits are either unquantified or unquantifiable. Dam repair costs may include long-term maintenance, dredging of silted-in reservoirs, impacts to fisheries and other ecosystem components, and hard-to-quantify flood risks. The costs of replacing the benefits formerly provided by a dam are also changing. New energy technologies are available to generate electricity; new water-use efficiency technologies may be able to reduce demands for local water withdrawals. Alternative flood protection policies can reduce flood risks without new structures.

When even some of these factors are included in economic assessments, new analyses can favor dam removal. For example, Woolen Mills Dam in the Milwaukee River in Wisconsin would have cost $3.3 million to repair and cost only $500,000 to remove. Sandstone Dam on the Kettle River in Minnesota would have cost $400,000 to repair but cost only half as much to remove. Savage Rapids Dam on the Rogue River in Oregon will cost an estimated $17 to 24 million to repair but only $11 million to remove, even including the cost of new irrigation pumps to permit growers to continue to take water for agriculture. The Condit Dam on Washington State's White Salmon River is to be removed by 2006 because removal is far less expensive than the $30 million that would be required to make it less harmful to the river's fisheries (Verhovek 1999).

Another barrier to fair economic assessments is differing perceptions of effects on property values. For example, "lakefront" property owners behind Salling Dam on the AuSable River in Michigan complained of potential property value losses associated with removing the dam. The Michigan Department of Natural Resources evaluated local property values and found that AuSable River frontage was at least equal to, if not more valuable than, "lake" or reservoir frontage (Haberman 1995). "Lakefront" property owners behind Manitowoc Rapids Dam on the Manitowoc River in Wisconsin were strongly opposed to removal of the dam and loss of their reservoir, even though the deteriorating dam was a major public safety hazard. The dam was removed in 1990, and residents now enjoy improved water quality and the restoration of 65 kilometers of excellent fisheries habitat.

Removing dams is still a relatively new strategy in river conservation. Almost every case sets a new precedent for legal, economic, or social issues. Writer Ted Williams described the Edwards Dam hydropower dam relicensing process as

> a case study of how Americans have looked on their rivers in the past and how they perceive them today. . . . The dam's removal would mark a profound shift in America's philosophy about river management and river ownership. . . . The notion that a dam doesn't belong just because it is there is revolutionary; yet it's catching on. (Williams 1993)

Many practical hurdles remain. Experts with experience in removing dams are few. Engineering and biological sciences are far more advanced in the areas of dam construction than removal, and many unknowns remain. Small dams are relatively easy to remove, but experience removing or decommissioning large dams is very limited. How should accumulated sediments be managed? What is the quality of that sediment? If sediment is released uncontrolled, will it degrade or restore habitat downstream? Will fish return to old habitat after decades or even centuries? These and other questions must be asked and answered as more knowledge and experience accumulate in the field of dam removal.

Following are case studies (the first of which is modified from Owens-Viani and Wong 1999) of dams that have been removed and others that are being seriously considered for removal.

Dam Removal Case Studies: Some Completed Removals

The Sacramento River Valley, United States

Natural flows in California rivers and streams have been declining since the late 1800s with the advent of hydraulic mining, urban water withdrawals, and agricultural diversions. In the past half century, the timing and magnitude of these flows have been further affected by the construction of large and small water projects that store water and transfer it from one part of the state to another. One consequence of altered flows has been the dramatic decline in the populations of California's once-abundant salmon. Changes in stream and river temperatures and other habitat conditions have also contributed to these declines. In recent years, numerous programs have been implemented in an effort to try to restore salmon runs. Those programs include the U.S. Fish and Wildlife Service's Anadromous Fish Restoration Program, the California Department of Fish and Game's 1993 plan for restoring Central Valley streams, the state Resources Agency's Upper Sacramento River Fisheries and Riparian Habitat Management Plan, the CALFED's Ecosystem Restoration Program, and a habitat improvement program established by the 1994 Bay–Delta Accord. Decade-long efforts to restore the Sacramento River winter-run Chinook salmon appear to be meeting with some success (McClurg 1998). Now spring-run Chinook, once so abundant they supported an extensive inland fishery, are the focus of intensive restoration efforts, in part in an attempt to forestall listings under the state and federal Endangered Species Acts.

Key habitat for the remaining wild, genetically isolated spring-run exists in only a few places, including the upper stretches of Butte Creek, a tributary to the Sacramento River. Parts of this tributary are eligible for protection under the National Wild and Scenic Rivers Act (FOTR 1997). Unfortunately, dams and water withdrawals along the lower stretches of this river are serious obstacles to the salmon. Figure 6.4 shows a map of this region of northern California.

In late August 1998, the California Fish and Game Commission decided to list the spring-run Chinook as a threatened species under the state Endangered Species Act; in September 1999, this species received protection as a threatened species under the federal Endangered Species Act (ESA) (U.S. NOAA 1999). These actions led to the proposal for major state and federal actions that would have affected interests as diverse as commercial fishermen, landowners, Sacramento Valley rice growers, and San Joaquin Valley farmers. These stakeholders, together with state, federal, and local agencies agreed in 1998 to join together to improve spring-run habitat.

Over the past three and a half decades, the number of spring-run salmon in Butte Creek dropped from a high of 8,700 (in 1960) to lows of a few hundred in some years (see Figure 6.5). In the Sacramento Valley reach of Butte Creek alone, eight diversion dams (primarily for rice irrigation) have been built; farther upstream are three small

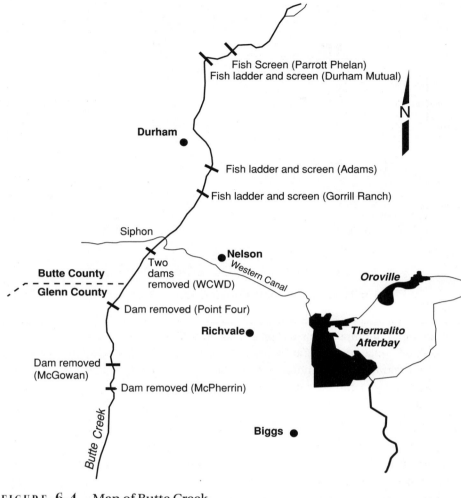

FIGURE 6.4   Map of Butte Creek

Butte Creek is a tributary of the Sacramento River, north of the City of Sacramento. This map shows the stretch where several dams were removed in a cooperative effort among farmers, regulatory agencies, ecologists, and local communities. *Source:* Owens-Viani and Wong 1999

hydropower dams. For salmon numbers to recover, fish need access to the more pristine habitat in the creek's upper reaches, at least to just below these hydro dams. Removing the diversion dams lower on the river, improving fish ladders, and screening diversions are among the projects underway to help fish in Butte Creek.

Most of the diversion dams on Butte Creek permit water to be delivered to rice farmers, altering or reducing spring flows and impeding fish migration. In recent years, growers have also taken water in the winter to flood fields for decomposing rice stubble, which also threatens fish by reducing flows during critical periods. Two particular dams on Butte Creek belonged to the Western Canal Water District (WCWD), which irrigates around 24,000 hectares of farmland in Butte and Glenn counties. The WCWD, together with other regional water districts, imports 750 million cubic meters of water from the Feather River into the Butte Basin every year. This water is used for irrigation, and part of it ultimately flows into Butte Creek (Reisner 1997).

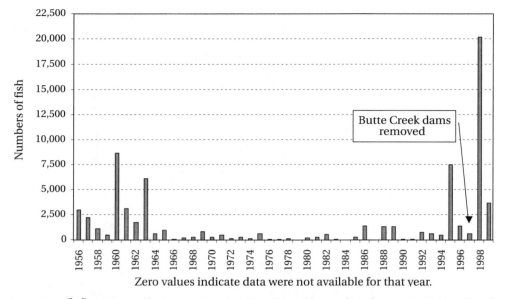

FIGURE 6.5   Naturally Spawning Spring-Run Chinook Salmon in Butte Creek, 1956–1999

The construction of multiple dams in Butte Creek in the 1950s and 1960s led to dramatic drops in salmon populations. Several dams were removed as part of an effort to restore populations in 1997 and 1998. Populations have now begun to rebound, though it is too early to tell how successful the project will ultimately be in helping salmon recovery efforts. *Sources:* Owens-Viani and Wong 1999; Paul Ward, personal communication.

Until November 1997, these two WCWD dams kept water from the Western Canal from flowing down Butte Creek and raised the water level in the creek high enough to allow pumps and gravity-driven diversions on the other side of the creek to suck the water back into the continuation of the Western Canal in order to continue delivering irrigation water to rice lands to the west. The dams, the subsequent alterations in flows, and the many unscreened diversions in the creek either killed or stranded large numbers of salmon each year.

As a solution, the Butte Creek siphon was built to carry the water that formerly passed through the creek via the canal. The siphon (and the associated new delivery canals) eliminated the need for at least 12 unscreened diversions as well as the two Western Canal dams (Owens-Viani and Wong 1999). It also enabled five downstream dams to be removed, including McPherrin, McGowan, and Point Four dams. These actions opened up 30 kilometers of Butte Creek to salmon migration and spawning (McClurg 1998).

The alternative to this project would have been to install fish screens on inflow and outflow pipes at a cost of between $6 million and $7 million. The irrigation pumps might also have had to be shut down during crucial fish migration periods (Owens-Viani and Wong 1999). Moreover, there was concern that the expensive fish screens might not have been effective, and that, after a substantial investment, the water district might find itself in the position of seeking yet another solution at even higher cost. The project to build a pipeline to move the water under Butte Creek and to remove four dams cost approximately $9 million. It was funded by a combination of state and federal agencies together with the local water district.

As a result of removing these diversions, the streambed is already returning to a more natural state, which also improves conditions for other wildlife, and in 1998 a record number of salmon spawned in the creek. The Butte Creek case shows that previously unthinkable acts—like removing dams—can become a reality when diverse interests with a common goal are able to think creatively about how to conserve and allocate scarce and valuable resources.

Edwards Dam on the Kennebec River, Maine, United States

Edwards Dam was built in 1837 on the Kennebec River in Maine in the northeastern United States and, in one of the most important efforts in the dam removal movement, was torn down in 1999. The dam was the first encountered moving upstream along the Kennebec River, approximately 65 kilometers from the river's mouth on the Atlantic. It measured 280 meters long and 7.6 meters high and created a reservoir covering more than 400 hectares (U.S. DoI 1998). Built primarily to provide mechanical power to Augusta's textile industry, it was fitted with turbines in 1913. The hydropower produced by this small dam was only about one-tenth of 1 percent of the entire power supply of Maine. The dam, however, effectively blocked access to the historical prime spawning grounds of Kennebec River's fisheries.

In the early 1990s, the license for the Edwards Dam came up for renewal before FERC. As part of the relicensing process, biologists from the U.S. Fish and Wildlife Service and other fisheries agencies, along with the Kennebec Coalition and other organizations and individuals, argued that removal of Edwards Dam would help Atlantic salmon, American shad, river herring, striped bass, shortnose sturgeon (listed as endangered under the federal ESA), Atlantic sturgeon, rainbow smelt, and the American eel. The Kennebec Coalition includes the Natural Resources Council of Maine, American Rivers, Atlantic Salmon Federation, and Trout Unlimited. It was formed in 1991 for the purpose of removing the Edwards Dam and restoring migrating fish populations in the river. The Interior Department, Environmental Protection Agency, former governor John McKernan, and current governor Angus King also supported removal of the dam (Cheever 1998). Republican U.S. senator Olympia Snowe also supported removing the dam and later urged FERC to deny the request to relicense the dam. Proponents of dam removal argued it would allow shortnosed sturgeon, Atlantic sturgeon, striped bass, and rainbow smelt to reach 30 kilometers of historic upstream spawning habitat and result in an overall increase in wetland habitat, recreational boating, and fishing benefits.

In an unusual decision, FERC determined that installing a $10 million fish ladder would be more costly than retiring and removing the dam and might not even achieve the desired goal. In late 1995 FERC voted not to reissue the operating license for the Edwards Dam. In November 1997, FERC recommended and ordered the decommissioning and removal of the dam. This was the first time that FERC determined that continued operation of a dam caused unacceptable environmental damage that could not be adequately addressed through any action short of dam removal. This was also the first time that FERC has denied an application for relicensing and the first time that FERC explicitly ordered that a dam be removed by an owner seeking relicensing (ENN 1997, EDF 1998a).

FERC Chairman James Hoecker argued that this decision was a special case where the economic benefits of allowing the dam to continue operating were outweighed by its environmental and social impact. FERC noted that the power produced at the dam could easily be replaced by other resources in the region at a lower cost. Electricity from the dam was twice as expensive as prevailing power rates (ENN 1997). The FERC decision sets a precedent, despite the statements of Hoecker that Edwards was a special case.

The original FERC ruling called for the Edwards Manufacturing Company to pay the cost of removing its dam. In an effort to avoid a long-term, expensive dispute, the dam owner, other corporate river users, state and local officials, and representatives of the Kennebec Coalition worked out a new solution in mid-1998 to remove the dam on an expedited schedule without using any public funds.

Elements of the settlement agreement included the following: Edwards Manufacturing transferred ownership of the dam at the beginning of 1999 to the state of Maine; all parties would seek government permits to remove the dam in 1999; $7.25 million of private funds would be committed to dam removal and fisheries restoration projects in the Kennebec River; the funds would be provided by a local shipbuilder, Bath Iron Works (BIW), and the Kennebec Hydro Developers group, a coalition of dam operators upstream of the Edwards Dam (ENN 1998); and the funds would be managed by the National Fish and Wildlife Foundation. BIW's contribution of $2.5 million of these funds is a central component of a mitigation plan required by state and federal agencies in connection with BIW's plans to expand a manufacturing facility along the river. Members of the Kennebec Hydro Developers Group were to contribute the remaining $4.75 million in exchange for being permitted to delay installation of fish-passage systems at their seven hydropower facilities further up the Kennebec River and its tributaries. The new deadlines in the agreement for the upstream dams will range from 2002 to 2014. The restoration funds also provide for a fisheries-restoration effort for river herring and Atlantic salmon. Finally, the agreement called for the dam to be dismantled between May and September 1999. In September 1998, an engineering firm was awarded the contract for removing Edwards Dam. Removal began on July 1, 1999, and the dam was breached shortly thereafter (www.state.me.us/spo/edwards/volume5.htm). Secretary of the Interior Bruce Babbitt, at a ceremony marking the agreement to remove the dam, said:

> Today, with the power of our pens, we are dismantling several myths: that hydrodams provide clean pollution-free energy; that hydropower is the main source of our electricity; that dams should last as long as the pyramids; and that making them friendlier for fisheries is expensive and time consuming. There are 75,000 large dams in this country, most built a long, long time ago. Many are useful but some are obsolete, expensive and unsafe. They were built with no consideration of the environmental costs. We must now examine those costs and act accordingly. This is not a call to remove all, most, or even many dams. But this is a challenge to dam owners and operators to defend themselves—to demonstrate by hard facts, not by

sentiment or myth, that the continued operation of a dam is in the public interest, economically and environmentally. Often the outcome will mean more environmentally friendly operating regimes, perhaps achieved through the installation of fish passages or other technological fixes. In some cases, like the one we are here to highlight today, it will mean actual removal. (U.S. DoI 1998)

Maisons-Rouge and Saint-Etienne-du-Vigan Dams in the Loire River Basin, France

The Maisons-Rouges and Saint-Etienne-du-Vigan Dams are located on tributaries of the Loire River in France. Between September 1998 and early 1999, the Maisons-Rouges Dam was torn down. Maisons-Rouges Dam was 5 meters high and eliminated all 700 hectares of spawning grounds of the Atlantic Salmon on the Vienne River, the second-most-important tributary of the Loire River. This dam also blocked other migratory fish, such as eel and shad, from roughly a fifth of the entire Loire basin (EDF 1998e). An earlier unsuccessful traditional restoration project was launched in the mid-1970s. This effort included fish ladders and other infrastructures. The latest fish-restoration program now includes the removal of this dam and the Saint-Etienne-du-Vigan Dam and the construction of an improved fish ladder on a third dam in the watershed (Epple 1998). Costs of demolition were about $1.6 million. Demolition was supervised by Electricité de France (Epple 1998).

Saint-Etienne-du-Vigan was a small hydro dam on the upper Allier River. This dam was also removed in 1998. The dam had an installed capacity of 35 MW, was 13 meters high, and blocked or flooded 30 hectares of prime spawning habitat. Prior to the construction of the dam, this area produced 10 tons of salmon a year for nearby villages. The Allier River is the only tributary in the Loire Basin where the Atlantic salmon still return to spawn, and only 67 returning salmon were counted in 1996. The Atlantic salmon have disappeared from all the large rivers on the European Atlantic coast. The regional managers of the rivers in the area stated that the dam "représentait une menace pour la sécurité publique et constituait un obstacle infranchissable pour de nombreuses espèces de poisson migrateur" (the dam represents a menace to public safety and constitutes an impassable obstacle for numerous species of migratory fish) (http://www.architectes.net/archinews/breves/1206-2.htm).

Newport No. 11 Dam, Clyde River

Newport No. 11 Dam—a 6-meter-high dam—was built on the Clyde River, Vermont, in 1957 by Citizens Utility. Its purpose was the generation of a modest amount of hydropower—the dam had an installed capacity of under two megawatts. Before construction of the dam, the runs of Atlantic salmon in the river caused *Vermont Life* magazine to brag "Detroit may boast of its autos, Pittsburgh of its steel mills, and Boston of its beans, but up Newport way it's the fabulous salmon which busts vest buttons and makes local chests puff out" (AR & TU 1999). After the dam was built, the fishery collapsed. In 1994, high water washed away the banks along the dam. In

1995, FERC staff recommended the removal of the dam as a condition for licensing other dams located on the Clyde River. Ultimately, the dam's owner reached agreement with resource management agencies and conservation groups. It was removed in late August 1996 before the fall spawning run of the Atlantic salmon, and salmon were seen moving past the site (EDF 1998d, http://www.flyshop.com/News/08-96damgone). The increased interest and willingness to consider dam removal at other sites in the northeastern United States have partly been attributed to the experience gained in removing the Newport No. 11 Dam on the Clyde River in Vermont.

Quaker Neck Dam, Upper Neuse River, North Carolina, United States.

The Quaker Neck Dam was built on the upper Neuse River in North Carolina in 1952 to provide cooling water for energy production. The dam was 2-meters high and nearly 100 meters long and blocked migratory fish species from reaching historic spawning grounds. It also affected two species now protected under the ESA—the shortnose sturgeon and the dwarf wedge mussel (AR & TU 1999). All together, this dam blocked off 125 kilometers of the river and over 1,400 kilometers of tributaries. The U.S. Fish and Wildlife Service officially classified the Quaker Neck Dam as a barrier to fish in 1989, but removal was delayed by the need to find an alternative source of cooling water for the Carolina Power and Light Company, which owned the dam. Between 1993 and 1996 an alternative was developed and an agreement worked out to remove the dam. Deconstruction began in December 1997 and was completed by August 1998 for a total cost of only $180,000 (AR & TU 1999). The Neuse is now open to the sea for nearly 350 kilometers.

   Removal of the dam was done as part of a voluntary watershed restoration project carried out by a public–private partnership involving federal and state agencies, fisheries groups, and the Carolina Power and Light Company. Fishing generated nearly $2 billion in revenue for North Carolina in 1996, and there are hopes that opening up the tributaries of the Neuse will generate significant new revenue in rural areas. Scientists are doing studies to determine whether fish will take advantage of their new range (U.S. Water News 1999a).

Some Proposed Dam Removals or Decommissionings

Scotts Peak Dam, Gordon River, Tasmania, Australia

Scotts Peak Dam on the Gordon River in Tasmania has been proposed for removal. The dam created a reservoir that flooded Lake Pedder, a lake in Tasmania famous for the beauty of its pink quartzite beach. Local environmentalists have proposed draining the reservoir and restoring Lake Pedder to its natural state. Scotts Peak Dam produces hydropower, but local activists believe that the hydropower from the dam can be replaced or simply eliminated because Tasmania has a substantial surplus of generating capacity. The Tasmanian government and the Opposition Party oppose the proposed removal. Further research and planning are needed before a final decision is made. Pedder 2000, a campaign to restore Lake Pedder, has been launched with the goal of draining the reservoir (EDF 1998b; (http://192.111.219.6/programs/International/Dams/AsiaOceania/t_ScottsPeak.html).

Elwha and Glines Canyon Dams, Elwha River, Washington State, United States

The Elwha Dam is a 32-meter-high, gravity/concrete–faced rockfill/earthfill dam, built in 1913. It is only 8 kilometers from the mouth of the Elwha River. The Glines Canyon Dam is located about 12 kilometers above Elwha Dam and is nearly three times higher at 82 meters. Glines Canyon Dam was built between 1925 and 1927 and created Lake Mills. Both dams have hydroelectric facilities and neither has fish ladders. Glines Dam is licensed by the Federal Power Commission and operates on annual licenses. The Elwha Dam has not been licensed. Most of the lands inundated by Lake Mills became part of Olympic National Park in 1940. The Olympic National Park is now a World Heritage Site (for more information see, http://www.nps.gov/olym/issues/isselwha2.htm).

The Elwha River drains part of the park, though both dams were built before the national park was created. Their combined installed electric capacity is only 19 MW. Both dams were built by a logging company to supply electrical power to a wood products plant at Port Angeles. The Elwha River system, about 80 percent within the boundaries of the Olympic National Park, was once a rich spawning ground for salmon and other anadromous fish. The Elwha S'Klallam Tribe had guaranteed fishing rights to these fisheries through a treaty with the U.S. government. At least one Elwha River salmon stock, the sockeye, is now thought to be extinct. Two others, the pink and spring chinook, are present only in small numbers.

In 1992, the U.S. Congress passed the Elwha River Ecosystem and Fisheries Restoration Act of 1992. This law authorizes the secretary of the interior to purchase and remove both dams to fully restore the ecosystem and native anadromous fisheries. The National Park Service completed two Environmental Impact Statements (EIS), which concluded that both dams must be removed to meet the goals of the law. The U.S. Department of Interior is now studying how the river can be completely restored. The plan after the decommissioning of the dams is to fully restore the ecosystem and the native fisheries (http://www.nps.gov/olym/issues/isselwha2.htm).

One significant problem with the proposal to dismantle these two dams is management of the sediment stored in the reservoirs. These sediments must be carefully redistributed in a fashion to minimize downstream effects, and there is little practical experience in this. Estimated costs for dismantling the dams and to physically remove the sediments are between $60 and $200 million (Collier et al. 1996, http://www.edf.org/programs/International/Dams/NAmerica/e_ElwhaGlines.html.)

In April 1998 additional pressure to remove the dam was applied when President Clinton included $154 million in his proposed budget toward demolishing both dams. In a political subterfuge, Senator Slade Gorton from Washington State proposed demolishing the Elwha Dam under the condition that a series of other proposals to help salmon on the larger Columbia River system be delayed or eliminated. The proposal would also have prohibited demolition of the Glines Canyon Dam for at least 12 years. Environmentalists denounced the proposal and no action was taken on it (http://www.oregonlive.com/todaysnews/9804/st04043.html).

Savage Rapids Dam, Rogue River, Oregon, United States

The U.S. Fish and Wildlife Service and the federal Bureau of Reclamation currently support a $12.6 million plan to remove the 77-year-old Savage Rapids Dam on the Rogue River in the northwestern United States and replace it with irrigation pumps to deliver water to users. A Fish and Wildlife Service study estimates that 22 percent more salmon and steelhead a year could be produced if the dam were not there. The Grants Pass Irrigation District (GPID), which owns the dam, disagrees. GPID estimates the loss of fish at more like 1 to 2 percent annually, and it supports a $15 million option to retrofit the dam. In mid-1998, Secretary of the Interior Bruce Babbitt supported evaluating the alternatives, including identifying funds that might be available for dam removal (Freeman 1998). In September 1998, the Oregon State Department of Water Resources Commission ruled that the Grants Pass Irrigation District had failed to diligently pursue efforts to remove the dam as part of a salmon restoration evaluation. The commission is pushing forward to remove the dam, as is the National Marine Fisheries Service (http://www.oregonlive.com/todaysnews/9809/st091310.html).

The Peterson Dam, Lamoille River, Vermont, United States

The Lamoille River, a tributary of the Connecticut River, has a variety of barriers to migratory fish. Over the past 50 years, dams on the Lamoille have affected water quality and fisheries, blocking migration and access to spawning habitat for salmon and sturgeon. In 1998, following more than a decade of negotiations, mediation, and regulatory battles, local environmental groups and other organizations (including Trout Unlimited and the Vermont Natural Resources Council) won an agreement that removal of the largest downstream dam—Peterson Dam—would be studied and considered as an alternative in both the Vermont water quality certification hearings and in FERC relicensing. Three other upstream dams will also be studied to see if flow modifications or fish ladders might be appropriate (Trout Unlimited 1998). Altogether, the four dams have an installed capacity of about 17 MW of power. The agreement requires that the owner of Peterson Dam must consult with Trout Unlimited and the state of Vermont in planning dam-removal studies and determining if dam removal is warranted.

Glen Canyon Dam, the Colorado River

In what was one of the greatest early battles of the environmental movement in the United States in the 1950s and 1960s, great public opposition prevented the construction of a series of massive dams in and around Grand Canyon National Park, one of the most phenomenal natural wonders on the planet. As part of that battle, however, a compromise that prevented flooding in the parks permitted the construction of a single huge dam above the park that destroyed a little-known but spectacularly beautiful site called Glen Canyon. The dam was authorized in the late 1950s and was completed in 1963 (http://www.glencanyon.org/FACTSDAM.HTM), creating the vast Lake Powell with more than 33 billion cubic meters of storage capacity. Today, 1 billion cubic meters of sediment have settled to the bottom of that reservoir, sand beaches in the stretches downstream of the dam have been deprived of the replenishing sediments that used to come down the river, two native fish species have been extirpated from Grand Canyon, and the dam has critically endangered a third. A wide range of other ecological impacts is felt all the way to the Colorado River delta in Mexico.

This dam is perhaps the most despised dam in the United States and has become a rallying point for many environmental activists and movements. A famous novel, *The Monkey Wrench Gang,* has been written about efforts by a mythical band of eco-warriors to destroy the dam (Abbey 1976). In recent years, real-life efforts have gained momentum to do the same thing with legislation and public opinion rather than explosives. An organization (the Glen Canyon Institute) has even been formed whose sole objective is the restoration of free flows of the Colorado River through Glen and Grand Canyons by decommissioning (not necessarily removing) Glen Canyon Dam. Richard Ingebretsen, a Salt Lake City physician who visited Glen Canyon as a Boy Scout before it was inundated, grew increasingly dismayed at having to accept its loss and founded the institute in 1995. An inspiration to this group is David Brower, who was largely responsible for protecting the Grand Canyon from the original dam proposals and who also feels that the compromise that led to the construction of Glen Canyon Dam is his greatest failure (Brower 1990).

The proposal to drain Lake Powell is supposedly based upon three factors: the biological impacts of the dam, economic considerations, and dam safety concerns. Probably just as important, however, is the desire to right a wrong—to attempt to restore a natural wonder destroyed in an era when such things had little or no economic or political value.

Dams have drastically altered the ecosystems of the Colorado River, from the headwaters all the way to the Sea of Cortez. Prior to the construction of several large dams along the river, the Colorado provided breeding and rearing grounds for many endemic species. The construction of Glen Canyon Dam has affected the ecosystem of Grand Canyon National Park more than any other human development. Warm spring floods that previously deposited millions of tons of vital sediment and nutrients have been replaced with nutrient-poor, cold, regulated flows. Native fish that evolved in the pre-dam environment have been unable to adapt to the new conditions, and they are increasingly displaced by nonnative, introduced species.

Evaluating the economic costs and benefits of Glen Canyon Dam, or conversely, removing Glen Canyon Dam, is extremely difficult. While economic estimates of the dam's hydroelectric and recreational benefits are readily available, ecological and environmental costs are notoriously difficult to evaluate. According to Dan Beard, former Commissioner of the Bureau of Reclamation, the long-term costs associated with maintaining the reservoir offer the most compelling argument in favor of decommissioning the dam (Hyde 1999). The electric power provided by Glen Canyon Dam is an asset to portions of several states in the region. But a study conducted by the Environmental Defense Fund determined that the power generated represents a mere 3 percent of the total power use for an area that currently has a surplus. With energy deregulation sweeping the nation, more flexible and competitive power sources might be available at even lower costs (Rosekrans 1999).

In the arid southwest, water is more valuable than power. The role of the dam and reservoir in meeting downstream water demands must be more clearly evaluated, together with the implications for meeting those demands without the facility. Complex upstream and downstream water management issues must be evaluated, including the impacts on the timing and amount of water that will be available to users in the seven states that border the Colorado River, as well as Mexico, which receives Colorado River water under treaty with the United States. The enormous reservoir behind Glen Canyon Dam also loses a tremendous amount of water to evaporation in the hot desert climate: an estimated 1 billion cubic meters of water are lost to evaporation and seepage annually. Insufficient information is available on how operations and water contracts would be met with a different set of infrastructure and operating rules than is currently in place.

An argument is also being made that the safety of the dam must be reevaluated. During the summer of 1983, high flows nearly resulted in catastrophic spillway failure. The flood subsided just in time, but only after inflicting significant and costly damage to the spillways, and the historical hydrologic record suggests far higher flows are possible.

Catastrophic failure of the dam would have enormous environmental and economic consequences, and could result in the loss of human life, yet evaluating the economic costs associated with these high-consequence, low-probability events has yet to be done.

Many other difficult issues must be addressed before any proposal to remove this dam can be approved. The reservoir behind the dam now provides recreational opportunities to millions of visitors annually. Proponents of dam decommissioning note that the water behind Glen Canyon Dam, or flood flows now trapped in wet years by the dam, might be useful in helping to restore ecosystems in both the Salton Sea and the Colorado River Delta in Mexico—both sites of ongoing discussion and controversy (Morrison et al. 1996, Cohen et al. 1999).

The decommissioning of the dam and the draining of Lake Powell are extremely unlikely to occur in the next few years. Nevertheless, even the idea of taking action on such a large scale has attracted serious public and scientific attention. No formal federal efforts are underway to evaluate the issues associated with removing Glen Canyon Dam, but using the process required of federal agencies under the National Environmental Policy Act, the Glen Canyon Institute has begun to facilitate a citizen-led environmental assessment of the proposal to restore a free-flowing Colorado River through Glen and Grand Canyons. As part of that effort, scientific and technical studies of a number of questions are being done, covering issues such as biological and cultural consequences, economic implications, dam safety, and social/policy issues.

Ultimately, federal agencies will have to become involved in any discussions involving eliminating the dam and restoring Colorado River ecosystems. In a remarkable acknowledgment of the possibility that valid arguments and information might be found to justify removing such a large dam, a few panicked elected officials have actually passed legislation banning the use of federal funds to even *study* these issues. Language in a bill passed by the U.S. House of Representatives on September 23, 1999, prohibits the Interior Department from using appropriated funds "to study or implement any plan to drain Lake Powell or to reduce the water level of the lake below the range of water levels required for the operation of the Glen Canyon Dam." This is the second consecutive year that Congress has included this type of ban in appropriations legislation (http://www.glencanyon.org). Dam removal advocates will continue to watch this particular project with great interest.

Ice Harbor, Lower Monumental, Little Goose, and Lower Granite Dams, Lower Snake River, Washington State, United States

A vast network of dams is in place in the Pacific Northwest of the United States. Some of the dams date back to the Great Depression, when massive public works projects helped to employ huge numbers of people, bring hydroelectricity to remote regions, and open shipping channels far from the ocean. Four major dams (Ice Harbor, Lower Monumental, Little Goose, and Lower Granite) were built by the U.S. Army Corps of Engineers on the Lower Snake River—a major tributary of the Columbia River—between 1962 and 1975. Each dam is about 30 meters high and generates electricity: the total hydropower generation capacity of the four dams is about 1,230 MW. These dams produce around 7.2 billion kWh of electric power per year, approximately 4 percent of the total electricity used in the Northwest. The four dams on the Lower Snake River are operated as "run of the river" dams, which means that they have minimal storage capacity and their operation does not significantly alter the timing of flows. As such, they provide no flood control. They do not store water for irrigation, though some irrigation withdrawals are made from the Ice Harbor pool. The dams are also operated in conjunction with four other major dams on the Lower Columbia River to allow barges to travel between Lewiston, Idaho, and Portland, Oregon, via navigation locks. Barges are able to ship goods at a slightly lower rate than shipping via rail, if the ecological costs of the dams and waterways are excluded from the calculation.

Two federal agencies, the U.S. Army Corps of Engineers and the National Marine Fisheries Service, with the support of a variety of local environmental and community groups, Native Americans, and fishermen, are now exploring proposals to remove or at least breach the four Lower Snake River dams in a process that is proving to be far more contentious than building them in the first place. Opposing the plan are farmers and other interests in the region that rely on the water, inexpensive electricity, and navigational benefits provided by the dams.

The discussion about removing these dams is the result of their devastating negative effect on Snake River salmon and on the economics of the dams themselves. When Lewis and Clark first saw the Snake nearly two hundred years ago, an estimated 2 million fish a year returned to their spawning grounds. Today, Snake River salmon must navigate a total of eight major dams from their hatching grounds in Idaho down 1,400 kilometers to the Pacific Ocean and back up again two years later to spawn. Major losses occur among juvenile fish heading downstream. Fisheries biologists calculate that the four Snake River dams kill vast numbers of young fish on their way to the ocean. Altogether, warm waters in reservoirs, nonnative predators, and spinning turbines kill more than 90 percent of the young salmon before they can even reach the ocean. The surviving juvenile salmon reach the Pacific where they are subject to other predatory and environmental pressures.

Over $3 billion have already been spent on salmon recovery efforts in the Columbia basin. These funds have paid for dam bypass systems, trucks and barges to transport young salmon around turbines, and the construction of hatcheries (Lammers 1998, Verhovek 1999). These efforts have not been successful. In 1998, the Idaho Department of Fish and Game counted 8,426 spring and summer Chinook, 306 fall Chinook, and 2 sockeye (Verhovek 1999). Snake River coho are now considered extinct, sockeye are severely depleted, and chinook and steelhead are listed as endangered species.

In the ongoing effort to help the fisheries, the U.S. Army Corps of Engineers (ACoE) was directed in the mid-1990s to study three options for salmon restoration, including dam removal, barging of fish around the dams, and surface "bypass collectors" to assist fish migration around the dams. In an interim report, the ACoE found that the bypass collectors are not feasible, that barging is a continuation of the status quo, and that dam removal warranted further investigation. The report concluded that removing the four dams across the lower Snake River would be the most effective—but may also be the most expensive—option for restoring the fisheries. A study that same year by scientists for the Northwest Power Planning Council reached similar conclusions: there is an 80 to 95 percent chance that wild Snake River salmon and steelhead runs would recover to 1960s levels if the dams were breached (Independent Science Group 1996).

Serious consideration is now being given to the idea of breaching the dams, and further investigation has been turned over to the Snake River Dam Removal Economics Working Group. Another ACoE review is scheduled to produce a draft by the end of 2000. The National Marine Fisheries Service (NMFS), the U.S. agency charged with protecting and restoring endangered species, is also examining changes in the operation of the entire river system, including breaching or bypassing the dams and augmenting flows for fish survival (EDF 1998c). The NMFS has produced a biological opinion about the impact of hydropower operations as part of their legal responsibility under the ESA. They concluded that drawdown of the reservoirs to natural levels is the long-term recovery option with the greatest degree of scientific certainty (http://research.nwfsc.noaa.gov/afis/index.html).

There is tremendous controversy over the net economic and ecological effects of removing the dams. Many stakeholders have conflicting interests in keeping or removing the dams. The dams were built with public funds, and fisheries advocates note that in addition to the huge expenses for salmon restoration efforts, taxpayers are subsidizing the dams by paying $18 million each year to pay off the debt for building the locks and $25 million a year to maintain and operate them. A modest fuel tax was levied on barges in 1979 in an effort to get them to share more of the costs, but these fees are not adequate. Water used to pass barges through the locks cannot be used to produce hydroelectric power. In the mid-1990s, this lost revenue was estimated at around $1.4 million annually (http://www.cyberlearn.com/remove.htm).

Advocates of removing or breaching the dams argue that the four Lower Snake River dams produce relatively little power compared to other dams in the watershed, that this power could be easily and inexpensively replaced, that only a few farmers depend on irrigation water from the reservoirs, and that cost-effective alternatives to barges are available for moving products from Lewiston to Portland (http://www.cyberlearn.com/remove.htm).

A local fishing organization, Idaho Rivers United, estimated that bypassing and mothballing the dams could be done for a one-time investment of $500 million and would take two years to fully implement (http://www.cyberlearn.com/remove.htm). They estimated that plans to maintain the current operation of the dams would cost $435 million each year. The western division of the American Fisheries Society has also come down on the side of removing the Lower Snake River dams. The Society resolved by a 115–47 vote in summer of 1999 that the dams must be removed if wild runs of Idaho salmon and steelhead are to be saved (http://www.amrivers.org/membersh.html).

A study commissioned by the Oregon Natural Resources Council (ONRC) estimated that, altogether, total operating costs for the dams, including subsidies, amounted to over $230 million a year (http://cnn.com/EARTH/9805/07/snake.river.dam). The report concluded that the region would see a net gain of between $87 and $183 million by eliminating subsidies for salmon-recovery programs and ongoing maintenance costs. The ONRC study includes some simple estimate of the benefits to local economies from recovered fisheries (http://www.onrc.org/wild_oregon/salmonriver98/salmonriver98.html).

No study adequately quantifies the potential economic benefits resulting from a restored salmon fishery. These economic benefits are difficult to quantify, but they are potentially very large and would certainly offset at least part of the costs of dam removal. Yet even the costs of dam removal are uncertain. Estimates for how much it would cost to remove the dams vary enormously, from hundreds of millions of dollars to as much as a billion dollars.

Some local industry and farmers argue that the dams provide low-cost electric power and that healthy salmon populations can coexist with the dams. The Columbia River Alliance (CRA), a coalition of navigation, agricultural, and timber interests, opposes decommissioning of these dams. They argue that drawdown of the lower Snake River to "natural river" levels by itself would likely not provide any greater benefits to salmon survival than the current or improved program of barging juvenile fish past the dams for release downstream. They argue that if the Snake River dams were removed, the fish would still encounter four more dams on the Columbia River. Their choice is an artificial transportation program that can move young fish past all eight dams. The CRA also argues that removing the dams would have a wide range of impacts on the people of the region. The social costs of removing the dams would be borne by navigation interests, as well as some growers, local communities, recreational interests, and electricity users.

These conflicting figures and positions illustrate the complex nature of this issue. A final decision is likely to be one of two options: natural river drawdown (dam removal or decommissioning) and surface collection technology, where young fish are collected and transported downstream via barge to improve survival rates sufficiently to achieve recovery. Congressional authorization to implement natural river drawdown would be requested if the surface-collection options proved insufficient. It should be noted that natural river drawdown would not restore the natural hydrograph, which is still controlled by water releases from upstream reservoirs. Some additional flow augmentation would be necessary even after dam decommissioning.

In August 1999, the Resources Committee of the U.S. House of Representatives approved a resolution opposing the removal of the four Snake River dams. This resolution lacks any legal authority, but underscores the opposition of some of the parties involved (U.S. Water News 1999b). Senator Slade Gorton (R-WA) has said that the breaching of the dams would be "an unmitigated disaster and an economic nightmare," but 107 members of Congress called the fisheries "an economic and environmental asset whose preservation is a national responsibility" in a letter to President Clinton urging consideration of the proposal (Verhovek 1999). They noted that if dam removal is done as a part of a comprehensive strategy toward restoring and recapturing the economic value of salmon harvests for the region, then at some point it becomes more economical to remove these dams than to maintain them. The final decision is likely to rest in the hands of Congress.

CONCLUSION

A remarkable change in perception and public opinion over the past few decades has created a new movement—efforts to remove dams that can no longer be justified because of safety, economic, or ecological reasons. Hundreds of small- and medium-size dams have already been destroyed around the world in efforts to restore ecosystems and protect human property and lives. And local or regional environmental and community groups have now begun to target some of the world's largest dams for removal. It may be many years before any of these massive dams are actually decommissioned, but water managers can no longer take for granted the continued, unexamined operation of dams that no longer provide adequate benefits compared to their cumulative environmental and social costs and impacts.

REFERENCES

Abbey, E. 1976. *The Monkey Wrench Gang.* Avon Books, New York

American Rivers and Trout Unlimited (AR & TU). 1999. *Dam Removal Success Stories.* Draft Report (June). http://www.amrivers.org

Brower, D. 1990. *For Earth's Sake: The Life and Times of David Brower.* Peregrine Smith Books, Salt Lake City.

Cheever, D. 1998. "It's official: Tide to pass through Augusta." *Kennebec Journal Morning Sentinel.* http://www.centralmaine.com/052798/news01.html. (May 27).

Cohen, M.J., J.I. Morrison, and E.P. Glenn. 1999. *Haven or Hazard: The Ecology and Future of the Salton Sea.* Pacific Institute for Studies in Development, Environment, and Security, Oakland, California (February).

Collier, M., R.H. Webb, and J.C. Schmidt. 1996. *Dams and Rivers: Primer on the Downstream Effects of Dams.* U.S. Geological Survey Circular 1126. Tucson, Arizona.

Environmental Defense Fund (EDF). 1998a. Edwards Dam. http://www.edf.org/programs/International/Dams/NAmerica/d_Edwards.html

Environmental Defense Fund (EDF). 1998b. Scotts Peak Dam at Lake Pedder/Gordon River, Australia Proposed for Removal. http://www.edf.org/programs/International/Dams/AsiaOceania/t_ScottsPeak.html

Environmental Defense Fund (EDF). 1998c. Lower Snake River Dams Proposed for Removal. http://www.edf.org/programs/International/Dams/NAmerica/f_Snake.html

Environmental Defense Fund (EDF). 1999d. Newport No. 11 Dam. http://www.edf.org/programs/International/Dams/NAmerica/g_Newport.html

Environmental Defense Fund (EDF). 1998e. Maisons-Rouge & Saint-Etienne-du-Vigan Dams.http://www.edf.org/programs/International/Dams/EuroMidE Russ/h_MaisonsRouges.html

Environmental News Network (ENN). 1997. "Maine's Edwards Dam to come down." http://www.enn.com/enn-news-archive/1997/11/112697/11269714.asp (November 26).

Environmental News Network (ENN). 1998. "Pact clears way for Edwards Dam removal." http://www.enn.com/enn-news-archive/1998/05/052798/edwards.asp. (May 27).

Epple, R. 1998. "Second dam demolished for salmon." *World Rivers Review,* Vol. 13, No. 6, p. 4.

Friends of the River (FOTR). 1997. "California rivers eligible for National Wild and Scenic Status." *FOTR Newsletter* Sacramento, California.

Freeman, M. 1998. "Babbitt: Fish can't wait forever." *Oregon Mail Tribune.* Available at http://www.mailtribune.com/archive/98/july98/71698n2.htm. (July 16).

Haberman, R. 1995. "Dam fights of the 1990s: Removals." *River Voices,* Vol. 5, No. 4 (Winter). Available at http://rivernetwork.org/dam.htm.

Hyde, P. 1999. Glen Canyon Institute, personal communication.

Independent Science Group. 1996. "Return to the river." Report to the Northwest Power Planning Council, ISG 96-6. Available at: http://www.nwppc.org/ftpfish.htm#I2.

Keefer, M. 1990. "Flowing free." *Milwaukee Journal* (June 10, p 27).

Lammers, O. 1998. "Undamming the Snake River to free salmon." *World Rivers Review,* Vol. 13, No. 6, p. 5

McClurg, S. 1998. *Western Water.* Water Education Foundation. Sacramento, CA. January/February, pp.4–13.

McPhee, J. 1999. "Farewell to the nineteenth century: Letting the country's rivers run free." *New Yorker,* Vol. 75, No. 28 (September 27), pp. 44–53.

Morrison, J.I., S.L. Postel, and P.H. Gleick. 1996. *The Sustainable Use of Water in the Lower Colorado River Basin.* Pacific Institute for Studies in Development, Environment, and Security, Oakland, California (November).

New York Times. 1999. "Rethinking Dams." *New York Times* (July 7), editorial page.

Owens-Viani, L., and A. Wong. 1999. "Improving passage for spring-run salmon: Cooperative efforts on Deer, Mill, and Butte Creeks." In (L. Owens-Viani, A.K. Wong, and P.H. Gleick eds.), *Sustainable Use of Water: California Success Stories,* Pacific Institute for Studies in Development, Environment, and Security, Oakland, California, pp. 241–254.

Reisner, Marc. 1997. "Deconstructing the age of dams." *High Country News,* Vol. 29, p. 20 (October 27). Available at: http://www.hcn.org/1997/oct27/dir/Feature_Deconstruc.html.

River Alliance of Wisconsin. 1999. "Restoring rivers through selective small dam removal." Available at: http://www.igc.apc.org/wisrivers/smldam.html.

Rosekrans, S. 1999. Environmental Defense Fund, personal communication.

State of North Carolina. 1967. Dam Safety Law. Available at http://www.ehnr. state.nc.us/ENR/DLR/MELL/dams.htm.

Trout Unlimited. 1998. "Lamoille could see salmon return." http://www.tu.org/ trout/newsstand/ll/October_1998/5/iart.sub, (October).

United States Army Corps of Engineers (U.S. ACoE). 1996. *National Inventory of Dams.* http://crunch.tec.army.mil

United States Department of the Interior (U.S. DoI). 1998. "Interior Secretary signs landmark conservation agreement." http://www.doi.gov/news/980526r. htm (May 26).

United States House of Representatives. 1994. *BPA at the Crossroads.* Congressional report. The Natural Resources Committee for the BPA Task Force, Peter DeFazio, Task Force Chairman, Washington, D.C. (May).

United States National Oceanic and Atmospheric Administration (U.S. NOAA). 1999. "Fisheries Service makes final decision for protection of Chinook salmon populations in California, Oregon." U.S. Department of Commerce, National Oceanic and Atmospheric Administration News Release, September 9. NOAA 99-R150. see also http://www.nwr.noaa.gov and http://swr.ucsd.edu.

U.S. Water News. 1999a. "Removal of dam clears way for spawning runs." *U.S. Water News,* Vol. 16, No. 6, p. 3.

U.S. Water News. 1999b. "House panel votes against removal of Snake River dams." *U.S. Water News,* Vol. 16, No. 9, p. 3.

Verhovek, S.H. 1999. "Returning river to salmon, and man to the drawing board." *New York Times* (September 26), p.1.

Williams, T. 1993. "Freeing the Kennebec." *Audubon Magazine* (September/October 1993), pp 36–42.

Water Reclamation and Reuse: Waste Not, Want Not

As human pressures on limited water resources increase worldwide, alternative sources of supply are attracting more and more attention. Desalination, addressed in Chapter 5 in this book, has long been considered a possible source of water. Another unconventional source is water that has been used before, recaptured, and cleaned for reuse. There is a growing conviction in many parts of the world that wastewater must not be treated as a liability, but as a resource capable of meeting many different needs. In some regions, treated wastewater is already being used for a wide range of purposes, from industrial and commercial production to agricultural irrigation. Highly treated wastewater is even being used directly for drinking water. As populations and water demands grow, this old waste product will increasingly be viewed as a new and valuable resource.

The history of water reuse stretches back 5,000 years to when it was used as a source of water for agricultural production in the Minoan civilization (Angelakis and Spyridakis 1995). Roman and Greek waterworks often had separate water supply and sewage systems, but most early civilizations failed to understand the links between contaminated water and human illness. As a result, they failed to separate wastewater from pristine water supplies, which often resulted in disastrous epidemics of water-related diseases, including typhoid and cholera (Gleick 1998). While major European and American cities built separate sewers in the early 1800s for disposal of human wastes, these wastes were often dumped right back into the very water bodies from which water supplies were then withdrawn. Not until the medical community figured out the links between contaminated water and human health were efforts made to separate and treat wastewater. These efforts began in the mid-1800s and included the development of independent water-supply systems, relocation of water intakes and water discharges, and the introduction of a range of water filtration and treatment technologies. At this point, water containing human or industrial wastes began to be treated only as a liability, not an asset.

A renewed look at planned use of wastewater began in the early twentieth century when water scarcity and the disposal of wastewater became problematic. In the United States, intentional use of reclaimed wastewater for irrigation of grain and field crops began in California as early as 1912, and the State of California issued reuse regulations in 1918 addressing the use of sewage for crop production. A wastewater reclamation plant was constructed by the U.S. government in 1926 to treat and reuse water for lawn irrigation and power-plant cooling water at Grand Canyon National Park in Arizona. In 1929 the City of Pomona, California, began providing treated wastewater for outdoor landscape use (Cologne 1998). Industrial wastewater reuse for steel processing began in the 1940s and urban water reuse systems began to be developed in the 1960s. The Israeli Ministry of Health issued regulations on the use of secondary effluents for crop irrigation in 1965 (Asano and Levine 1998). Extensive research on direct potable reuse began in Windhoek, Namibia, in 1968, and that city continues to have one of the highest levels of direct potable reuse worldwide (van der Merwe 1999)

Treated wastewater can be used for nearly any purpose, though at present it is most economical for large users located close to water-treatment facilities, such as large parks and playing fields, industries with large cooling water needs, golf courses, and certain types of farms. While extensive use of reclaimed wastewater directly for drinking water is unlikely to gain wide acceptance, a wide range of alternative nonpotable uses can be satisfied. Even today, because of its flexibility, wastewater is increasingly attractive as a reliable source of supply where water systems must meet greater and greater demands. Box 7.1 presents some definitions of terms used in this chapter.

Wastewater can be treated to meet a wide range of standards and water-quality objectives, with the primary objectives being the protection of public health and the quality of ecosystems. Wastewater treatment consists of a combination of physical, chemical, and biological processes, depending on the final desired quality of the water. The level of treatment, the processes involved, and their applications are described here and in Table 7.1.

Primary treatment typically refers to the simple removal of solids and particulate matter from wastewater. Primary treatment processes remove about 50 percent of suspended solids, 25 to 50 percent of biological oxygen demand, 10 to 20 percent of organic nitrogen, and about 10 percent of phosphorus. Primary treatment does not provide adequate treatment to meet most water-quality objectives. As a result, most industrialized countries rely on a wide range of secondary treatment methods to further improve water quality.

Secondary treatment also separates liquids and solids and includes biological processes to remove microbiological contaminants. Many different combinations of secondary processes are available to remove suspended solids and nutrients and to eliminate organic constituents. Examples include aerobic processes to encourage biological metabolism of organics by microorganisms, and oxidation ponds to break down bacteria and ammonia.

Tertiary treatment typically occurs downstream of biological processes and includes processes for the additional removal of solids and nutrients, and the inactivation of pathogenic organisms. A wide range of processes is available for tertiary treatment of wastewater, including chemical coagulation and flocculation, filtration, and disinfection. Advanced tertiary treatment techniques are increasingly being applied where higher quality water is needed or where wastewater may eventually end up back in potable supplies. Advanced treatment involves highly specialized techniques for filtering, disinfecting, or otherwise treating water to meet particular strict standards.

BOX 7.1

Definitions

Wastewater reclamation is the treatment of wastewater to make it reusable.

Wastewater reuse is the beneficial use of treated water.

Water recycling or wastewater recycling is the use, capture, and reuse of water by a single user.

Nonpotable water reuse refers to the use of reclaimed water for applications other than drinking water.

Potable water reuse refers to the use of reclaimed water for drinking water.

Direct reuse refers to a direct connection between a water-treatment system and a reuse application.

Direct potable reuse involves the direct conveyance of treated wastewater to a potable-water distribution system.

Indirect reuse involves mixing, dilution, and dispersion of reclaimed wastewater into another water system before additional use. Indirect reuse is not normally defined as planned reuse.

Incidental, or unplanned indirect potable reuse is the discharge of treated wastewater into a surface or underground source of water that is subsequently used for drinking.

Sources: Modified from Asano and Levine 1998, McEwen 1998, NAS 1998.

Wastewater Uses

Wastewater can be reused for any purpose. What varies is the level of treatment required to bring it to a quality appropriate for a particular need. Effective integration of wastewater reuse into overall water planning thus requires careful assessment of both water needs and water quality. In regions where agricultural water demands are high and other sources of supply are limited, like the Middle East and North Africa, the use of reclaimed water for irrigation can be an important application. Because of water scarcity, South Africa treats wastewater as too valuable for use in irrigation, with the exception of use for parks and playing fields where wastewater can replace water taken from high-quality municipal sources, or where wastewater would be lost to a saline sink such as the ocean (Odendaal et al. 1998). In industrialized nations, water withdrawals for power-plant cooling often make up nearly half of all water withdrawals, but only a small fraction of total consumptive use. Where industrial and commercial water demands are high, as in Japan, reuse

TABLE 7.1 Wastewater Reclamation Processes

Process	Description	Application
Separation of Solids and Liquids		
Sedimentation	Gravity sedimentation, chemical flocculation, precipitation.	Removes particles larger than about 30 micrometers (1 μm = 10^{-6} m). Typically used for primary treatment.
Filtration	Particle removal by passing water through porous medium.	Removes particles larger than about 3 μm. Typically used following sedimentation.
Biological Treatment		
Aerobic treatment	Biological metabolism by microorganisms.	Removal of dissolved and suspended organic matter.
Oxidation pond	Ponds with approximately 1-meter depth for mixing and sunlight penetration.	Reduction of suspended solids, BOD, pathogenic bacteria, and ammonia.
Biological nutrient removal	Combination of aerobic, anoxic, and anaerobic processes.	Conversion of organic and ammonia nitrogen to molecular nitrogen; removal of phosphorus.
Disinfection		
Oxidizing or caustic chemicals	Inactivation of pathogenic organisms.	Protection of public health by removal of pathogenic organisms.
Ultraviolet light	Inactivation of pathogenic organisms.	Protection of public health by removal of pathogenic organisms.
Separation processes	Physical separation of pathogenic organisms.	Protection of public health by removal of pathogenic organisms.
Advanced Treatment		
Activated carbon	Contaminants are physically absorbed onto surface of activated carbon.	Removal of hydrophobic organic compounds.
Air stripping	Transfer of ammonia and other volatile constituents from water to air.	Removal of ammonia nitrogen and volatile organics from wastewater.
Ion exchange	Use of exchange resins to separate ions.	Effective for removal of cations such as calcium, magnesium, iron, ammonium, and anions such as nitrate.
Chemical coagulation/ precipitation	Use of salts, polyelectrolytes, or ozone to destabilize colloidal particles from wastewater and precipitation of phosphorus.	Formation of phosphorus precipitates and flocculation of particles for removal by sedimentation and filtration.
Lime treatment	Use of lime to precipitate cations and metals from solution.	Used to reduce phosphorus, scale-forming minerals, and to modify pH.
Membrane filtration	Micro-, nano-, and ultrafiltration.	Removal of particles and microorganisms.
Reverse osmosis	Membrane filtration to separate ions from solution.	Removal of pathogens and dissolved salts and minerals.

Source: Modified from Asano and Levine 1998.

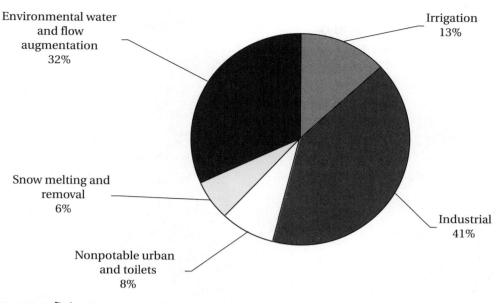

Environmental water
and flow
augmentation
32%

Irrigation
13%

Snow melting and
removal
6%

Industrial
41%

Nonpotable urban
and toilets
8%

FIGURE 7.1   Wastewater Reuse in Japan, 1990s

Wastewater is reused in Japan for a variety of purposes. Over 40 percent goes to meet industrial water needs, and over 30 percent is used to supplement natural flows in rivers and streams. Less than 15 percent is used for irrigation of crops. *Source:* Asano et al. 1996.

TABLE 7.2   Use of Recycled Water in California by Category for Survey Years (amounts in million cubic meters)

	1987	1989	1993	1995	2000[a]
Agricultural irrigation	207	213	99	179	201
Groundwater recharge	48	86	228	150	167
Landscape irrigation	49	67	58	94	141
Industrial uses	7	7	9	36	40
Environmental uses	12	22	36	19	20
Other	5	5	44	76	77
Total	328	400	474	554	646

Sources: Wong 1999b, MacLaggan 1999.

[a]Year 2000 figures from 1996 survey of "existing" facilities.

applications in these sectors predominate. Figure 7.1 shows that 41 percent of all wastewater reuse in Japan occurs in the industrial sector. California currently reuses more than 640 million cubic meters of wastewater annually for a wide-range of purposes (Table 7.2 and Figure 7.2), primarily groundwater recharge and irrigation. Because of the growing interest in this source of supply, considerable effort was made several years ago to identify different kinds of wastewater uses and to set criteria for the kinds of wastewater treatment required for each use. Table 7.3 shows the most recent California regulations set in this area.

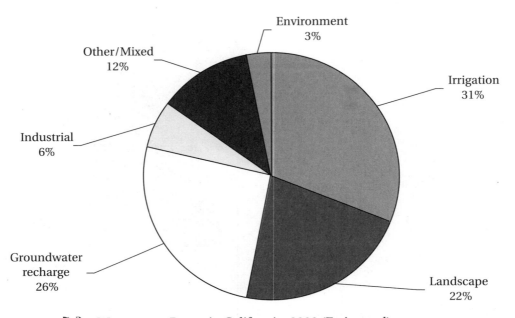

FIGURE 7.2   Wastewater Reuse in California, 2000 (Estimated)

The three principal uses of reclaimed water in California are crop irrigation, groundwater recharge, and outdoor landscape irrigation. A total of 646 million cubic meters of wastewater is currently reused in California. *Sources:* Wong 1999b, MacLaggan 1999.

Agricultural Water Use

The largest single use of water worldwide is agricultural irrigation, which accounts for approximately 80 percent of consumptive water use. Total irrigated area has just about doubled since 1960 (see Table 10 in the Data Section) and some analysts project that large increases will be needed in the future, with concomitant increases in agricultural water needs (see Chapter 4). Where shortages of water exist, the use of reclaimed water for irrigation offers a reliable and valuable resource; it can increase crop yields, decrease reliance on chemical fertilizers, and provide a relatively drought-proof source of supply. In parts of Saudi Arabia, Tunisia, California, Israel, and Egypt, wastewater reuse for agriculture is already an essential water source. Studies demonstrate that even uncooked foods grown with carefully treated wastewater can be eaten without adverse health effects (Asano and Levine 1995).

In Israel, treated wastewater has been used for irrigating a variety of field crops, orchards, and edible vegetables. Some analysts expect that because of Israel's severe water constraints, agricultural water demands there will increasingly be supplied solely from wastewater (Shuval 1994). In Portugal, the volume of treated wastewater discharged in 2000 is estimated to be enough to cover about 10 percent of irrigation water needs in a dry year (Marecos do Monte 1998), and experimental studies on various crops are now underway. In Tunisia, about 35 million m³/yr of reclaimed water go to irrigation (Bahri 1998). Wastewater from Tunis has been used since the 1960s to irrigate citrus and olive orchards near the city. More recently, wastewater reuse has been expanded in Tunisia to other commercial and industrial crops, golf courses, and hotel gardens.

California has substantial experience with the use of reclaimed water for

TABLE 7.3 Uses of Recycled Water in California by Treatment Level[a] (Summary of Allowed Uses According to Draft of March 1997 Proposed Revisions for California Department of Health and Safety Title 22 Regulations)

USE	TREATMENT LEVEL			
	Disinfected Tertiary Recycled Water	Disinfected Secondary–2.2 Recycled Water[b]	Disinfected Secondary–23 Recycled Water[c]	Undisinfected Secondary Recycled Water
Irrigation of:				
Food crops where recycled water contacts the edible portion of the crop, including all root crops	Allowed	Not allowed	Not allowed	Not allowed
Parks and playgrounds	Allowed	Not allowed	Not allowed	Not allowed
School yards	Allowed	Not allowed	Not allowed	Not allowed
Residential landscaping	Allowed	Not allowed	Not allowed	Not allowed
Unrestricted-access golf courses	Allowed	Not allowed	Not allowed	Not allowed
Any other irrigation uses not prohibited by other provisions of the California Code of Regulations	Allowed	Not allowed	Not allowed	Not allowed
Food crops where edible portion is produced above ground and not contacted by recycled water	Allowed	Allowed	Not allowed	Not allowed
Cemeteries	Allowed	Allowed	Allowed	Not allowed
Freeway landscaping	Allowed	Allowed	Allowed	Not allowed
Restricted access golf courses	Allowed	Allowed	Allowed	Not allowed
Ornamental nursery stock and sod farms	Allowed	Allowed	Allowed	Not allowed
Pasture for milk animals	Allowed	Allowed	Allowed	Not allowed
Nonedible vegetation with access control to prevent use as a park, playground, or school yard	Allowed	Allowed	Allowed	Not allowed
Orchards with no contact between edible portion and recycled water	Allowed	Allowed	Allowed	Allowed
Vineyards with no contact between edible portion and recycled water	Allowed	Allowed	Allowed	Allowed
Non-food-bearing trees, including Christmas trees not irrigated less than 14 days before harvest	Allowed	Allowed	Allowed	Allowed

(continues)

TABLE 7.3 *Continued*

USE	TREATMENT LEVEL			
	Disinfected Tertiary Recycled Water	Disinfected Secondary–2.2 Recycled Water [b]	Disinfected Secondary–23 Recycled Water [c]	Undisinfected Secondary Recycled Water
Fodder crops (e.g., alfalfa) and fiber crops (e.g., cotton)	Allowed	Allowed	Allowed	Allowed
Seed crops not eaten by humans	Allowed	Allowed	Allowed	Allowed
Food crops that undergo commercial pathogen-destroying processing before consumption by humans (e.g., sugar beets)	Allowed	Allowed	Allowed	Allowed
Ornamental nursery stock, sod farms not irrigated less than 14 days before harvest	Allowed	Allowed	Allowed	Allowed
Supply for impoundment:				
Nonrestricted recreational impoundments, with supplemental monitoring for pathogenic organisms	Allowed [d]	Not allowed	Not allowed	Not allowed
Restricted recreational impoundments and fish hatcheries	Allowed	Allowed	Not allowed	Not allowed
Landscape impoundments without decorative fountains	Allowed	Allowed	Allowed	Not allowed
Supply for cooling or air conditioning:				
Industrial or commercial cooling or air conditioning with cooling tower, evaporative condenser, or spraying that creates a mist	Allowed [e]	Not allowed	Not allowed	Not allowed
Industrial or commercial cooling or air conditioning without cooling tower, evaporative condenser, or spraying that creates a mist	Allowed	Allowed	Allowed	Not allowed
Other uses:				
Groundwater recharge	Allowed under special case-by-case permits by RWQCBs [f]			
Flushing toilets and urinals	Allowed	Not allowed	Not allowed	Not allowed
Priming drain traps	Allowed	Not allowed	Not allowed	Not allowed

TABLE 7.3 *Continued*

USE	TREATMENT LEVEL			
	Disinfected Tertiary Recycled Water	Disinfected Secondary–2.2 Recycled Water [b]	Disinfected Secondary–23 Recycled Water [c]	Undisinfected Secondary Recycled Water
Industrial process water that may contact workers	Allowed	Not allowed	Not allowed	Not allowed
Structural fire fighting	Allowed	Not allowed	Not allowed	Not allowed
Decorative fountains	Allowed	Not allowed	Not allowed	Not allowed
Commercial laundries	Allowed	Not allowed	Not allowed	Not allowed
Consolidation of backfill material around potable water pipelines	Allowed	Not allowed	Not allowed	Not allowed
Artificial snow making for commercial outdoor uses	Allowed	Not allowed	Not allowed	Not allowed
Commercial car washes excluding the general public from washing process	Allowed	Not allowed	Not allowed	Not allowed
Industrial boiler feed	Allowed	Allowed	Allowed	Not allowed
Nonstructural fire fighting	Allowed	Allowed	Allowed	Not allowed
Backfill consolidations around nonpotable piping	Allowed	Allowed	Allowed	Not allowed
Soil compaction	Allowed	Allowed	Allowed	Not allowed
Mixing concrete	Allowed	Allowed	Allowed	Not allowed
Dust control on roads and streets	Allowed	Allowed	Allowed	Not allowed
Cleaning roads, sidewalks, and outdoor work areas	Allowed	Allowed	Allowed	Not allowed
Flushing sanitary sewers	Allowed	Allowed	Allowed	Allowed

Source: WateReuse Association of California 1997.

[a] Refer to the full text of the latest version of Title-22: California Water Recycling criteria.

[b] Median of 2.2 coliform per 100 ml.

[c] Maximum does not exceed 23 coliform per 100 ml.

[d] With "conventional tertiary treatment." Additional monitoring for two years or more is necessary with direct filtration.

[e] Drift eliminators and/or biocides are required if public or employees can be exposed to mist.

[f] Regional Water Quality Control Boards.

irrigation. In 1995, nearly 200 million m^3 of wastewater were used in the agricultural sector, while 94 million m^3 were used for landscape irrigation (Table 7.2). Some innovative projects involve the collaboration of cities with local farms. The city of Santa Rosa, California, for example, has implemented several agricultural wastewater reuse projects, in part to relieve a wastewater disposal problem. The city produces tertiary-treated water with additional UV disinfection. Part of the water is used on 1,660 hectares of fodder, sod, and pasture; 202 hectares of urban landscaping; 283 hectares of vineyard; 101 hectares of row crops; and 2.8 hectares of organic vegetables. The row crops are primarily several varieties of squash; they are started with recycled water, and then switched to well water when the fruit sets. Agricultural water users take the recycled water free of charge. In fact, some of the earlier contracts were written with incentives—farmers were paid to take a specified amount of recycled water for irrigation (Fidell and Wong 1999, Wong and Gleick 1999).

The city also provides reclaimed water to a local farm (Left Field Farm), which uses the water to grow organic vegetables for direct human consumption. As part of the agreement with the farmers, the city delivers pressurized reclaimed water directly to a sprinkler irrigation system that waters roughly 4 hectares of land. The farm's neighbors are dairies, a cattle farm, and a poultry plant that also receive tertiary-treated water from the city. Left Field Farm grows 47 varieties of vegetables with recycled water. The local climate allows them to grow "spring" crops throughout the year, and the recycled water supply is sufficient to grow broccoli and cauliflower that nearby growers can't produce (Fidell and Wong 1999).

Nonpotable Urban Reuses of Wastewater

There are substantial opportunities for the reuse of treated water in urban settings. Three factors favor this kind of reuse. Wastewater treatment plants are typically located in urban areas; there are many different kinds of water uses where wastewater use may be appropriate; and urban water users pay much more for water than do irrigators. Urban wastewater use options include water for outdoor irrigation of landscapes such as golf courses, municipal lawns, freeway medians, and parks; toilet flushing; small-scale industrial purposes; and creation or enhancement of wetlands. As with all reuse applications, the appropriate uses must be tied to water of appropriate quality.

One approach being taken in some communities is the development of dual distribution systems to deliver both potable and nonpotable water supplies. This permits water agencies to deliver a high-quality supply for drinking water and in-building applications, while nonpotable water needs can be met with reclaimed water. Retrofitting existing systems with a dual system is usually prohibitively expensive, but such systems have been installed where new housing or commercial uses are being developed. In such circumstances the costs of dual systems are more favorable.

Several U.S. urban areas have developed dual systems to provide separate reclaimed water for residential outdoor irrigation, car washing, fire protection, or commercial uses. The first U.S. system was installed in 1926 in Grand Canyon Village. Altamonte Springs, Florida, installed a reclaimed water-supply system for certain outdoor uses. This system can deliver a peak of 1.1 m^3 per second of high-quality reclaimed water. With the exception of nitrogen and phosphorous, which are considered beneficial for irrigation systems, the water meets drinking water standards (Asano and Levine 1998). A more comprehensive and ambitious program of urban reclaimed water use has been developed in the St. Petersburg, Florida, area, with a program to provide reclaimed water for selected residential areas in a dual-piped system. This program has halted the growth of potable water demand, despite substantial population growth (Figure 7.3). By 1995, nearly 8,600 customers

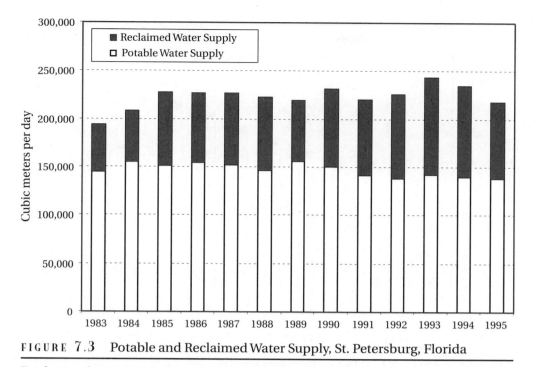

FIGURE 7.3   Potable and Reclaimed Water Supply, St. Petersburg, Florida

Total water demand in St. Petersburg, Florida has leveled off, with an increasing proportion of supply being provided by reclaimed water. *Source:* Johnson and Parnell 1998.

were receiving more than 80,000 m³/d of reclaimed water to meet part of their needs; 8,200 of these were residential customers (Johnson and Parnell 1998) (see Figure 7.4).

The Irvine Ranch Water District (IRWD) in Southern California also has a long history of providing reclaimed water for use. They initiated a reclaimed water program in 1963 and today can provide about 49,000 m³/day for diverse purposes (Young et al. 1998). The earliest uses were for irrigation and outdoor landscaping. More recently, efforts have included construction of a separate dual system for use in high-rise office buildings and in all new developments within the district. The district maintains about 320 kilometers of dual distribution pipelines, serves approximately 400 hectares of agricultural and nursery crops, and provides reclaimed water for parks, schools, golf courses, and other outdoor landscapes.

In 1991, IRWD became the first U.S. water district to obtain permission to use reclaimed water indoors. Seven facilities now have reclaimed water available for toilet flushing, urinals, and landscape irrigation. Demands for potable quality supplies in these buildings have dropped by as much as 75 percent due to the use of reclaimed water (Young et al. 1998). In the mid-1990s IRWD began to convert all landscape connections to reclaimed water. Because their water was treated to tertiary standards, and because the price was lower than regular domestic supply, customer acceptance was usually easy. Domestic water was offered at around $0.23/m³; reclaimed water for only $0.20/m³ (Young et al. 1998).

Some of the largest and highest-value wastewater uses are for industrial applications. Water recycling and wastewater reuse have been implemented successful in a

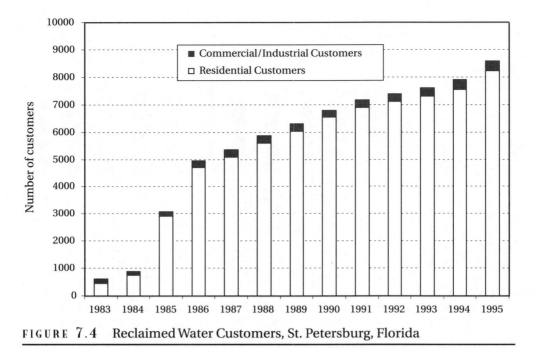

FIGURE 7.4   Reclaimed Water Customers, St. Petersburg, Florida

The number of customers using reclaimed water in St. Petersburg, Florida has risen substantially since the program began in 1983. Most of the increase has been in the residential sector. *Source:* Johnson and Parnell 1998.

wide range of industrial settings, and great potential exists. Among the most important uses are for process water and cooling water. There is a wide range of industrial cooling systems. Once-through-cooling is often used where large volumes of water are available, such as coastal power plants. Such use is typically nonconsumptive, though water used for direct contact cooling, such as in the primary minerals industry, may be consumed. Recirculating systems take the heat added to water and transfer it to the air through cooling towers via evaporation. A fraction of the water evaporates and is lost, but most is collected for reuse.

A major industrial reuse project is underway in the West Basin Municipal Water District in southern California. As part of a drought-response plan, the district initiated a U.S.\$200 million water-recycling project that could ultimately provide approximately 125 million m^3/yr of recycled water, or nearly half of the basin's water needs. Treatment facilities were dedicated just four years after project conception in late 1994. The first customer was connected in March 1995. Phase 1 of the project, completed in June 1996, provides 18.5 million m^3/yr of tertiary-treated water annually to over 100 industrial and landscape customers. Another 6.2 million m^3/yr of water undergoes additional treatment and is provided to a seawater intrusion barrier operation. Phase 2 is expanding the tertiary-treatment plant capacity and will provide for another 27 million m^3/yr of recycled water deliveries annually, as well as an additional 3 million m^3/yr of advanced-treated water to the seawater intrusion barrier project (Wong and Gleick 1999).

Two levels of treatment are planned. First, tertiary treatment meets the state's standards for the broadest level of 40 specified nonpotable uses. Two nitrification plants were required to meet additional treatment needs for industrial use at two of

TABLE 7.4   Sites Connected to West Basin Water
Recycling Project, February 1998

Type of Use	Number of Sites	Cubic Meters per Year (m³/yr)
Landscaping		
Public	88	2,611,000
Private	10	1,173,000
Industrial	2	12,335,000
Seawater intrusion Barrier	1	9,251,000
Other	1	2,500
Total	102	25,373,000

Source: Wong and Gleick 1999.

Notes: Totals may not add due to rounding. The two large industrial users are oil refineries.

the district's oil refineries. Because some of the water is directly injected into groundwater basins that provide some potable supplies, an advanced treatment facility treats the tertiary water, using reverse osmosis membranes, to drinking water standards. Table 7.4 summarizes the uses of the West Basin reclaimed water.

Environmental and Ecosystem Restoration

In some regions of the world, wastewater or return flows from a prior use of water constitute the only reliable sources of instream flows for natural ecosystems. As interest in ecosystem restoration grows, wastewater is increasingly being considered as a potential source of new supply for the environment, if it can be treated to an appropriate level. A demonstration of the use of wastewater for ecosystem restoration can be seen at the Kelly Farm demonstration marsh in northern California. In 1989, nearly 5 hectares of marsh in a 10-hectare setting were established as a test site to study additional beneficial uses for tertiary-treated reclaimed water. Reclaimed water is provided to create over 3.2 hectares of freshwater marsh, 1.2 hectares of open water, and about 0.5 hectare of seasonal wetland. The demonstration area also includes riparian woodland and shrub, grassland, and oak woodland. Water quality is monitored every two weeks. The tertiary-treated water meets all state and federal standards and is not expected to adversely affect wildlife at the wetland. It was feared that the recycled water contained constituents that could be toxic, accumulate in organisms, or overstimulate the growth of aquatic life, but preliminary results of the chemical monitoring of water, sediment, and biological quality suggest no adverse effects. Toxic substances (primarily heavy metals) have not been found in elevated levels in the plants, animals, or sediments in the wetland (Fidell and Wong 1999).

Another California water-recycling project was also designed with explicit environmental and ecosystem benefits. The San Jose/Santa Clara Wastewater Pollution Control Plant serves over 1.2 million residents, businesses, and industries in Santa Clara Valley, just south of San Francisco. In 1997, this wastewater plant discharged about 510,000 m³/d of tertiary-treated effluent into the southern end of San Francisco Bay. The southern part of the bay is considered an environmentally sensitive area, and the area of discharge, a salt marsh, is habitat for the endangered salt marsh harvest mouse and California clapper rail. According to studies done in the area, freshwater effluent from the plant was converting the salt marsh into a brackish or freshwater marsh, destroying the natural habitat. In 1989, the state required that the wastewater plant reduce freshwater discharges to the marsh during dry weather months, create or enhance salt marsh, relocate the plant discharge, or find other means to mitigate for impacts of plant discharge on salt-marsh habitat in South San Francisco Bay.

Faced with this requirement, the plant owners developed an action plan, with three main elements:

- The purchase and restoration of South Bay marsh properties to mitigate past conversion
- The development of potable water conservation programs to reduce influent flows
- Programs for the use of treated wastewater that could reduce effluent discharge to the Bay

Phase 1 will divert 57,000 m³/d of treated effluent to various water users in the cities of San Jose, Santa Clara, and Milpitas. Deliveries will expand to 114,000 m³/d in Phase 2 and up to 190,000 m³/d in the future (Wong 1999a), greatly relieving the ecological problems in the marsh.

Groundwater Recharge

Groundwater use in some parts of the world makes up a significant, and often primary, part of water supply. In regions where groundwater use exceeds rates of natural recharge, groundwater depletion can lead to higher economic costs for water, problems with reliability, and ultimately, constraints on development. Table 7.5 shows major regions of the world where annual pumping rates are near or even exceed natural recharge rates.

In some regions with severe groundwater overdraft, reclaimed water is increasingly being used to augment natural recharge. Artificial recharge can slow or halt the depletion of groundwater aquifers, protect coastal aquifers from saltwater intrusion due to overpumping, and provide storage of water for later use. Reclaimed water is used to recharge groundwater either by surface spreading and subsequent

TABLE 7.5   Heavily Exploited Aquifers of the World

Region	Aquifer	Average Annual Recharge (km³/yr)	Average Annual Use (km³/yr)
Algeria/Tunisia	Saharan basin	0.58	0.74
Saudi Arabia	Saq	0.3	1.43
China	Hebei Plain	35	19
Canary Islands	Tenerife	0.22	0.22
Gaza Strip	Coastal	0.31	0.50
United States	Ogallala	6 to 8	22.2
United States	selected Arizona	0.37	3.78

Source: Margat 1996.

percolation through soils, or through direct injection of treated wastewater into a groundwater aquifer. Surface spreading has the benefits of providing soil filtering and treatment. Direct injection is particularly useful for protecting coastal aquifers from saltwater contamination by creating freshwater barriers.

Good examples of these uses of reclaimed water can be found in California. Agencies in the southern part of the state have turned to recycled water to replenish and recharge overdrafted groundwater basins. This water displaces the need to buy imported surface supplies for this purpose. The County Sanitation Districts of Los Angeles County, Los Angeles County Department of Public Works, and Water Replenishment District of Southern California have been recharging the Central Basin groundwater aquifers with treated effluent since the 1960s. Orange County Water District has used recycled water in its seawater barrier injection operation since 1976, and the district is developing a proposal for a groundwater recharge project that will use up more than a million m^3 of recycled water annually. In Monterey County, the Monterey Regional Water Pollution Control Agency and Monterey County Water Resources Agency have partnered on a $75 million regional water-recycling project to ultimately provide around 200,000 m^3 of water annually for nearly 5,000 hectares of farmland, reducing pressure on the coastal aquifers there (Wong and Gleick 1999).

One of the greatest concerns with using reclaimed water for recharging groundwater is contamination of an aquifer. Water extracted from a groundwater basin must be free of chemical, microbiological, and radiological contaminants (or the water must be treatable at reasonable cost). As a result, ensuring that wastewater is free of these contaminants helps prevent contamination of a groundwater basin. The actual level of wastewater treatment necessary depends upon the characteristics of the basin, the quality of the water to be mixed with it, soil conditions, and the length of time water is retained in the aquifer before withdrawal. Specific legal requirements have rarely been set for groundwater recharge using reclaimed water (Asano 1994).

Direct and Indirect Potable Water Reuse

Wastewater treatment technologies are available to provide water of almost any desired quality. As a result, it is possible to treat, for a price, wastewater of nearly any initial quality to a standard high enough to use directly for drinking. Indirect potable reuse typically refers to wastewater that is intentionally (planned) or unintentionally (unplanned) discharged into a lake, stream, or aquifer where it is then reused for drinking water, often after further treatment. Direct potable reuse is the intentional and immediate use of treated wastewater as part of domestic drinking-water supply without intervening storage.

The vast majority of intentional wastewater reuse is indirect and nonpotable, as described earlier. But almost everywhere, some form of treated (or untreated) wastewater is discharged to streams, lakes, or groundwater basins where it is later withdrawn for uses that include drinking water. Indeed, very few large communities are situated in locations where their drinking water comes entirely from pristine sources, and most communities now use water that has been used at some time by someone else. London receives 20 percent of its drinking water from a tributary of the Thames that itself receives treated wastewater (Asano and Levine 1998). All the cities along the Rhine, Danube, Ohio, Mississippi, Colorado, and other major rivers use water that has been used upstream, put back in the river, and then reused, sometimes many times.

Despite these findings, major social barriers to potable reuse, particularly direct reuse, still exist. More and more communities, however, are developing and implementing specific plans for the indirect potable reuse of treated wastewater for drinking. These include systems where wastewater is injected into groundwater aquifers also used for water supply, placed in large reservoirs, or returned to rivers. As a result, the distinction between direct and indirect potable reuse often blurs. A study from the U.S. Environmental Protection Agency 20 years ago noted that 25 million people in U.S. cities with populations greater than 25,000 withdrew their water supplies from sources that contain from 5 to 100 percent wastewater during low-flow periods (McEwen 1998). Four million of these people were supplied by sources of water that contained 100 percent wastewater during low-flow conditions (U.S. EPA 1980). These numbers are likely to be much higher today. This study concluded that the indirect, unplanned reuse of wastewater for domestic purposes was widespread, supporting the argument that far better intentional planning for this kind of reuse would be important for protecting human health.

In 1956 and 1957, severe drought conditions forced the city of Chanute, Kansas, to treat wastewater for direct potable reuse. No adverse health impacts were observed, though the water became aesthetically unappealing after several cycles (McEwen 1998). The Orange County Water District in California began operation of Water Factory 21 in 1976, which treats secondary effluent to drinking water quality and uses it to recharge a heavily used groundwater aquifer to prevent saltwater intrusion, freeing up potable water for other uses. In 1978, the Upper Occoquan Sewage Authority in Virginia began reclaiming wastewater for release into the Occoquan Reservoir, which serves as a source of drinking water for a million people. In drought years, the plant discharge can account for as much as 90 percent of total inflow to the system. In the same year, the Tahoe-Truckee Sanitation Agency near the California–Nevada border began operation of an advanced treatment plant that releases water to the Truckee River, which is then used downstream as water supply by Reno, Nevada. These examples are increasingly the rule, not the exception.

Integrating wastewater systems with drinking water supplies, however, still requires careful examination of health concerns, public perceptions, water planning goals, and economic issues. The direct use of treated wastewater for drinking water is, at present, limited to a small number of extreme situations. Perhaps the greatest experience with direct potable reuse technologies has occurred in the city of Windhoek, the capital of Namibia in southern Africa (described in detail in upcoming text).

Health Issues

The health risks of untreated municipal wastewater are real. Table 7.6 lists common infectious agents potentially present in untreated wastewater. Table 7.7 shows typical concentrations of microorganisms in untreated wastewater. Primary treatment

TABLE 7.6 Common Infectious Agents Potentially Present in Untreated Wastewater

Classification	Agent	Disease
Protozoa		
	Entamoeba histolytica	Amebiasis (amebic dysentery)
	Giardia lamblia	Giardiasis
	Balantidium coli	Balantidiasis (dysentery)
	Cryptosporidium	Cryptosporidiosis, diarrhea, fever
Helminths		
	Ascaris (roundworm)	Ascariasis
	Trichuris (whipworm)	Trichuriasis
	Taenia (tapeworm)	Taeniasis
Bacteria		
	Shigella (four spp.)	Shigellosis (dysentery)
	Salmonella typhi	Typhoid fever
	Salmonella (1,700 serotypes)	Salmonellosis
	Vibrio cholerae	Cholera
	Escherichia coli (enteropathogenic)	Gastroenteritis
	E. coli 0157:h7 (enterohemorrhagic)	Bloody diarrhea
	Yersinia enterocolitica	Yersiniosis
	Leptospira (spp.)	Leptospirosis
	Legionella pneumophilia	Legionnaire's disease, Pontiac fever
	Camphylobacter jejuni	Gastroenteritis
Enteroviruses		
	Poliovirus	Paralysis, aseptic meningitis
	Echovirus	Fever, rash, respiratory illness, aseptic meningitis, gastroenteritis, heart disease
	Coxsackie A	Herpangina, aseptic meningitis, respiratory illness
	Coxsackie B	Fever, paralysis, respiratory, heart, and kidney disease
	Norwalk	Gastroenteritis
Viruses		
	Hepatitis A virus	Infectious hepatitis
	Adenovirus (47 types)	Respiratory disease, eye infections
	Rotavirus (4 types)	Gastroenteritis
	Parvovirus (3 types)	Gastroenteritis
	Reovirus (3 types)	Not clearly established
	Astrovirus (7 types)	Gastroenteritis
	Calicivirus (2–3 types)	Gastroenteritis
	Coronavirus	Gastroenteritis

Source: NAS 1998.

does little to remove dangerous disease agents. Even secondary treatment can leave high concentrations of microorganisms. Table 7.8 shows typical removal percentages by primary and secondary treatment processes. These percentages are too low for potable reuse, but they permit the kinds of reuse described in Table 7.3. During the past two decades, various studies have been done on the potential health effects of potable and nonpotable reuse of treated wastewater. These assessments

TABLE 7.7 Microorganism Concentrations
in Untreated Municipal Wastewater

Microorganism	Concentration (number per 100 ml)
Fecal coliforms	10^5–10^7
Fecal streptococci	10^4–10^6
Shigella	1–10^3
Salmonella	10^2–10^4
Pseudomonas aeruginosa	10^3–10^4
Clostridium perfringens	10^3–10^5
Helminth ova	1–10^3
Giardia lamblia cysts	10–10^4
Cryptosporidium oocysts	10^2–10^5
Entamoeba histolytica cysts	10^2–10^5
Enteric viruses	10^3–10^4

Source: NAS 1998.

TABLE 7.8 Typical Removal of Microorganisms by Conventional Primary
and Secondary Wastewater Treatment (Percentage Removal)

Microorganism	Primary Treatment	Secondary Activated Sludge	Secondary Trickling Filter
Fecal coliforms	<10	0–99	85–99
Salmonella	0–15	70–99	85–99+
Mycobacterium tuberculosis	40–60	5–90	65–99
Shigella	15	80–90	85–99
Entamoeba histolytica	0–50	Limited	Limited
Helminth ova	50–98	Limited	60–75
Enteric viruses	Limited	75–99	0–85

Source: NAS 1998.

demonstrated that clean and safe effluents could be produced from municipal wastewater treatment through a combination of treatment approaches (Asano and Levine 1995, NAS 1998). The direct and indirect reuse of treated wastewater for potable uses can thus be completely safe.

The purpose for which wastewater is to be used must define the level of treatment that water receives. Even for irrigation systems, a higher degree of treatment is required for irrigation of crops that are consumed directly or uncooked or where humans may have direct contact with the water. Several efforts have been made to define water quality guidelines to govern the use of reclaimed water. Both the World Health Organization and the State of California have comprehensive microbiological quality regulations (Table 7.9). The World Health Organization standards recommend use of stabilization ponds; the California criteria stipulate conventional biological wastewater treatment followed by tertiary treatment including filtration and chlorine disinfection (Asano and Levine 1998).

A critical concern is whether public health would be threatened by use of treated

TABLE 7.9   Microbiological Quality Regulations and Restrictions

Category	Water Reuse Conditions	Fecal or Total Coliforms[a]	Intestinal Nematodes[b]	Wastewater Treatment Process
WHO	Irrigation of crops likely to be eaten raw; sports fields, public parks	<1000/100 ml	< 1/liter	A series of stabilization ponds or equivalent
WHO	Landscape irrigation with public access	<200/100 ml	<1/liter	Secondary treatment and disinfection
WHO	Irrigation of cereal, industrial, fodder crops, pasture, trees	No standard recommended	< 1/liter	Stabilization ponds with 8- to 10-day retention or equivalent removal
California	Spray and surface irrigation of food crops, high exposure landscape irrigation	<2.2/100 ml	No standard recommended	Secondary treatment followed by filtration and disinfection
California	Irrigation of pasture for milking animals, landscape impoundment	<23/100 ml	No standard recommended	Secondary treatment followed by disinfection

Source: Asano and Levine 1998.

[a]Median number of total coliforms per 100 ml over past seven days for which analyses have been completed.

[b]Arithmetic mean number of eggs per liter for *Ascaris* and *Trichuris* species and hookworms.

wastewater. Two studies conducted in California—the Pomona Virus Study and the Monterey Wastewater Reclamation Study for Agriculture—demonstrated conclusively that virtually pathogen-free water could be produced through adequate treatment and that even food crops consumed uncooked could be successfully irrigated with reclaimed municipal wastewater without adverse health or environmental consequences (Asano and Levine 1995, Sheikh et al. 1998).

It is important to note that there is a potential health risk from little-known or unknown pathogens that might be found in wastewater and may not be eliminated by treatment. For some outbreaks of waterborne diseases, no etiological agent was ever found (NAS 1998). In the past two decades, several agents have been identified as "emerging" diseases, with increased threats to humans. Cryptosporidium, for example, was only first described in the past 10 to 20 years, but it has recently caused several major outbreaks of disease. The prevalence and health significance of many of these agents in both drinking water and treated wastewater are not well known.

A recent study by the National Academy of Sciences in the United States offered a series of recommendations to operators of water-supply systems considering potable reuse of treated wastewater. The most important of these related to human health are summarized here.

- Potable reuse systems should employ a combination of advanced physical treatment processes and strong chemical disinfectants as the principal line of defense against most microbial contaminants.

- Current and future facilities should assess and report the effectiveness of their treatment processes in removing microbial pathogens.
- Research should be conducted on methods for detecting emerging pathogens in environmental samples. Research should also be conducted on the effectiveness of various water or wastewater treatment processes and disinfectants in removing or inactivating these pathogens.
- The industry and research communities need to establish the performance and reliability of individual barriers to microorganisms within treatment trains and to develop performance goals appropriate to planned potable reuse.

Wastewater Reuse in Namibia

Namibia is one of the most arid countries in the world, with no perennial rivers entirely within its borders. The capital, Windhoek, is over 700 kilometers from the Okavango River—the nearest perennial river. Average rainfall in the country is 360 millimeters/yr, while average potential evaporation is 3,400 millimeters/yr. Windhoek's traditional water is supplied from three surface reservoirs on ephemeral rivers and local groundwater aquifers.

Most groundwater and surface-water sources near Windhoek had been tapped by the end of the 1960s. Rapid population growth—on the order of 6 to 8 percent annually in the 1960s and 1970s—led to strong water conservation programs as well as an aggressive wastewater reuse program that includes direct potable reuse. Even today, urban population growth rates are around 5 percent per year. Total residential water use in the city is only 150 liters per person per day at present (compared with 500 or more in some U.S. cities) and plans are to reduce this to 100 liters per person per day.

In 1968, the City of Windhoek inaugurated a pilot-scale potable reclamation plant with a capacity of 4,800 m^3/d. After several expansions and upgrades, current capacity now exceeds 8,000 m^3/d and a new plant with a capacity of 21,000 m^3/day is scheduled to be completed in November 2000. The cost of production for this water, including capital costs, is around N\$2.80 (Namibian dollars)/m^3, compared to the cost of bulk water at N\$3.17/m^3 and the cost of developing new supply from the Okavango River nearly 700 kilometers away at N\$7.10/m^3 (WTC 1997). In 1999, N\$6.00 were approximately equal to US\$1.00.

The system includes primary, secondary, and advanced treatments. Primary treatment includes settling and solids separation. Secondary treatment includes biofiltration, an activated sludge module, maturation ponds, and biological phosphate removal. Several forms of advanced treatment have been used since the first plant was commissioned in 1968. These include additional ammonia removal, dissolved air flotation, filtering, chlorine and lime treatments, carbon and sand filtration, and most recently experimentation with membrane and ozonation systems.

As a fundamental part of Namibia's wastewater system, poor-quality industrial wastewater that contains toxic chemicals or heavy metals and water with high organic loads are diverted to separate wastewater streams. Wastewater from the heavily industrial areas of the city go to a separate, small treatment plant, and industrial siting decisions are at least partly made on the basis of the wastewater composition (Haarhoff and van der Merwe 1996).

Comprehensive monitoring of both water quality and human health was initiated early. Water produced by the reclamation plant was initially monitored by four independent laboratories, which evaluated chemical quality, bacteriological content, toxicity, mutagenicity, and the presence of viruses. An independent technical committee initially reviewed water quality, technological and operating improvements, and health issues. Since 1988, expert reviews have been invited on an ad hoc basis (Odendaal et al. 1998). Monitoring and testing of the system were scaled down after 1990, but still account for about 17 percent of the total production costs (Odendaal et al 1998, van der Merwe and Menge 1996). During plant operation, samples are drawn at up to 13 monitoring points and 170 samples are analyzed weekly for the main chemical constituents. *Cryptosporidium* and *Giardia* are known to exist in the effluent water and recent experimentation with filtration and membrane systems is partly an effort to address this problem.

A more comprehensive set of tests is performed twice a month, including a metals assessment. Mutagenicity and the presence of viruses are monitored at three sampling points twice a month, with additional virological monitoring being done more frequently at certain locations. Bacteria are monitored twice a week at three points. Chlorophyll levels are determined weekly at five points. When compared with the water quality from two other primary water supply sources, reclaimed water is at least equal or even better in quality.

Associated with the water-quality monitoring, every patient who visited a hospital, clinic, or general practitioner's office was monitored between 1968 and the mid-1980s. An epidemiological study was conducted between 1976 and 1983 of a population of 75,000 to 100,000. No apparent patterns of morbidity or mortality associated with reclamation were found, though more analysis is needed.

One of the greatest barriers to direct potable reuse is public opposition. In Windhoek, such opposition has always existed, but a public-education program combined with more than 27 years of operating experience has led to overall public acceptance of direct potable reuse for augmenting supply (du Toit and Sguazzin 1995). In 1998, around 1.25 million m³ of purified effluent was used through a dual-pipe system on sports fields, gardens, nurseries, and parks. This water use is metered and customers are charged monthly (van der Merwe 1999).

Reclaimed wastewater is not the principal source of supply for Windhoek. First priority is given to the use of water in surface reservoirs in order to minimize evaporation losses in the very arid region. In 1997, the three surface reservoirs supplied 15.7 million m³ of water, while evaporation losses were over 35 million m³. In addition, the costs of production of reclaimed water were higher than the subsidized costs of the state-run conventional sources. As a result of these factors, average production at the reclamation plant between 1968 and 1995 was only 27 percent of total capacity, and the contribution of reclaimed water to the total Windhoek supply was only 4 percent. During the droughts of 1982 and 1995, however, the reclamation plant produced 80 percent of its maximum capacity. From April 1996 to

March 1997, reclaimed water provided over 30 percent of total water supply (van der Merwe and Menge 1996).

A comprehensive water-use efficiency and conservation program is integrated into the City of Windhoek's water plans. As an example of the progress made in this area, the local brewery in Windhoek consumes only 4 liters of water to produce a liter of beer, compared with 5 to 7 liters of water per liter of beer in Europe and between 5.5 and 8 liters of water per liter of beer in neighboring South Africa (van der Merwe 1999). Gray water systems are encouraged in residential areas, where some water used in the home is captured for reuse in outdoor gardens

Continued implementation of an aggressive water-use efficiency program and expansion of the use of nonconventional water sources like gray water and reclaimed water will ensure that Windhoek needs only 33 percent of its total water demand from conventional sources by 2005. This experience should be of significant value to other arid regions of the world.

Wastewater Reclamation and Reuse in Japan

Freshwater availability in Japan is relatively high compared with many other countries, but its large population and limited storage capacity have created some difficult water-management challenges in both rural areas and densely populated urban centers. As a result, Japanese water managers and agencies have begun to invest heavily in a range of wastewater systems for treatment and reuse. Following World War II, large-scale construction, industrial redevelopment, and rapid growth were accompanied by an explicit effort to build water-treatment systems, and by 1995, half the population was served by central sewage systems (JSWA 1993).

Early wastewater reclamation and reuse in Japan included the treatment of water for an industrial paper plant that could no longer use the heavily contaminated river water that was available nearby—an ironic example of having to clean up water because a basic supply source had not been adequately protected (Maeda et al. 1996). In recent years, most reclaimed water use has been limited to direct non-potable uses by industry or large commercial facilities. By the early 1990s, publicly owned treatment plants discharged approximately 11 billion m^3 of secondary treated effluents. Figures 7.1 and 7.2 show the different breakdown of uses in Japan and California (Asano et al. 1996).

A very small fraction of total water effluent is used beneficially in Japan, though some municipalities have proportionally higher use. About 0.2 percent of total municipal water supply is reclaimed water or around 85 million m^3/yr (Asano and Levine 1995). Tokyo Metropolitan Districts, for example, have promoted wastewater reuse because of highly concentrated populations, increasing demands for limited supplies of water, and constraints on finding new sources of supply. To promote improved water-use efficiency, the local government has imposed increasing block rate structures and supported the construction of dual distribution systems for large office buildings. The largest reuses in the Tokyo area are for stream augmentation and industrial use (see Table 7.10) followed by toilet flushing. The largest single water use in Tokyo commercial buildings is toilet flushing (Asano et al. 1996).

TABLE 7.10   Wastewater Reclamation and Reuse in Tokyo Metropolitan Districts

Wastewater Plant	Reclaimed Water Use	Volume of Use (1000 m³/yr)
Shibaura	Passenger train washing	111
Sunamachi	Dust control	6
Morigasaki	Refuse incineration plant	386
Mikawashima	Industrial water	8,835
Ochiai	Toilet flushing	970
Tamagawa-Joryu	Stream augmentation	12,370

Source: Asano et al. 1996.

In the dense urban area of Shinjuku, 19 high-rise buildings use a combined average of about 2,700 m³/day for toilet flushing. Reclaimed water is priced about 20 percent lower than normal domestic supply, but the cost of dual plumbing is the responsibility of the building owner.

Wastewater Costs

The cost-effectiveness of using reclaimed water depends on several factors, including the processes used to produce it, the distance to the point of use, the quality of the water needed, and the marginal cost of new supplies. Table 7.11 offers a wide range of costs associated with various wastewater reuse and treatment categories.

The costs of production are lowest for reclaimed water used to irrigate fodder, fiber, seed crops, and orchards and vineyards where the water doesn't come in direct contact with the fruit. The estimated range of costs is from $0.09 to $2.65 per m³. As the quality required increases, so do the costs up to the point where water suitable for potable reuse can be as high as $1/m³ or more. Actual costs in different regions will, of course, vary, depending on the many factors raised in this chapter, the regional costs of capital and labor, and a wide range of other factors.

Developing countries need reliable, low-cost, low-technology methods for collecting wastewater and treating it to a level suitable for meeting many different kinds of demands. Complex wastewater treatment and reuse systems will be harder to implement in developing countries because of constraints on financial resources available for public works. It is worth noting, however, that in many regions, sewage collection systems and wastewater treatment are nonexistent, leading to major costs associated with water-related diseases.

SUMMARY

High-quality, reliable water supplies can be produced by careful treatment of wastewater. This water is being viewed as an increasingly valuable resource to be used, not thrown away. It is technically feasible to produce water of any quality

Table 7.11   Estimated Water Reclamation Treatment Process
Life-Cycle Costs (1996 dollars)

Reuse Category	Treatment Process	Annual Cost $/m³ (Low)[a,b]	Annual Cost $/m³ (High)
Agricultural irrigation	Activated sludge	$0.20	$0.55
Livestock and wildlife	Trickling filter	$0.22	$0.58
Power plant and industrial cooling: once-through	Rotating biological contactors	$0.31	$0.59
Landscape irrigation	Activated sludge, filtration	$0.24	$0.73
Power plant and industrial cooling: recirculation	Tertiary lime treatment	$0.33	$1.08
Industrial boiler make up: low pressure	Tertiary lime treatment, nitrified effluent	$0.33	$1.19
Industrial supply: chemicals and allied products; food	Activated sludge, filtered secondary effluent, carbon adsorption	$0.31	$0.97
Industrial boiler make up: intermediate pressure	Tertiary lime treatment, carbon adsorption, ion exchange	$0.51	$1.59
Groundwater recharge: spreading basins	Infiltration— percolation	$0.09	$0.21
Groundwater recharge: injection wells	Activated sludge, filtration, carbon adsorption, reverse osmosis	$0.95	$2.65

Sources: Modifed from Richard 1998, Culp et al. 1980.

[a]Ranges are for facilities from 1 to 50 million gallons per day. Lower costs are for the larger plant.

[b]Annual normalized costs include amortized capital costs with a facility life of 20 years and a return rate of 7 percent.

desired. By carefully matching water quality with specific end-use requirements and by addressing the economics of water supply, a wide range of reuse applications could become available as part of integrated water systems. Such applications can already be found in many locations around the world. Most attention has been focused on nonpotable reuse of water for agricultural and landscape irrigation, industrial cooling, and limited commercial indoor use. But indirect potable reuse, long practiced unintentionally in any city downstream of any other, is increasingly receiving explicit attention in an effort to develop consistent standards for water quality and treatment. In the final analysis, while significant social and economic barriers to wastewater reuse still remain in many regions, water managers can ill afford to ignore any plausible source of cost-effective and reliable water supply, even reclaimed water.

REFERENCES

Angelakis, A.N., and S.V. Spyridakis. 1995. "The status of water resources in Minoan times: A preliminary study." In A.N. Angelakis, A. Issar, and O.K. Davis (eds.), *Diachronic Climatic Impacts on Water Resources in the Mediterranean Region.* Springer-Verlag, Heidelberg, Germany.

Asano, T. 1994. "Groundwater recharge with reclaimed municipal wastewater and waters of impaired quality." *Water Down Under '94,* 21–25 November, Adelaide, Australia.

Asano, T., and A.D. Levine. 1995. "Wastewater reuse: A valuable link in water resources management." *Water Quality International (WQI),* No. 4, pp. 20–24.

Asano, T., and A.D. Levine. 1998. "Water reclamation, recycling, and reuse: An introduction." In T. Asano (ed.) *Wastewater Reclamation and Reuse,* Vol. 10. Water Quality Management Library, Technomic Publishing Co. Inc. Lancaster, Pennsylvania, pp. 1–56.

Asano, T., M. Maeda, and M. Takaki. 1996. "Wastewater reclamation and reuse in Japan: Overview and implementation examples." *Water Science Technology,* Vol. 34, pp. 219–226.

Bahri, A. 1998. "Wastewater reclamation and reuse in Tunisia." In T. Asano (ed.) *Wastewater Reclamation and Reuse.* Vol. 10. Water Quality Management Library, Technomic Publishing Co. Inc. Lancaster, Pennsylvania, pp. 877–916.

Cologne, G. 1998. "Legal aspects of water reclamation." In T. Asano (ed.), *Wastewater Reclamation and Reuse,* Vol. 10. Water Quality Management Library, Technomic Publishing Co. Inc. Lancaster, Pennsylvania, pp. 1397–1416.

Culp, G., G.M. Wesner, R. Williams, and M. Hughes. 1980. *Wastewater Reuse and Recycling Technology.* Noyes Data Corporation, Park Ridge, New Jersey. Office of Water Research and Technology OWRT/RU-79-1,2.

du Toit, D., and T. Sguazzin. 1995. *Sink or Swim: Water and the Namibian Environment.* Envirotech, Swakopmund, Namibia.

Fidell, M., and A.K. Wong. 1999. "Using recycled water for agricultural irrigation: City of Visalia and City of Santa Rosa." In L. Owens-Viani, A.K. Wong, and P.H. Gleick (eds.), *Sustainable Use of Water: California Success Stories.* Pacific Institute for Studies in Development, Environment, and Security, Oakland, California, pp.141–148.

Gleick, P.H. 1998. "Water and human health." In P.H. Gleick, *The World's Water 1998-1999.* Island Press, Washington, D.C., pp. 39–67.

Haarhoff, J., and B. van der Merwe. 1996. "Twenty-five years of wastewater reclamation in Windhoek, Namibia." Water Science and Technology, also in City of Windhoek, *Recent Papers and Publication on Water Reclamation in Windhoek, Namibia.* Department of the City Engineer, Windhoek, Namibia.

Japan Sewage Works Association (JSWA). 1993. *Sewerage in Japan—Status and Plans.* JSWA, Tokyo, Japan.

Johnson, W.D., and J.R. Parnell. 1998. "Wastewater reclamation and reuse in the City of St. Petersburg, Florida." In T. Asano (ed.), *Wastewater Reclamation and Reuse,* Vol. 10. Water Quality Management Library, Technomic Publishing Co. Inc. Lancaster, Pennsylvania, pp. 1037–1104.

MacLaggan, P. 1999. Personal communication, Oct. 11.

Maeda, M. K. Nakada, K. Kawamoto, and M. Ikeda. 1996. "Area-wide use of reclaimed water in Tokyo, Japan." *Water Science Technology,* Vol. 33, pp. 10–11.

Marecos do Monte, M.H.F. 1998. "Agricultural irrigation with treated wastewater in Portugal." In T. Asano (ed.), *Wastewater Reclamation and Reuse,* Vol. 10. Water Quality Management Library, Technomic Publishing Co. Inc. Lancaster, Pennsylvania, pp. 827–875.

Margat, J. 1996. "Comprehensive assessment of the freshwater resources of the world: Groundwater component." Contribution to Chapter 2 of the *Comprehensive Global Freshwater Assessment,* United Nations.

McEwen, B. 1998. "Indirect potable reuse of reclaimed water." In T. Asano (ed.), *Wastewater Reclamation and Reuse,* Vol. 10. Water Quality Management Library, Technomic Publishing Co. Inc. Lancaster, Pennsylvania, pp. 1211–1268.

National Academy of Sciences (NAS). 1998. *Issues in Potable Reuse: The Viability of Augmenting Drinking Water Supplies with Reclaimed Water.* National Research Council, National Academy Press, Washington, D.C.

Odendaal, P.E., J.L.J. Van der Westhuizen, and G.J. Grobler 1998. "Wastewater reuse in South Africa." In T. Asano (ed.), *Wastewater Reclamation and Reuse,* Vol. 10. Water Quality Management Library, Technomic Publishing Co. Inc. Lancaster, Pennsylvania, pp. 1163–1192.

Richard, D. 1998. "The cost of wastewater reclamation and reuse." In T. Asano (ed.), *Wastewater Reclamation and Reuse,* Vol. 10. Water Quality Management Library, Technomic Publishing Co. Inc. Lancaster, Pennsylvania, pp. 1335–1395.

Sheikh, B., R. Cort, R.C. Cooper, and R.S. Jaques. 1998. "Tertiary-treated reclaimed water for irrigation of raw-eaten vegetables." In T. Asano (ed.), *Wastewater Reclamation and Reuse,* Vol. 10. Water Quality Management Library, Technomic Publishing Co. Inc. Lancaster, Pennsylvania, pp. 779–825.

Shuval, H. 1994. "Proposed principles and methodology for the equitable allocation of the water resources shared by the Israelis, Palestinians, Jordanians, Lebanese, and Syrians." In J. Isaac and H. Shuval (eds.), *Water and Peace in the Middle East.* Elsevier, Amsterdam, pp. 487–496.

United States Environmental Protection Agency (U.S. EPA). 1980. *Wastewater in Receiving Waters at Water Supply Abstraction Points.* U.S. EPA-600/2-80-044, Cincinnati, Ohio.

Van der Merwe, B., and J. Menge. 1996. "Water reclamation for potable reuse in Windhoek, Namibia." In City of Windhoek, *Recent Papers and Publication on Water Reclamation in Windhoek Namibia* Department of the City Engineer, Windhoek, Namibia.

Van der Merwe, B. 1999. "Reuse of water in Windhoek, Namibia." Presented at the Stockholm Water Symposium, 9-12 August. Stockholm, Sweden.

Water Reuse Association of California, 1997. "Recycling water to meet California's needs." Sacramento, California.

Water Transfer Consultants (WTC). 1997. "Feasibility study on the Okavango River to Grootfontein link of the Eastern National Water Carrier." Ministry of Agriculture, Water and Rural Development, Department of Water Affairs, File No. 13/2/2/2. Windhoek, Namibia.

Wong, A.K. 1999a. "Using recycled water in urban settings: West Basin recycling project and South Bay water recycling program." In L. Owens-Viani, A.K. Wong, and P.H. Gleick (eds.) *Sustainable Use of Water: California Success Stories.* Pacific Institute for Studies in Development, Environment, and Security, Oakland, California, pp. 127–140.

Wong, A.K. 1999b. "An overview to water recycling in California." In L. Owens-Viani,

A.K. Wong, and P.H. Gleick (eds.), *Sustainable Use of Water: California Success Stories*. Pacific Institute for Studies in Development, Environment, and Security, Oakland, California, pp. 121–126.

Wong, A.K., and P.H. Gleick. 1999. "Overview to Water Recycling in California: Success Stories." Presented at the 1999 Stockholm Water Symposium, Stockholm, Sweden. Available from the Pacific Institute for Studies in Development, Environment, and Security, Oakland, California.

Young, R.E., K.A. Thompson, R.R. McVicker, R.A. Diamond, M.B. Gingras, D. Ferguson, J. Johannessen, G.K. Herr, J.J. Parons, V. Seyde, E. Akiyoushi, J. Hyde, C. Kinner, and L. Oldewage. 1998. "Irvine Ranch Water District's reuse today meets tomorrow's conservation needs." In T. Asano (ed.), *Wastewater Reclamation and Reuse*. Vol. 10. Water Quality Management Library, Technomic Publishing Co. Inc. Lancaster, Pennsylvania, pp. 941–1036.

Water Briefs

Arsenic in the Groundwater of Bangladesh and West Bengal, India

It took a long time to get people to use tube well water, and it will take a long time to get them to change.

KHUSHI KABIR, BANGLADESHI NGO DIRECTOR

If I die, I will die, but I will not go to fetch water from another man's house.

SALIM UDDIN MONDAL, BANGLADESHI VILLAGER

Background

Bangladesh is a small country with a huge population—approximately 120 million people—making it the most densely populated country in the world, with 890 people per square kilometer. It also ranks very low on the Human Development Index—number 147 out of 174 (UNDP 1998).[1] Water problems are common and diverse in Bangladesh. Flooding is a major risk during parts of the year when monsoon rains combine with runoff from the Himalayas in the low-lying delta of the Ganges/Brahmaputra/Meghna river systems before passing into the Bay of Bengal. Damaging cyclones cause massive hardship in the delta—severe ones killed 500,000 people in 1970 and 200,000 in 1991. During other parts of the year, scarcity of clean water can be a problem. The major rivers are too contaminated to drink without treatment, and the low-lying nature of the landscape prevents the economic construction of any significant storage facilities. In the past few years, another water problem of crisis proportion has appeared—the large-scale contamination of drinking water supplies with naturally occurring arsenic.

Prior to the 1970s, most Bangladeshis relied almost exclusively on surface water for their daily requirements. Water was taken directly from ponds and shallow hand-pumped wells. But industrial effluents and poor or nonexistent sewage systems increasingly polluted this water, leading to widespread water-related

165

diseases, particularly diarrheas and cholera. Because of the geological characteristics of the region and the abundant seasonal rainfall and groundwater, international aid agencies, primarily the World Bank and the United Nations International Children's and Educational Fund (UNICEF) advocated tapping the country's extensive groundwater reserves as a source of clean water.

The proposed solution was to put in shallow and deep tubewells to provide protected drinking water sources. Since that time nearly 4 million such tubewells have been installed and the incidence of water-related diseases in the region has dropped tremendously, saving millions of lives. The development of tubewells is believed to be responsible for the reduction of infant mortality from diarrhoeal diseases, and the achievement of food-grain self-sufficiency using groundwater irrigation. It is estimated that 95 percent or more of Bangladeshis now use groundwater for drinking. Because of these results, Bangladesh has long been held up as one of the few major successes in the effort to provide basic water services to unserved poor populations around the world.

The possibility that this groundwater could be contaminated was overlooked. In a public-health crisis of potentially catastrophic proportions, elevated concentrations of arsenic have now been found in the groundwater of Bangladesh and neighboring West Bengal, India. Millions of people live in areas where drinking water is now known to have arsenic concentrations above—often far above—acceptable levels (see Box WB-1). Thousands of people have already been diagnosed with symptoms of arsenic toxicity and the scope of the problem is only slowly being understood.

Some information and concern about arsenic contamination of groundwater in Asia has circulated for nearly 20 years, but international water and public-health experts have only recently become involved.

BOX WB.1

About Arsenic and Standards for Protection

Arsenic is a metalloid element known for its toxicity. It is relatively soluble in water and occurs naturally in the environment in both organic and inorganic forms. Humans can be exposed to arsenic through many pathways, including air, food, and water. The World Health Organization (WHO) set a recommended upper limit of 0.05 mg/l for drinking water. Both Bangladesh and India adopted this standard, as did the U.S. Environmental Protection Agency (EPA). In 1993, the WHO lowered their guideline value for arsenic to 0.01 mg/l.

In 1999, the U.S. National Academy of Sciences (NAS) released a study calling on the EPA to strengthen its arsenic standards. The NAS study concluded that 0.05 mg/l "does not sufficiently protect public health" and that the standard should be lowered "as promptly as possible" (Hebert 1999, U.S. NAS 1999). According to the NAS study, males who drink water daily at or near the maximum allowable arsenic level have a risk of developing bladder cancer of 1 in 1,000, with the possibility of a combined cancer risk as high as 1 in 100 over a lifetime. This is far in excess of EPA's general goal of limiting lifetime cancer risks to 1 in 10,000.

The Scope of the Problem

Reports of contamination of water supplies with arsenic first appeared in West Bengal, India, in 1983 when patients showing symptoms of arsenic poisoning were identified (Tsushima 1998). In 1988, Dr. Dipankar Chakraborti, now at the School of Environmental Studies at Jadavpur University in Calcutta (SOES), collected water samples from wells in West Bengal, India. Testing in the laboratories at the University of Antwerp in Belgium revealed high levels of arsenic (Bearak 1998). Hints of similar arsenic poisoning in Bangladesh were reported in 1992 when Indian surveys revealed arsenic patients who had come from Bangladesh and reported that friends and family had similar symptoms (Tsushima 1998). Groundwater contamination by arsenic was formally discovered during testing in the west of Bangladesh in 1993. In March 1994, Chakraborti wrote letters to the Ministry of Public Health Engineering of the Government of Bangladesh and to the representatives of the World Health Organization(WHO) and UNICEF in Dhaka to draw attention to the problem. In February 1995, SOES convened an international conference on Arsenic in Ground Water: Cause, Effect, and Remedy. This brought the scale of the arsenic problem in West Bengal, India to a wider audience. It became evident at that meeting that there was serious risk of comparable contamination in neighboring Bangladesh. Further testing in the mid- and late-1990s revealed that contamination extended across large parts of southern Bangladesh. It is the most serious instance of arsenic groundwater contamination ever discovered.

The contamination occurs in groundwater in alluvial and deltaic sediments, with large variability at both local and regional scales. The arsenic is of geological, not human, origin and has probably been present in the groundwater for thousands of years (BGS 1999). It is apparent now because it is only in the last few decades that groundwater has been extensively used to meet drinking water needs.

The source of the arsenic problem appears to be the main, shallow aquifer that has been most extensively exploited. This aquifer extends from less than 10 meters to around 100 meters from the surface. Groundwater from deep aquifers at depths of more than 150–200 meters appears to be essentially arsenic-free. Shallow hand-dug wells at the top of the shallow aquifer, at depths of less than 10 meters, also appear to be less contaminated than deeper wells, though shallow wells face the highest risk of microbiological contamination (BGS 1999).

Early studies by the National Institute of Preventative and Social Medicine highlighted the problem but were too limited to provide an overall sense of the risks. WHO helped provide laboratories of the Department of Public Health and Engineering (DPHE) with equipment to detect arsenic, and several thousand analyses have now been carried out in these laboratories. Other early data came from the Dutch-funded Eighteen District Towns project of DPHE. This project was also significant in instituting regular monitoring of wells (BGS 1999).

More extensive surveys by the Dhaka Community Hospital in association with SOES between 1995 and 1997 confirmed the scope of the problem and further helped to raise public awareness. The SOES reported in 1995 that 312 villages in six districts of Bangladesh were affected. By January of 1996 this had increased to 560 villages in seven districts, and then to 830 villages in eight districts, as more and more monitoring was finally undertaken (SOES 1996, 1997). At that time it was estimated that 1.1 million people were drinking well water with arsenic concentrations exceeding the maximum permissible level of 0.05 mg/1. One and a half million people were estimated to be drinking water with arsenic concentrations over the WHO guideline of 0.01 mg/1.

Galvanized by growing concerns among scientists, the public, and health professionals, a serious effort at evaluating the nature and scope of the problem began in 1997, with an increasing number of studies carried out by governments, university scientists, nongovernmental organizations (NGOs), and international agencies. Information collected in 1997 included high-quality, multi-element data from the DPHE for a range of Bangladesh groundwaters. Of 63 samples taken, 60 percent had arsenic concentrations greater than the Bangladesh standard of 0.05 mg/l. A random survey of wells in six districts in northeast Bangladesh was undertaken by the Bangladesh University of Engineering and Technology for the North-East Minor Irrigation Project, a region away from the perceived center of the problem. Yet even here, 61 percent of the 1,210 samples tested had arsenic levels above 0.01 mg/l; 33 percent were above 0.05 mg/l. Another 751 samples from the same region were analyzed by the Bangladesh Council for Scientific Research and showed 42 percent of samples above 0.05 mg/l. A British Geological Survey study of Chapai Nawabganj in early 1997 confirmed some drinking water sources with extremely high concentrations of arsenic—up to 2.4 mg/l—which is 240 times the WHO recommended limit.

Around this time various inexpensive kits that could be used in the field became available to NGOs and other users. These kits were evaluated to determine their practicality and accuracy. Five different kinds of field kits were tested and it was concluded that they can greatly improve knowledge about the extent of contamination (BGS 1999). Controlled field and laboratory testing in India and Bangladesh showed that

- field kits reliably identify highly contaminated water containing above about 0.20 mg/l of arsenic;
- field kits do not falsely indicate the presence of arsenic in wells where laboratory tests show the arsenic concentration is below 0.05 mg/l;
- field kits do not reliably identify the presence of arsenic in groundwater containing between 0.05 and 0.20 mg/l of arsenic.

While these kits lack the precision of expensive laboratory analyses, their low cost permits wide sampling to be done. The National Institute of Preventative Medicine analyzed nearly 3,500 samples from various regions of Bangladesh and found 28 percent with concentrations above 0.05 mg/l of arsenic. An extensive survey of about 23,000 wells was carried out by DPHE with assistance from UNICEF using simple field-test ("yes/no") kits. These surveys demonstrated that the most serious arsenic contamination was in southeastern Bangladesh.

The Bangladesh Rural Advancement Committee (BRAC), a major NGO working on development problems, is among the groups to begin arsenic monitoring and mitigation programs. In 1998 they tested 802 tubewells at their field offices and another 11,954 wells in one area (Hajiganj thana). Local women who had earlier been trained to treat common illnesses in their villages did the testing. The testing program was implemented with the collaboration of the Department of Public Health and Engineering of the Government of Bangladesh.

Out of the total of 11,954 tubewells tested in Hajiganj thana, 11,096 tubewells (93 percent) were found to be contaminated with arsenic concentrations greater than the acceptable limit. Fifty-three percent of the 156 villages had contamination in all their wells and hence had no access to safe water. A total of 193 randomly selected water samples from Hajiganj were analyzed further to determine the validity of the field kit results. They agreed in 178 (92 percent) of the cases. Sixty-four tubewells (33 percent) had contamination of over 0.25 mg/l—five times higher than the standard (Table WB.1). In addition to providing information on the scope of the problem, this survey also demonstrated the potential for successful community involvement in testing programs.

The British Geological Survey (BGS) undertook another study in 1998 with SOES and the Dhaka Community Hospital in 41 of the 64 districts of Bangladesh (BGS 1999, Source 1999). They focused on the regions thought to be most severely contaminated, including most of southern Bangladesh (except the Chittagong Hill Tracts) and the northeastern districts. More than two thousand samples provided uniform spatial coverage and a representative range of well types and depths. Duplicate samples were collected at each well and analyzed at the DPHE in Bangladesh and the BGS laboratory in the United Kingdom. The results of the 2,022 samples analyzed in the UK are summarized as follows:

- 51 percent of the samples had arsenic above the WHO Guideline Value of 0.01 mg/l.
- 35 percent were above the Bangladesh Drinking Water Standard of 0.05 mg/l.
- 25 percent were above 0.10 mg/l.
- 8.4 percent were above 0.30 mg/l.
- 0.1 percent was above 1.0 mg/l.
- The maximum concentration found was 1.67 mg/l.
- The minimum concentration was below the detection limit of 0.0005 mg/l.
- About 20 percent of the samples were considered essentially arsenic-free.

Combining the available testing data to date on 50,000 wells in Bangladesh indicate that around 40 percent are too contaminated with arsenic to provide safe drinking water. Further tests on another 30,000 wells carried out by the UNDP–World Bank Water and Sanitation Program also showed arsenic in nearly 40 percent of the wells (Lockwood 1999).

TABLE WB.1   Summary of Arsenic Test Results from Hajiganj Thana

Total population of the thana	254,057
Number of unions	11
Number of villages	156
Number of tubewells tested by field kit	11,954
Number of tubewells contaminated with arsenic	11,095 (93 percent)
Number of villages with all tubewells contaminated	83 (53 percent)
Number of samples tested by spectrophotometer	193
Field test results confirmed by spectrophotometer (% cases)	92

Source: British Geological Survey 1999.

The extensive testing of groundwater prompted by concerns over arsenic presented an opportunity to test for other contaminants. The BGS survey also found other minerals of potential health significance:

- Boron exceeding the WHO Guideline Value was found in 13 percent of wells.
- Manganese exceeded the WHO Guideline Value in 31 percent of wells.
- Barium and chromium each exceeded the WHO Guideline Value in three wells.
- Ammonium frequently exceeded the WHO Guideline Value.

Effects on Health: Cases of Arsenic Poisoning

Arsenic is both toxic and carcinogenic. Inorganic forms of arsenic dissolved in drinking water are the most significant forms of natural exposure. Organic forms of arsenic that may be present in food are much less toxic to humans. Clinical manifestations of arsenic poisoning include various forms of skin disease, damage to internal organs, and, in some cases, cancer and death. Arsenic may also cause neurological damage over time after exposure to concentrations in drinking water as low as 0.1 mg/l (http://www.ce.kth.se/aom/AMOV/PEOPLE/Prosun/garg.h747tm). Some clinical features of arsenic poisoning include weakness, muscle aches, tingling and numbness, of hands and feet, cough, changing skin pigmentation, thickening of palms and soles of feet, anemia, hepatomegaly, splenomegaly, cirrhosis, hypertension, and abdominal pain (Guha Mazumder et al. 1992).

Of great concern to health specialists is that symptoms of chronic arsenic poisoning may take between 5 and 15 years—even longer in the case of cancers—to reveal themselves. The period depends on the amount of arsenic ingested, the length of exposure, and the susceptibility of the person. It is thus possible that the vast numbers of tubewells installed in Bangladesh in the past two decades have been slowly poisoning their users since they began operation. It is also possible that many of the worst health impacts may not appear for a number of years.

People with poor socioeconomic and nutritional status appear to be more vulnerable to arsenic poisoning because of their greater reliance on water sources of unknown or poor quality and because they have fewer options for shifting to better supplies. Arsenicosis, the disease caused by arsenic contamination, is not infectious, contagious, or hereditary, but patients with arsenicosis suffer psychologically, economically, and socially as well as physically. They lose jobs and income and become a burden to their family.

One of the earliest health studies was conducted in West Bengal in the late 1980s. Patients who consumed large amounts of arsenic were found to be suffering from a wide range of health effects. These patients all drank water from shallow tube wells with arsenic concentrations higher than the WHO recommended levels (Guha Mazumder et al. 1988).

A survey run by the Dhaka Community Hospital examined 920 patients in the villages of Pabna, Pakshi, Bera, and Kushitia in October 1996. Water was also collected from 41 tubewells. Sixty-six percent of the tubewells studied yielded concentrations of arsenic of more than 0.01 mg/l. Some of the samples had arsenic levels up to 900 times the standard. Analyses of the data from patients showed that nail, hair, skin, and urine samples were heavily impregnated with arsenic and that various other signs of arsenic poisoning were appearing.

Based on early surveys of the areal extent of the contamination, SOES estimated in 1996 that there could be as many as 220,000 people with identifiable signs of arsenic poisoning (SOES 1996). Chakraborti now estimates that as many as six million Indians in West Bengal are drinking contaminated water and that 300,000 people already show signs of arsenic poisoning (Bearak 1998). As of early 1999, 4,600 patients had been detected with severe arsenicosis, though much of the at-risk population had still not been assessed for arsenic-related health problems (Source 1999). SOES researchers have recently found symptoms of arsenic poisoning in patients in 22 of the 23 districts where they did focused health studies. One-third of the 7,600 people investigated were found with skin lesions (Source 1999). Experts predict that tens of millions of people might be affected within the next decade and one World Bank hydrologist believes that the number of people consuming dangerous quantities of arsenic could be as high as 18 million (Bearak 1998). Another assessment of the number of people living in areas where water (but not necessarily drinking water) has arsenic concentrations above the Bangladesh standard (0.05 mg/l) is about 21 million people (BGS 1999). This number would be roughly doubled if the WHO Guideline value of 0.01 mg/l were used as the standard. The highest density of people coinciding with high arsenic concentrations occurs southeast of Dhaka in the region of Chandpur.

The Source of Arsenic

Several hypotheses have been offered about the sources of arsenic contamination in groundwater of the Ganges delta region, including both natural and anthropogenic causes. One early theory—studied and quickly rejected—was the leaching of wood preservatives containing arsenic from treated electricity pylons. Soil samples collected at selected sites did not show extensive arsenic, and this hypothesis was eliminated when the true scope of the problem was determined (BGS 1999).

Other more serious theories noted the apparent natural geologic sources of the problem. The arsenic content of alluvial sediments in Bangladesh is usually in the range of 2 to 10 milligrams of arsenic per kilogram of soil (mg/kg); only slightly greater than typical sediments (2 to 6 mg/kg). However, it appears that an unusually large proportion of the arsenic is present in a potentially soluble form. The first major hypothesis noted that high levels of arsenic could be found in some pyrite rock in alluvial regions of Bangladesh. This theory proposed that when groundwater is extracted, pyrite is exposed to air and the arsenic is freed by oxidation to contaminate the water table. Thus, a high rate of water withdrawal by humans that caused groundwater levels to drop permitted oxygen to reach the minerals (Das et al. 1995, Chowdhury et al. 1998). In 1997, however, more intensive investigation revealed that arsenic concentrations were correlated with concentrations of dissolved iron and bicarbonate and that they are low, not high, in waters with oxygen (Nickson et al. 1998). The "pyrite oxidation" hypothesis is therefore now thought unlikely to play much of a role.

An alternative analysis now suggests that the contamination of groundwater with arsenic appears to be caused by the "reductive dissolution of arsenic-rich iron oxyhydroxides which in turn are derived from weathering of base-metal sulphides" (Nickson et al. 1998). In simpler terms, iron-rich minerals containing arsenic are common in the region. When these minerals come into contact with groundwater lacking oxygen—also common in the region—the arsenic is released into the water. Thus the groundwater arsenic problem in Bangladesh appears to arise because of an unfortunate combination of three factors: a ready source of arsenic (the sediments), mobilization (arsenic is released from the sediments to groundwater), and transport (arsenic is brought to the surface by groundwater pumps).

This finding also offers guidance for identifying where new water wells can be placed and for treating contaminated water. Soil surveys that identify sediments with arsenic-rich minerals can be used to avoid contaminated groundwater withdrawals. Furthermore, because oxidation of dissolved iron oxyhydroxide scavenges arsenic from water, the simple aeration of oxygen-poor groundwater, followed by settling, should remove a considerable amount of arsenic from solution (Nickson et al. 1998). In reality, no single, simple solution is likely for technical, social, and economic reasons.

Responses

Various responses to the problem have been taken or suggested. The government of Bangladesh has formed a National Arsenic Steering Committee involving the Ministry of Health and Family Welfare, the Ministry of Local Government and Rural Development, local NGOs, and various development partners. India and Bangladesh signed an agreement in March 1997 to work in close cooperation to address the situation. The national program in Bangladesh includes a data analysis and dissemination office called the National Arsenic Mitigation Information Centre.

UNICEF was the first UN agency to fund research on the extent of the arsenic problem. Together with DPHE, UNICEF is assisting in preparing alternate sources of safe water, such as pond sand filters and rainwater harvesting. With government ministries, UNICEF is preparing a program to train doctors and health workers to be able to diagnose arsenic patients and give them proper advice. UNICEF is also working with NGOs, such as BRAC and the Grameen Bank, on water testing and community-based mitigation.

In 1997, UNDP initiated a $1.7 million emergency assistance program at Bangladesh's request. The UNDP program consisted of five components: field water-quality testing to determine the level of arsenic contamination; identification and marking of tubewells to indicate the degree of contamination; identification of patients and provision of information, medical advice, and medicine; collection of data on the location of affected tubewells; and field testing of new technologies for separating arsenic from water.

WHO has supported studies on arsenic contamination and its effects on human health and has provided specific epidemiological expertise. WHO also organized a conference in New Delhi in 1997 on "Arsenic in Drinking Water and Resulting Toxicity in India and Bangladesh." Together with UNICEF, WHO has evaluated and supported the further development of field test kits for arsenic detection.

In August 1998, the World Bank announced a $32.4 million credit program with both short- and long-term measures to support the Bangladesh Arsenic Mitigation/Water Supply Project. World Bank funds will help provide comprehensive national screening of all the 4 million tubewells in Bangladesh. Funds will also provide on-site mitigation in both rural villages and urban areas. In regions with the most severe contamination, funds will be used to identify and provide alternative sources of supply (World Bank 1998). The Bank estimates that Bangladesh will require US$ 275 million in the next 10 to 12 years to fight the arsenic problem.

In February 1999, the Bangladesh minister of Local Government and Rural Development announced a plan to study the feasibility of keeping one pond in each of the country's 68,000 villages reserved for drinking water, with purification facilities. Later in 1999, UNICEF allocated $2 million to DPHE for installation of some 5,500 water points such as deep tube wells, rainwater collection jars, and pond sand filters for families affected by arsenic contamination.

Because individuals cannot remove arsenic from water by boiling or the use of normal filters, the most important action is to provide safe drinking water. This water is available, but only if the arsenic concentrations in wells are actually monitored and only if people use just those wells with low concentrations. As of mid-1999, only a small percentage of all wells had had their arsenic concentrations measured. Because of the vast number of wells, priorities have been set for monitoring wells thought to be at greatest risk of contamination, but the overall rate of monitoring must increase.

Other factors that are relevant for designing solutions include the following:

- Some health effects of acute arsenic poisoning can be reversed by drinking arsenic-free water and by eating nutritious and vitamin-rich food, but there is insufficient information at present about the reversibility of longer-term exposure or chronic effects.
- Surface water from rivers, ponds, canals, lakes, and rainwater is most likely to be arsenic-free. When this is the case, such water can be used for drinking after proper boiling.

To mitigate the problem, rapid detection of arsenic contaminated tubewells, provision of safe water, treatment of ill people, and health education are all essential. Until easy and inexpensive methods of arsenic removal become available, early detection and provision of alternative sources of water are vital. Ultimately, however, effective solutions to this problem will be complicated by the social and cultural factors at play. In a stunning example of the complex human dynamics that complicate addressing this problem, a Bangladeshi villager whose well has been found to be contaminated and who is already showing signs of advanced arsenic poisoning could not be persuaded to go to a neighboring clean well for water, saying "If I die, I will die, but I will not go to fetch water from another man's house" (Bearak 1998).

NOTE

1. The HDI is an indicator that measures life expectancy at birth, literacy, and per capita income. It is considered a far better measure of overall well-being than the traditional use of purely economic values such as gross national produce.

REFERENCES

For information from the West Bengal and Bangladesh Arsenic Crisis Information Centre, the online focal point for the environmental health disaster in Bangladesh and West Bengal, India, visit: http://bicn.com/acic/.

Bearak, B. 1998. "New Bangladesh disaster: Wells that pump poison." *New York Times* (November) 10, p. 1.

British Geological Survey (BGS). 1999. *Phase I, Groundwater Studies of Arsenic Contamination in Bangladesh*. Executive Summary, Main Report. (With Mott MacDonald) Department for International Development (DFID), United Kingdom. http://www.bgs.ac.uk/arsenic

Chowdhury, M., A. Jakariya, A.H. Tareq, J. Ahmed. 1998. "Village health workers can test tubewell water for arsenic." Bangladesh Rural Advancement Committee (BRAC). http://bicn.com/acic/infobank/brac1.html.

Das, D., G. Basu, T.R. Chowdhury, and D. Chakraborty. 1995. "Possible sources of arsenic contamination in groundwater." Proceedings of the International Conference on Arsenic in Groundwater. Calcutta, pp. 44–45.

Guha Mazumder, D.N., A.K. Chakraborty, A. Ghose, J. Das Gupta, D.P. Chakraborti, S.B. Dey, N. Chattopadhyay. 1988. "Chronic arsenic toxicity from drinking tubewell water in rural West Bengal." *Bulletin of the World Health Organization,* Vol. 66, No. 4, pp. 499–506.

Guha Mazumder, D.N., J. Das Gupta, A.K. Chakraborty, A. Chatterjee, D. Das, D. Chakraborti. 1992. "Environmental pollution and chronic arsenicosis in South Calcutta." *Bulletin of the World Health Organization.* Vol. 70, No. 4, pp. 481–485.

Hebert, H.J. 1999. "Panel warns about arsenic in water." Associated Press (March 24). http://dailynews.yahoo.com/headline...ap/19990324/hl/water_arsenic_4.html.

Lockwood, D.E. 1999. Statement at the National Conference on Coordinated Action for Arsenic Mitigation, February 27. http://www.un-bd.org/undpb/pr/arsenic.html

Nickson, R., J. McArthur, W. Burgess, K.M. Ahmed, P. Ravenscroft, and M. Rahman. 1998. "Arsenic poisoning of Bangladesh groundwater." *Nature,* Vol. 395 (24 September) http://bicn.com/acic/infobank/nature98-1.htm.

School of Environmental Studies (SOES). 1996. "A preliminary status report on arsenic problems in groundwater of Bangladesh." School of Environmental Studies, Jadavpur University, Calcutta, India, May.

School of Environmental Studies (SOES). 1997. (drafts by Ratan Kumar Dhar, et al.) "Groundwater arsenic calamity in Bangladesh" School of Environmental Studies, Jadavpur University, Calcutta, India, March.

Source. 1999. "Bangladesh: One pond in every village reserved for safe water." *Source: Water and Sanitation News.* No. 4 (April). http://www.wsscc.org/source/bulletin/sb04.html#bangladesh

Tsushima, S. 1998. "Arsenic contamination in ground water in Bangladesh: An overview." http://www.kfunigraz.ac.at/fwiwww/aan/newsl2/contamin.html

United Nations Development Programme (UNDP). 1998. *Human Development Report 1998.* Oxford University Press, New York.

United States National Academy of Sciences (U.S. NAS). 1999. Arsenic study, see: http://www.nap.edu/readingroom/enter2.cgi?0309063337.html

World Bank. 1998. "World Bank approves fast-track project to assist Bangladesh in adressing arsenic crisis." New Release No. 99/1927/SAS. At http://www.worldbank.org/html/extdr/extme/1927.htm

Fog Collection as a Source of Water

There went up a mist from the Earth, and watered the whole face of the ground.

GENESIS, CHAPTER 2, VERSE 6

There would be water in those clouds. We could collect some of that water.

ROBERT SCHEMENAUER

The collection or harvesting of water directly from the atmosphere, specifically dense fogs, is an unusual approach to water supply. Nevertheless, in regions with few alternatives, fog collection has proven to be both possible and effective. This section describes the concept of fog collection and experience to date with a number of specific projects in developing countries to produce water supplies from fog. Fog collection can also be used to provide water where the greatest constraint is not water supply, but water quality. In areas where surface or groundwater is contaminated, fog collection offers a source of potable drinking water suitable for direct human consumption.

The ability to use fog as a water source results from a combination of climatological and technical factors. Persistent fogs often occur in coastal areas or in interior mountains. In locations where persistent fogs occur naturally, trees or other vegetation have long been observed to concentrate and precipitate water. The collection of fog droplets by vegetation supports local ecosystems and contributes to recharge of aquifers.

Artificial collectors can also collect this water. Schemenauer and Cereceda (1991), two of the leading analysts of this approach, have identified more than 20 countries on six continents where research supports the idea that fog collection by artificial arrays could produce reliable water for local use. These include the Sudan, Kenya, South Africa, Namibia, Angola, Ascension Island, the Cape Verde Islands, and the Canary Islands. Based on local conditions, evaluation programs would also be appropriate for parts of Eritrea, Ethiopia, Somalia, Tanzania, Madagascar, and Morocco, among others. Several of these are countries with considerable constraints on water supply. Even in California, coastal fogs have been shown to produce substantial amounts of water to support redwood ecosystems, but regions like California have a far wider range of alternatives that can be called upon.

A standard collection system consists of two posts mounted in a hole and anchored with cables. Cables support the mesh, and pipes for collecting water are suspended from the lower edge. Local materials and construction practices can be used but only certain kinds of mesh materials are effective. Most projects use a polypropylene mesh with fibers about a millimeter wide and a lifetime of around 10 years. Standard collectors are usually 12 meters long and 6 meters high, with the mesh covering the upper 4 meters, giving a total mesh area of 48 m² per collector (Schemenauer and Cereceda 1994, 1997). The cost of a system depends on the site, the distance water has to be moved, storage costs, and the uses of the water. Experience to date suggests that a collector with a surface area of 50 m² costs in the range of U.S. $300 to $500. The cost of 100 large fog collectors, suitable for a village, would thus be on the order of U.S. $40,000, inexpensive compared to many other water-supply systems for remote sites.

The Nature of Fog

Raindrops have diameters from 500 μm (0.5 mm) to approximately 5,000 μm and fall at velocities from 2 to 9 meters per second. Drizzle has diameters from 40 μm to 500 μm and fall at velocities from 5 cm per second to 2 meters per second. Fog droplets have diameters from about 1 μm to 40 μm and fall at speeds ranging from less than 1 cm per second to approximately 5 cm per second. This speed is so slow that modest winds cause fog droplets to travel almost horizontally, permitting fog collectors to use vertical surfaces (Schemenauer and Cereceda 1994).

The amount of water present in clouds can range from 0.05 g/m^3 in thin clouds to over 3 g/m^3 in thunderstorms. Typical coastal fogs contain around 0.2 g/m^3 (Schemenauer and Joe 1989).

Fog forms on the surface of the ocean, in low-lying areas, through advection of clouds over terrain, or from orographic effects on mountains. They often lack sufficient liquid water or sufficient wind speeds for substantial water collection. Most of the research and demonstration programs for fog collecting have focused on dense fogs with high enough water content for producing good quantities of water.

Persistent winds from a single direction are ideal for fog collection. These situations occur where the driving forces are global in scale (Cereceda et al. 1990, Schemenauer et al. 1988). Wind and atmospheric pressure patterns, for example, produce onshore winds in northern Chile most of the year and southerly winds along the coast of Peru. The trade winds are another example of persistent winds in regions of interest.

It is also helpful to have a mountain range that intercepts the clouds that are formed. On a continental scale, these include the coastal mountains of Chile, Peru, and Ecuador. On a local scale it can be an isolated hill. For coastal mountain ranges, it is important that the orientation of the range is perpendicular to the direction of the winds. The topography of the ridgeline or mountain influences site selection. In regions of complex terrain, it is vital to test different sites before making choices of where to locate collectors. Studies of wind flow over mountains show that fog-collection locations along ridges, or just slightly upwind of the crest, are optimal (Schemenauer and Cereceda 1994).

The amount of water that can actually be collected from fog depends on many factors, including the area and capture efficiency of the collector, the moisture available in the fog, wind speed, and temperature. A review of natural fog collection by vegetation, particularly trees, has shown that the vertical cross section of a tree can collect water at a rate of about 10 liters per square meter of tree area per day (Schemenauer and Cereceda 1992a).

The collection of fog water from forests or fields of collectors is more difficult to quantify than from individual trees (or collectors), and depends on deposition rates, evapotranspiration factors, wind flows over complex terrain, and other variables. A simple calculation can, however, demonstrate that significant amounts of water can be produced (see Box WB-2).

Experience with Regional Fog Collection Programs

Successful fog collection programs have been developed in several countries, including Chile, Peru, Ecuador, and Oman. The largest provides nearly 11,000 liters of water per day to a village in the arid coastal desert of northern Chile. A number of these regional projects are described here.

Chungungo, Chile

The largest fog collection project currently operating serves the fishing village of Chungungo on the north central coast of Chile. Up until the 1970s, Chungungo received water from a mine, but when the mine closed, that supply was cut off. Arrangements were then made to bring water to the village by truck from a well 40 kilometers away. The water delivery was irregular and expensive and there were concerns about the water quality.

The ridgeline above the village of Chungungo is frequently covered in fog. Beginning in 1987, this ridgeline was the site of a large pilot project to harvest fog for water supply. The incoming clouds are thin, 100 to 300 m, and rarely produce drizzle or rain. Their tops vary in altitude from perhaps 500 to 1,200 m depending on the height of the temperature inversion that persists throughout the year. Research at the site has included surveys of the local meteorology, interactions of topography and wind, fog microphysics, fog water chemistry, water production rates, and water costs.

In late 1987, 50 large fog collectors, each 48 m² of a double layer of polypropylene mesh, were constructed by the Corporacion Nacional Forestal with funding from

---

BOX WB.2

How Much Water Might a Forest Collect from Fog?

R. Schemenauer and P. Cereceda, two leading fog collection experts, offer the following example of the amount of water a forest might collect (or conversely, lose if the forest is cut) from fog. Assuming a fog-covered watershed with 100,000 trees planted for good exposure in a 500-meter-wide band roughly 10 meters apart. They estimate that each tree can collect around 250 liters per day during the fog season. If the season is half a year, the total water collected in a year is 45 m³ per tree, or 4.5×10^6 m³ in total. Assuming that three quarters of this water is used directly by the trees or evaporates, around a million m³ is still available for use. This is equivalent to having an extra 100 mm of precipitation fall on a 10-km² area in a year. In arid regions, this can equal or exceed the annual precipitation (Schemenauer and Cereceda 1992a). Collectors should have spaces between them to permit the wind to flow. Studies suggest that lengths of 0.5 to 1 kilometer may be needed for 50 collectors of 12 to 24 meters. Parallel rows of collectors may be built in some locations.

the Canadian International Development Research Centre (IDRC) (Schemenauer and Joe 1989). A survey was done in 1988 of the actual water use in the village. This study covered all the households and found that domestic water use was only 14 liters per person per day (lpcd), with a range from 3 to 57 lpcd, depending on income (Cereceda et al. 1992). The same study determined that the villagers considered 27 lpcd to be necessary for most domestic uses. The cost of the trucked water supply was high, averaging $1.80 per m^3. This is well above the costs paid by domestic water users in industrialized nations for high-quality piped water supply. Given local incomes, paying for this water often took up to 10 percent of total family income. Even at this price, the municipality heavily subsidized the water. The 1988 study estimated that the true cost of providing trucked water was over $7 per m^3. The situation in Chungungo was typical of other isolated villages in the coastal deserts of Chile and Peru (Cereceda et al. 1992).

In 1992, 25 additional collectors were added and a 6.2-km pipeline was laid to the village. A 100-m^3 storage tank was placed above the village and water flows through pipes laid to houses in the village. The system became operational in March 1992. Additional support for the project came from the Pontifical Catholic University of Chile, the University of Chile, Environment Canada, and the Canadian embassy in Santiago (Schemenauer and Cereceda 1994, 1997).

This fog collection project showed that the collectors could generate more than 3 liters per square meter of collector area per day, or an average of 10,800 liters per day. The range varied from zero on fogless days to a maximum of about 100,000 liters per day. On average, each of the 330 villagers now receives about 33 liters of water per day from the system—more than double their previous average water use.

Water in the incoming fog and from the fog collectors is generally of very high quality. The concentrations of ions and trace elements in the fog water have been studied in detail and meet Chilean and WHO drinking water standards (Schemenauer and Cereceda 1992b). It contains some marine salts and soil dust but little contamination from anthropogenic sources. The water also receives chlorination treatment required by law for domestic water supplies in Chile.

The unconventional nature of the water-supply system led to an unconventional management system. The system is run by a "Potable Water Committee" of villagers with elected, unpaid members from the village and a paid administrator. The administrator maintains the storage tank and distribution system, monitors water use in each home, and collects a monthly fee based on the volume of use. The fee pays the administrator's salary and minor maintenance costs. A portion of the fees is reserved for future expenses. The committee also regulates water use and monitors the amount of water in the reservoir. During periods of low supply, the Committee alerts villagers to reduce consumption; during periods of high production, surplus water is diverted to a larger 400-m^3 open reservoir for agricultural purposes. As a result of the fog-collection system the overall costs of water have decreased for both the families and the municipality (Schemenauer and Cereceda 1994).

Peru and Ecuador

Other collection projects have been developed in Latin America, including projects in Peru and Ecuador. Along the coastline of Peru are areas of dry desert with small rivers carrying water from the Andes. The rural villagers are often extremely poor and suffer from both overall scarcity and from contaminated water (Pinche-Laurre 1986).

As described in Schemenauer and Cereceda (1994), in 1990 the Canadian International Development Agency provided support for an assessment project near Lima, with the assistance of the Canadian Embassy and the Servicio Nacional de Meterologia e Hidrologia. A site at Cerro Orara (at 430 m above sea-level) was selected just north of Lima, about 3.5 km from the coast. Surveys revealed that even more water might be produced per square meter of collector here than at the site in Chile because of higher wind speed and a light drizzle that could be collected. Subsequently, two private companies installed 1,200 m^2 of mesh to provide fog water to a school and 500 m^2 of mesh to produce fog water for reforestation purposes.

As described in Schemenauer and Cereceda (1994), in 1993, IDRC initiated funding for a separate large agricultural and forestry project in a region of Peru with only 5 mm of annual precipitation for the desert community of Collanac. As part of the project, the hillsides above the town are being reforested using fog water for irrigating the trees. In the mid-1990s, the Commission of European Communities sponsored a project with the Universidad de San Augustin in Arequipa in the southern part of Peru to irrigate trees in a reforestation effort on coastal hills.

Ecuador also has stretches of coastline where the people experience water shortages and high water costs. An evaluation of fog collection potential began in late 1992 near Puerto Lopez, a region of coastal mountains seasonally covered with fog. Two evaluation projects were also started in the high Andes at elevations of 2,830 m at Pululahua near Quito and 2,000 m near Celica in the south of the country. The first operational fog-collection project in Ecuador was at the Puluahua site and was a joint effort between the Ecuadorean nongovernmental organization Centro de Investigaciones Sociales Alternativas and the Fondo Ecuatoriano Canadiense de Desarrollo (Schemenauer and Cereceda 1997). Another project is operating at Pachamama Grande, an indigenous community in the south of Ecuador at an elevation of 3,700 m (Schemenauer and Cereceda 1997, Perera 1998).

The Sultanate of Oman

A major fog-collection evaluation project was undertaken in the Sultanate of Oman in 1989 and 1990, funded initially by the United Nations Development Programme, the World Meteorological Organization, and the government of Oman. During the southwest monsoon (the Khareef) from around mid-June to mid-September, local mountains are covered with heavy fog and drizzle (Schemenauer and Cereceda 1994).

A range of collectors was tested. At elevations of 900 to 1,000 m, collection rates averaged 30 liters per m^2 per day (Barros and Whitcombe 1989, COWI 1990) over the three-month wet period. This is around ten times the average volume collected at the Chilean site, but occurs only during limited periods of time. Because of the extended dry period between collection seasons, the most likely application would be reforestation rather than local water supply, unless storage can be provided and water quality maintained.

Table WB.2 shows a summary of the production at demonstration sites in three countries, as measured with standard collectors. The average water-collection rates

TABLE WB.2   Fog Water Production and
Fog Season Length at Three Sites

Country	Average Water Production (l/m²/d)	Length of Fog Season (d/yr)	Annual Water Production (l/m²/yr)
Chile	3	365	1,095
Peru	9	210	1,890
Oman	30	75	2,250

Source: Schemenauer and Cereceda 1994.

Notes: Production volume rates are liters of water per square meter of collector area per unit time.

during the fog seasons in Chile, Peru, and Oman were 3, 9, and 30 liters per m² of collector area per day, with significant differences in the length of the fog season and hence in total production of water. In Chile, with a low daily production but a reliable long-term collection period, the use of the water for domestic purposes is quite good. In Peru, with a moderately high production for seven months, possible uses include agriculture or domestic uses if storage is available to even out supply and demand. In Oman, the short wet season with high production rates could be suited to forestry applications where native trees could be irrigated for about three months.

CONCLUSIONS

Fog collection by artificial collectors is an unconventional but proven source of water for small-scale local needs. In some regions, fogs bring an essentially unlimited amount of water to mountain sites. Actual installations now exist in several countries where conventional methods do not provide an adequate supply of water. The water produced by these systems can be delivered in large quantities, it is potable, and the costs are comparable to, or lower than, the cost of other water systems in rural arid regions. In principle, the amount of water that can be collected is limited only by the number of collectors installed. Even with much larger arrays than have been installed so far, the amount of water removed from the incoming clouds is modest, and downwind effects will be negligible. Moreover, this source of water is sustainable because the driving forces for the formation of fogs are part of the natural hydrologic cycle.

It is also important to note that in the humid tropics, cloud forests owe their existence to the input of water from both precipitation and fog. This means that deforestation in tropical mountains can lead to reduced fog water inputs, less water in aquifers, and reduced streamflows. This, coupled with erosion from deforested lands, can lead to significant changes in local hydrology and natural ecosystems. At the same time, fogs are often found in the humid tropics outside of the rainy season in regions such as the Philippines, India, Kenya, Hawaii, Central America, and the

Caribbean. This suggests that it might be possible to collect fog water during dry periods for projects such as reforestation.

New evaluation efforts are underway in Namibia, Nepal, and Mexico, with a focus on domestic water supplies. While the total amount of water likely to come from fog collection systems may never be large, it has already proven itself to be a potentially important local source of reliable, high-quality water.

REFERENCES

Barros, J., and R.P. Whitcombe. 1989. "Fog and rain water collection in Southern Region 1989 research programme." Draft report to the Technical Secretariat of the Planning Committee for Development and Environment in the Southern Region, Salalah, Oman, 100 pp.

Cereceda, P., J. Barros, and R.S. Schemenauer. 1990. "Las nieblas costeras de Chile y Oman similitudes y diferencias." *Revista Geogr fica de Chile Terra Australis*, Vol. 33, pp. 49–60.

Cereceda,P., R. S. Schemenauer, and M. Suit. 1992. "An alternative water supply for Chilean coastal desert villages." *International Journal of Water Resources Development*, Vol. 8, pp. 53–59.

COWI. 1990. "Dhofar Khareef studies feasibility of fog rain water collection, and guidelines for pilot projects." Draft report to the Technical Secretariat of the Planning Committee for Development and Environment in the Southern Region, COWI Consulting. November, Salalah, Oman, 100 pp.

Perera, J. 1998. "Chilean village gets drinking water from fog." *Interpress Service (IPS)*. (March) 1. http://www.oneworld.org/ips2/mar98/fog.html.

Pinche-Laurre, C. 1986. "Estudio de las condiciones clim ticas y de la niebla en la costa norte de Lima." Thesis, Universidad Nacional Agraria La Molina, Dept. of Physics and Meteorology, Lima, Peru.

Schemenauer, R.S., and P. Cereceda. 1991. "Fog water collection in arid coastal locations." *Ambio*, Vol. 20, pp. 303–308.

Schemenauer, R.S., and P. Cereceda. 1992a. "The use of fog for groundwater recharge in arid regions." International Seminar on Groundwater and the Environment in Arid and Semiarid Areas, Beijing, China, pp. 84–91.

Schemenauer, R.S., and P. Cereceda. 1992b. "The quality of fog water collected for domestic and agricultural use in Chile." *Journal of Applied Meteorology*, Vol. 31, pp. 275–290.

Schemenauer, R.S., and P. Cereceda. 1994. "Fog collection's role in water planning for developing countries." *Natural Resources Forum*, Vol. 18, pp. 91–100. United Nations, New York. http://www1.tor.ec.gc.ca/armp/fog/old/unnrf.html.

Schemenauer, R.S., and P. Cereceda. 1997. "Fog collection." *Tiempo*, Vol. 26 (December), pp. 17–21.

Schemenauer, R.S., and P.I. Joe. 1989. "The collection efficiency of a massive fog collector." *Atmospheric Research*, Vol. 24, pp. 53–69.

Schemenauer, R.S., H. Fuenzalida, and P. Cereceda. 1988. "A neglected water resource: The camanchaca of South America." *Bulletin of the American Meteorological Society*, Vol. 69, pp. 138–147.

Environment and Security: Water Conflict Chronology—Version 2000

Water resources have rarely, if ever, been the sole source of violent conflict or war. But this fact has led some international security "experts" to ignore or belittle the complex and real relationships between water and security. It is easy for an academic approach to draw a narrow definition of "security" in a way that excludes water (or other resources) from the debate over international security, or to require that security threats be narrow, single-issue factors. But this approach both misunderstands the connections between water and security and misleads policymakers and the public seeking ways of reducing tensions and violence. In fact, there is a long and highly informative history of conflicts and tensions over water resources, the use of water systems as weapons during war, and the targeting of water systems during conflicts caused by other factors.

In an ongoing effort to understand the connections between water resources, water systems, and international security and conflict, the Pacific Institute for Studies in Development, Environment, and Security initiated a project in the late 1980s to track and categorize events related to water and conflict. In one of the more esoteric products from that effort, a list of water-related conflicts in the myths, legends, and history of the ancient Middle East was published in *Environment* magazine (Gleick 1994). This chronology included information from between 5,000 and 2,300 years before the present. During the effort to develop that timeline, it became clear that modern history was even richer in examples of the connections between water and conflicts. The first edition of *The World's Water*, published two years ago, included an updated version of the chronology from the ancient Middle East, together with a more modern chronology listing events from 1500 to 1998 where water, conflict, and security intersected (Gleick 1998).

The interest in this chronology has been strong. New information has been sent to me by various historians, water experts, and readers to update, correct, and expand the current chronology. Recent world events in the Balkans, East Timor, and other regions have, unfortunately, also added several new entries. As a result, a new chronology is presented here, with new entries and a range of corrections and modifications to the older ones. In addition, I have made changes in how several of these entries are categorized. The heading "Basis of Conflict" now offers, I think, a more clear set of categories than the previous listing. The current categories, or types of conflict, now include:

Control of Water Resources (state and nonstate actors): where water supplies or access to water is at the root of tensions.

Military Tool (state actors): where water resources, or water systems themselves, are used by a nation or state as a weapon during a military action.

Political Tool (state and nonstate actors): where water resources, or water systems themselves, are used by a nation, state, or nonstate actor for a political goal.

Terrorism (nonstate actors): where water resources, or water systems, are either targets or tools of violence or coercion by nonstate actors. This is a new definition and will be further clarified for the next version of the chronology in a new project on "Environmental Terrorism" at the Pacific Institute.

Military Target (state actors): where water resource systems are targets of military actions by nations or states.

Development Disputes (state and nonstate actors): where water resources or water systems are a major source of contention and dispute in the context of economic and social development.

It will be clear to even the casual reader that these definitions are imprecise and that single events can fall into more than one category, depending on perception and definitions. For example, intentional military attacks on water-supply systems can fall into both the Targets and Tools categories, depending on one's point of view. Disputes over control of water resources may reflect either political power disputes or disagreements over approaches to economic development, or both. I believe this is inevitable and even desirable—international security is not a clean, precise field of study and analysis. It is evolving as international and regional politics evolves and as new factors become increasingly, or decreasingly, important in the affairs of humanity. In all this, however, one factor remains constant: the importance of water to life means that providing for water needs and demands will never be free of politics. As social and political systems change and evolve, this chronology and the kinds of entries and categories will change and evolve. And I continue to look forward to contributions and comments from readers.

Date	Parties Involved	Basis of Conflict[1]	Violent Conflict or in the Context of Violence?	Description	Sources
1503	Florence and Pisa warring states	Military tool	Yes	Leonardo da Vinci and Machiavelli plan to divert Arno River away from Pisa during conflict between Pisa and Florence.	Honan 1996
1642	China, Ming Dynasty	Military tool	Yes	The Huang He's dikes have been breached for military purposes. In 1642, "toward the end of the Ming dynasty (1368–1644), General Gao Mingheng used the tactic near Kaifeng in an attempt to suppress a peasant uprising."	Hillel 1991
1863	United States Civil War	Military tool	Yes	General U.S. Grant, during the Civil War campaign against Vicksburg, cut levees in the battle against the Confederates.	Grant 1885, Barry 1997
1898	Egypt, France, Britain	Military and political tool, control of water resources	Military maneuvers	Military conflict nearly ensues between Britain and France in 1898 when a French expedition attempted to gain control of headwaters of the White Nile. While the parties ultimately negotiated a settlement of the dispute, the incident has been characterized as having "dramatized Egypt's vulnerable dependence on the Nile, and fixed the attitude of Egyptian policy-makers ever since."	Moorehead 1960
1924	Owens Valley, Los Angeles, California	Political tool, control of water resources, terrorism, and development dispute	Yes	The Los Angeles Valley aqueduct/pipeline suffers repeated bombings in an effort to prevent diversions of water from the Owens Valley to Los Angeles.	Reisner 1986, 1993
1935	California, Arizona	Political tool, development dispute	Military maneuvers	Arizona calls out the National Guard and militia units to the border with California to protest the construction of Parker Dam and diversions from the Colorado River; dispute ultimately is settled in court.	Reisner 1986, 1993
1938	China and Japan	Military tool, military target	Yes	Chiang Kai-shek orders the destruction of flood-control dikes of the Huayuankou section of the Huang He (Yellow) River to flood areas threatened by the Japanese army. West of Kaifeng, dikes are destroyed with dynamite, spilling water across the flat plain. The flood destroyed part of the invading army and its heavy equipment was mired in thick mud, though Wuhan, the headquarters of the Nationalist government was taken in October. The waters flooded an area variously estimated as between 3,000 and 50,000 square kilometers, and killed Chinese estimated in numbers between "tens of thousands" and "one million."	Hillel 1991, Yang Lang 1989/1994

Date	Parties Involved	Basis of Conflict[1]	Violent Conflict or in the Context of Violence?	Description	Sources
1940–1945	Multiple parties	Military target	Yes	Hydroelectric dams routinely bombed as strategic targets during World War II.	Gleick 1993
1943	Britain, Germany	Military target	Yes	British Royal Air Force bombs dams on the Mohne, Sorpe, and Eder Rivers, Germany (May 16, 17). Mohne Dam breech killed 1,200, destroyed all downstream dams for 50 km.	Kirschner 1949
1944	Germany, Italy, Britain, United States	Military tool	Yes	German forces use waters from the Isoletta Dam (Liri River) in January and February to successfully destroy British assault forces crossing the Garigliano River (downstream of Liri River). The German Army then dams the Rapido River, flooding a valley occupied by the American Army.	Corps of Engineers 1953
1944	Germany, Italy, Britain, United States	Military tool	Yes	German Army floods the Pontine Marches by destroying drainage pumps to contain the Anzio beachhead established by the Allied landings in 1944. Over 40 square miles of land were flooded; a 30-mile stretch of landing beaches was rendered unusable for amphibious support forces.	Corps of Engineers 1953
1944	Germany, Allied forces	Military tool	Yes	Germans flood the Ay River, France (July) creating a lake 2 meters deep and several kilometers wide, slowing an advance on Saint Lo, a German communications center in Normandy.	Corps of Engineers 1953
1944	Germany, Allied forces	Military tool	Yes	Germans flood the Ill River valley during the Battle of the Bulge (winter 1944–45) creating a lake 16 kilometers long, 3–6 kilometers wide, and 1–2 meters deep, greatly delaying the American Army's advance toward the Rhine.	Corps of Engineers 1953
1947 onwards	Bangladesh, India	Development disputes, control of water resources	No	Partition divides the Ganges River between Bangladesh and India; construction of the Farakka barrage by India, beginning in 1962, increases tension; short-term agreements settle dispute in 1977–82, 1982–84, and 1985–88, and 30-year treaty is signed in 1996.	Butts 1997, Samson and Charrier 1997
1947–1960s	India, Pakistan	Development disputes, control of water resources, and political tool	No	Partition leaves Indus basin divided between India and Pakistan; disputes over irrigation water ensue, during which India stems flow of water into irrigation canals in Pakistan; Indus Waters Agreement reached in 1960 after 12 years of World Bank–led negotiations.	Bingham et al. 1994, Wolf 1997
1948	Arabs, Israelis	Military tool	Yes	Arab forces cut off West Jerusalem's water supply in first Arab–Israeli war.	Wolf 1995, 1997
1950s	Korea, United States, others	Military target	Yes	Centralized dams on the Yalu River serving North Korea and China are attacked during Korean War.	Gleick 1993
1951	Korea, United Nations	Military tool and military target	Yes	North Korea releases flood waves from the Hwachon Dam damaging floating bridges operated by UN troops in the Pukhan Valley. U.S. Navy planes are then sent to destroy spillway crest gates.	Corps of Engineers 1953

(continues)

Date	Parties Involved	Basis of Conflict[1]	Violent Conflict or in the Context of Violence?	Description	Sources
1951	Israel, Jordan, Syria	Political tool, military tool, development disputes	Yes	Jordan makes public its plans to irrigate the Jordan Valley by tapping the Yarmouk River; Israel responds by commencing drainage of the Huleh swamps located in the demilitarized zone between Israel and Syria; border skirmishes ensue between Israel and Syria.	Wolf 1997, Samson and Charrier 1997
1953	Israel, Jordan, Syria	Development dispute, military target, political tool	Yes	Israel begins construction of its National Water Carrier to transfer water from the north of the Sea of Galilee out of the Jordan basin to the Negev Desert for irrigation. Syrian military actions along the border and international disapproval lead Israel to move its intake to the Sea of Galilee.	Samson and Charrier 1997
1958	Egypt, Sudan	Military tool, political tool, control of water resources	Yes	Egypt sends an unsuccessful military expedition into disputed territory amidst pending negotiations over the Nile waters, Sudanese general elections, and an Egyptian vote on Sudan–Egypt unification; Nile Water Treaty signed when pro-Egyptian government elected in Sudan.	Wolf 1997
1960s	North Vietnam, United States	Military target	Yes	Irrigation water supply systems in North Vietnam are bombed during Vietnam War.	Gleick 1993
1962 to 1967	Brazil, Paraguay	Military tool, political tool, control of water resources	Military maneuvers	Negotiations between Brazil and Paraguay over the development of the Paraná River are interrupted by a unilateral show of military force by Brazil in 1962, which invades the area and claims control over the Guaira Falls site. Military forces are withdrawn in 1967 following an agreement for a joint commission to examine development in the region.	Murphy and Sabadell 1986
1963–1964	Ethiopia, Somalia	Development dispute, military tool, political tool	Yes	Creation of boundaries in 1948 leaves Somali nomads under Ethiopian rule; border skirmishes occur over disputed territory in Ogaden desert where critical water and oil resources are located; cease-fire is negotiated only after several hundred are killed.	Wolf 1997
1965–1966	Israel, Syria	Military tool, political tool, control of water resources, development dispute	Yes	Fire is exchanged over "all-Arab" plan to divert the Jordan River headwaters and presumably preempt Israeli National Water Carrier; Syria halts construction of its diversion in July 1966.	Wolf 1995, 1997
1967	Israel, Syria	Military target and tool	Yes	Israel destroys the Arab diversion works on the Jordan River headwaters. During Arab–Israeli War Israel occupies Golan Heights, with Banias tributary to the Jordan; Israel occupies West Bank.	Gleick 1993, Wolf 1995, 1997, Wallenstein and Swain 1997
1969	Israel, Jordan	Military target and tool	Yes	Israel, suspicious that Jordan is overdiverting the Yarmouk, leads two raids to destroy the newly built East Ghor Canal; secret negotiations, mediated by the United States, lead to an agreement in 1970.	Samson and Charrier 1997

Date	Parties Involved	Basis of Conflict[1]	Violent Conflict or in the Context of Violence?	Description	Sources
1970s	Argentina, Brazil, Paraguay	Political goal, development dispute	No	Brazil and Paraguay announce plans to construct a dam at Itaipu on the Paraná River, causing Argentina concern about downstream environmental repercussions and the efficacy of its own planned dam project downstream. Argentina demands to be consulted during the planning of Itaipu but Brazil refuses. An agreement is reached in 1979 that provides for the construction of both Brazil and Paraguay's dam at Itaipu and Argentina's Yacyreta Dam.	Wallenstein and Swain 1997
1974	Iraq, Syria	Military target, military tool, political tool, development dispute	Military maneuvers	Iraq threatens to bomb the al-Thawra dam in Syria and masses troops along the border, alleging that the dam has reduced the flow of Euphrates River water to Iraq.	Gleick 1994
1975	Iraq, Syria	Development dispute, military tool, political tool	Military maneuvers	As upstream dams are filled during a low-flow year on the Euphrates, Iraq claims that flow reaching its territory is "intolerable" and asks the Arab League to intervene. Syrians claim they are receiving less than half the river's normal flow and pull out of an Arab League technical committee formed to mediate the conflict. In May Syria closes its airspace to Iraqi flights and both Syria and Iraq reportedly transfer troops to their mutual border. Saudi Arabia successfully mediates the conflict.	Gleick 1993, 1994, Wolf 1997
1978 onwards	Egypt, Ethiopia	Development dispute, political tool	No	Ethiopia's proposed construction of dams on the headwaters of the Blue Nile leads Egypt to repeatedly declare the vital importance of water. President Anwar Sadat states "The only matter that could take Egypt to war again is water" (Starr 1991). Egypt's Foreign Minister, Boutrous Ghali states, "The next war in our region will be over the waters of the Nile, not politics" (Walker 1998).	Gleick 1991, 1994
1981	Iran, Iraq	Military target and tool	Yes	Iran claims to have bombed a hydroelectric facility in Kurdistan, thereby blacking out large portions of Iraq, during the Iran–Iraq War.	Gleick 1993
1982	Israel, Lebanon, Syria	Military tool	Yes	Israel cuts off the water supply of Beirut during siege.	Wolf 1997
1986	North Korea, South Korea	Military tool	No	North Korea's announcement of its plans to build the Kumgansan hydroelectric dam on a tributary of the Han River upstream of Seoul raises concerns in South Korea that the dam could be used as a tool for ecological destruction or war.	Gleick 1993
1990	South Africa	Development dispute, control of water resources	No	Proapartheid council cuts off water to the Wesselton township of 50,000 blacks following protests over miserable sanitation and living conditions.	Gleick 1993

(continues)

Date	Parties Involved	Basis of Conflict[1]	Violent Conflict or in the Context of Violence?	Description	Sources
1990	Iraq, Syria, Turkey	Development dispute, military tool, political tool	No	The flow of the Euphrates is interrupted for a month as Turkey finishes construction of the Ataturk Dam, part of the Grand Anatolia Project. Syria and Iraq protest that Turkey now has a weapon of war. In mid-1990 Turkish president Turgut Ozal threatens to restrict water flow to Syria to force it to withdraw support for Kurdish rebels operating in southern Turkey.	Gleick 1993, 1995
1991	Karnataka, Tamil Nadu (India)	Development dispute, control of water resources	Yes	Violence erupts when Karnataka reacts to an Interim Order handed down by the Cauvery Waters Tribunal. The Tribunal had been established in 1990 to settle two decades of dispute between Karnataka and Tamil Nadu over irrigation rights to the Cauvery River.	Gleick 1993, Butts 1997
1991	Iraq, Kuwait, United States	Military target	Yes	During the Gulf War, Iraq destroys much of Kuwait's desalination capacity during retreat.	Gleick 1993
1991	Iraq, Turkey, United Nations	Military tool	Yes	Discussions are held at the United Nations about using the Ataturk Dam in Turkey to cut off flows of the Euphrates to Iraq.	Gleick 1993
1991	Iraq, Kuwait, United States	Military target	Yes	Baghdad's modern water supply and sanitation system are intentionally targeted by Allied coalition.	Gleick 1993
1992	Czechoslovakia, Hungary	Political tool, development dispute	Potential	Hungary abrogates a 1977 treaty with Czechoslovakia concerning construction of the Gabcikovo/Nagymaros project based on environmental concerns. Slovakia continues construction unilaterally, completes the dam, and diverts the Danube into a canal inside Slovakian republic. Massive public protest and movement of military to the border ensue; issue taken to the International Court of Justice.	Gleick 1993
1992	Bosnia, Bosnian Serbs	Military tool	Yes	The Serbian siege of Sarajevo, Bosnia–Herzegovina, includes a cutoff of all electrical power and the water feeding the city from the surrounding mountains. The lack of power cuts the two main pumping stations inside the city despite pledges from Serbian nationalist leaders to United Nations officials that they would not use their control of Sarajevo's utilities as a weapon. Bosnian Serbs take control of water valves regulating flow from wells that provide more than 80 percent of water to Sarajevo; reduced water flow to city is used to "smoke out" Bosnians.	Burns 1992, Husarska 1995
1993	Iraq	Military tool	No	To quell opposition to his government, Saddam Hussein reportedly poisons and drains the water supplies of southern Shiite Muslims.	Gleick 1993
1993	Yugoslavia	Military target and tool	Yes	Peruca Dam intentionally destroyed during war.	Gleick 1993

Date	Parties Involved	Basis of Conflict[1]	Violent Conflict or in the Context of Violence?	Description	Sources
1995	Ecuador, Peru	Military and political tool	Yes	Armed skirmishes arise in part because of disagreement over the control of the headwaters of the Cenepa River. Wolf argues that this is primarily a border dispute simply coinciding with location of a water resource.	Samson and Charrier 1997, Wolf 1997
1997	Singapore, Malaysia	Political tool	No	Malaysia supplies about half of Singapore's water and in 1997 threatened to cut off that supply in retribution for criticisms by Singapore of policy in Malaysia.	Zachary 1997
1998	Tajikistan	Terrorism, political tool	Potential	On November 6, a guerrilla commander threatens to blow up a dam on the Kairakkhum channel if political demands are not met. Col. Makhmud Khudoberdyev made the threat, reported by the ITAR-Tass News Agency.	WRR 1998
1999	Lusaka, Zambia	Terrorism, political tool	Yes	Bomb blast destroys the main water pipeline, cutting off water for the city of Lusaka, population 3 million.	FTGWR 1999
1999	Yugoslavia	Military target	Yes	Belgrade reports that NATO planes have targeted a hydroelectric plant during the Kosovo campaign.	Reuters 1999a
1999	Bangladesh	Development dispute, political tool	Yes	Fifty injured during strikes called to protest power and water shortages. Protest led by former Prime Minister Begum Khaleda Zia over deterioration of public services and law and order.	Ahmed 1999
1999	Yugoslavia	Military target	Yes	NATO targets utilities and shuts down water supplies in Belgrade. NATO bombs bridges on Danube, disrupting navigation.	Reuters 1999b
1999	Yugoslavia	Political tool	Yes	Yugoslavia refuses to clear war debris on Danube (downed bridges) unless financial aid for reconstruction is provided; European countries on Danube fear flooding due to winter ice dams will result. Diplomats decry environmental blackmail.	Simons 1999
1999	Kosovo	Political tool	Yes	Serbian engineers shut down water system in Pristina prior to occupation by NATO.	Reuters 1999c
1999	Kosovo	Terrorism/ political tool	Yes	Contamination of water supplies/wells by Serbs disposing of bodies of Kosovar Albanians in local wells.	CNN 1999
1999	East Timor	Military tool, political tool, terrorism	Yes	Militia opposing East Timor independence kill pro-independence supporters and throw bodies in water well.	BBC 1999

[1]Conflicts may stem from the drive to possess or control another nation's water resources, thus making water systems and resources a "political" or "military goal." Inequitable distribution and use of water resources, sometimes arising from a water development, may lead to "development disputes," heighten the importance of water as a strategic goal, or lead to a degradation of another's source of water. Conflicts may also arise when water systems are used as instruments of war, either as "targets" or "tools." These distinctions are described in detail in Gleick (1993, 1998).

REFERENCES

Ahmed, A. 1999. "Fifty hurt in Bangladesh strike violence." Reuters News Service, Dhaka, April 18, 1999. http://biz.yahoo.com/rf/990418/3.html

Barry J.M. 1997. *Rising Tide: The Great Mississippi Flood of 1927 and How It Changed America.* Simon and Schuster, New York, p. 67.

Bingham, G., A. Wolf, and T. Wohlegenant. 1994. "Resolving water disputes: Conflict and cooperation in the United States, the Near East, and Asia." U.S. Agency for International Development (USAID). Bureau for Asia and the Near East, Washington, D.C.

British Broadcasting Company (BBC). 1999. "World: Asia-Pacific Timor atrocities unearthed." September 22, 1999. http://news.bbc.co.uk/hi/english/world/asia-pacific/newsid_455000/455030.stm

Burns, J.F. 1992. "Tactics of the Sarajevo Siege: Cut Off the Power and Water," *New York Times* (September 25), p. A1.

Butts, K. (ed.). 1997. *Environmental Change and Regional Security.* Asia-Pacific Center for Security Studies, Center for Strategic Leadership, U.S. Army War College, Carlisle, Pennsylvania.

Cable News Network (CNN). 1999. "U.S.: Serbs destroying bodies of Kosovo victims." CNN. May 5, www.cnn.com/WORLD/europe/9905/05/kosovo.bodies

Corps of Engineers. 1953. "Applications of Hydrology in Military Planning and Operations and Subject Classification Index for Military Hydrology Data." Military Hydrology R&D Branch, Engineering Division, Corps of Engineers, Department of the Army, Washington.

Financial Times Global Water Report. 1999. "Zambia: Water Cutoff." *Financial Times Global Water Report,* Issue 68 (March 19), p. 15.

Gleick, P.H. 1991. "Environment and security: The clear connections." *Bulletin of the Atomic Scientists* (April), pp. 17–21.

Gleick, P.H. 1993. "Water and conflict: Fresh water resources and international security." *International Security* Vol. 18, No. 1, pp. 79–112.

Gleick, P.H. 1994. "Water, war, and peace in the Middle East." *Environment,* Vol. 36, No. 3, pp. 6–42. Heldref Publishers, Washington, D.C.

Gleick, P.H. 1995. "Water and Conflict: Critical Issues." Presented to the 45th Pugwash Conference on Science and World Affairs, Hiroshima, Japan, 23–29 July.

Gleick, P.H. 1998. "Water and conflict." In *The World's Water 1998–1999.* Island Press, Washington, D.C.

Grant, U.S. 1885. *Personal Memoirs of U.S. Grant.* C.L. Webster, New York. "On the second of February, [1863] this dam, or levee, was cut, . . . The river being high the rush of water through the cut was so great that in a very short time the entire obstruction was washed away. . . . As a consequence the country was covered with water."

Hillel, D. 1991. "Lash of the Dragon." *Natural History* (August), pp. 28–37.

Honan, W.H. 1996. "Scholar sees Leonardo's influence on Machiavelli." *New York Times* (December 8), p. 18.

Husarska, A. 1995. "Running dry in Sarajevo: Water fight." *New Republic* (July 17 & 24).

Kirschner, O. 1949. "Destruction and protection of dams and levees." Military Hydrology, Research and Development Branch, U.S. Corps of Engineers, Department of the Army, Washington District. From Schweizerische Bauzeitung 14 March 1949, translated by H.E. Schwarz, Washington, D.C.

Moorehead, A. 1960. *The White Nile.* Penguin Books, Harmondsworth, England.

Murphy, I.L., and J.E. Sabadell. 1986. "International river basins: A policy model for conflict resolution." *Resources Policy,* Vol. 12, No. 1, pp. 133–144. Butterworth and Co. Ltd., United Kingdom.

Reisner, M. 1986, 1993 (ed.). *Cadillac Desert: The American West and Its Disappearing Water.* Penguin Books, New York.

Reuters. 1999a. "Serbs say NATO hit refugee convoys." April 14, 1999. http://dailynews.yahoo.com/headlines/ts/story.html?s=v/nm/19990414/ts/yugoslavia_192.html

Reuters. 1999b. "NATO keeps up strikes but Belgrade quiet." June 5, 1999. http://dailynews.yahoo.com/headlines/wl/story.html?s=v/nm/19990605/wl/yugoslavia_strikes_129.html

Reuters. 1999c. "NATO builds evidence of Kosovo atrocities." June 17, 1999. http://dailynews.yahoo.com/headlines/ts/story.html?s=v/nm/19990617/ts/yugoslavia_leadall_171.html

Samson, P., and B. Charrier. 1997. "International freshwater conflict: Issues and prevention strategies." Green Cross International. http://www.dns.gci.ch/water/gcwater/study.html

Simons, M. 1999. "Serbs refuse to clean bomb-littered river." Foreign desk. *New York Times* (October 24).

Starr, J.R. 1991. "Water wars." *Foreign Policy,* Vol. 82 (Spring), pp. 17–36.

Wallenstein, P., and A. Swain. 1997. "International freshwater resources—conflict or cooperation?" Comprehensive Assessment of the Freshwater Resources of the World, Stockholm Environment Institute.

Wolf, A.T. 1995. *Hydropolitics along the Jordan River: Scarce Water and Its Impact on the Arab–Israeli Conflict.* United Nations University Press, Tokyo.

Wolf, A.T. 1997. "'Water wars' and water reality: Conflict and cooperation along international waterways." NATO Advanced Research Workshop on Environmental Change, Adaptation, and Human Security, 9–12 October, Budapest, Hungary.

World Rivers Review (WRR). 1998. "Dangerous dams: Tajikistan" *World Rivers Review,* Vol. 13, No. 6, p. 13 (December).

Yang Lang. 1989/1994. "High dam: The sword of Damocles." In Dai Qing (ed.), *Yangtze! Yangtze!* Probe International, Earthscan Publications, London, pp. 229–240.

Zachary, G.P. 1997. "Water pressure: Nations scramble to defuse fights over supplies." *Wall Street Journal* (December 4), p. A17.

Water-Related Web Sites

In the first edition of *The World's Water,* released in 1999, I included a list of water-related Web sites. Even then, there were many dozens of simple and sophisticated Web sites addressing water concerns or issues for nongovernmental organizations, government agencies, universities, and even individuals. There are many more today. For those interested in any aspect of freshwater resources, the Internet has become an indispensable tool for finding references, data, and information about publications, organizations, and individuals, as well as for educators interested in developing and presenting useful information at all levels. In that first edition, I issued a warning to Internet users that bears repeating here:

> **Much that is useful is not on the Internet. And much that is on the Internet is not useful.**

Below, I offer an updated and expanded list of more than 150 different Web sites that the reader may find of use. In this edition, the sites are presented alphabetically because of the difficulty of categorizing sites that serve multiple purposes or that come from combinations of agencies and organizations. A modest effort to add approximate categories to this list is shown in the first column. Any such list changes rapidly: Web sites move, some are created but never updated, and new resources are added all the time. By the time this book is published, some of the following Internet addresses will no longer point to an active site. For that reason, the Pacific Institute will continue to maintain a more up-to-date listing on the Web itself, at www.worldwater.org. Please come visit.

WATER-RELATED WEB SITES

Key	Organization	URL Address
E	African Water Page	http://www.sn.apc.org/afwater/index.htm
N	American Desalting Association	http://www.desalting-ada.org
N	American Institute of Hydrology	http://www.aihydro.org/
✓ N	American Water Resources Association (US)	http://www.awra.org/
N	American Water Works Association (US)	http://www.awwa.org/
N	American Water Works Association Research Foundation (US)	http://www.awwarf.com/
N	Amigos Bravos: Friends of the Wild Rivers	http://www.newmex.com/amigosbravos
G	Aquastat: UN FAO Water Information System	http://www.fao.org/waicent/faoinfo/agricult/ agl/aglw/aquastat/aquastat.htm
U	Arizona Water Resources Research Center, University of Arizona	http://ag.arizona.edu/azwater/
G	Australia Centre for Groundwater Studies	http://www.dwr.csiro.au/CGS/
G	Australian CSIRO Division of Water Resources (Australia)	http://www.dwr.csiro.au/
G	British Hydrological Society (United Kingdom)	http://www.salford.ac.uk/civils/BHS/
G	California Department of Water Resources	http://wwwdwr.water.ca.gov/
G	California Environmental Resources Evaluation System (CERES): Water Resources Links	http://ceres.ca.gov/cgi-bin/theme?keyword= Water%20resources
✓ N	Canadian Water Resources Association (Canada)	http://www.cwra.org
E	Canberra Cooperative Research Centre for Freshwater Ecology, Australia	http://enterprise.canberra.edu.au/WWW/ www-crcfe.nsf
E	Climate Change and Water Bibliography	http://www.pacinst.org/CCBib.html
N	Consejo para la Conservacion de Humedales de Norteamerica	http://uib.gym.itesm.mx/ine-nawcc/
C	Denver Water Company (US)	http://www.water.denver.co.gov/
E	Desalination Directory	http://www.desline.com
E	Encyclopedia of Water Terms: Texas Environmental Center	http://www.tec.org/tec/terms2.html
✓ G	Environment Canada (Canada): Water	http://www.ec.gc.ca/water/index.htm
G	EPA Center for Environmental Statistics (surface water quality study US/Mexico border)	http://www.epa.gov/ceis
N	European Desalination Society	http://www.eds.it
E	Flood Related Links	http://maligne.civil.ualberta.ca/water/misc/ floodlinks.html
N	Glen Canyon Institute	http://www.glencanyon.org
G	Global Applied Research Network (GARNET)	http://info.lut.ac.uk/departments/cv/wedc/ garnet/grntback.html
E	Global Energy and Water Cycle Experiment	http://www.cais.com/gewex/gewex.html
G	Global Environment Monitoring System (GEMS), Freshwater Quality Programme, UNEP	http://www.cciw.ca/gems/
E	Global Runoff Data Centre, Federal Institute of Hydrology, Germany	http://www.bafg.de/grdc.htm
✓ N	Global Water	http://www.globalwater.org
N	Global Water Partnership (GWP)	http://www.gwp.sida.se/
N	Global Water Partnership Forum	http://www.gwpforum.org
E	Great Lakes Information Network	http://www.great-lakes.net/
E	Groundwater and the Internet	http://gwrp.cciw.ca/internet/
G	Groundwater Atlas of the United States, U.S. Geological Survey	http://wwwcapp.er.usgs.gov/publicdocs/gwa/ index.html
N	Groundwater Foundation	http://www.groundwater.org

(*continues*)

WATER-RELATED WEB SITES *Continued*

Key	Organization	URL Address
G	Hydrology and Water Resources Program, WMO	http://www.wmo.ch/web/homs/hwrphome.html
E	Hydrology Web	http://terrassa.pnl.gov:2080/EESC/resourcelist/hydrology.html
C	Hydrology Web Directory	http://www.webdirectory.com/Science/Hydrology/
E	Hydrology Web: interface to many other sites	http://etd.pnl.gov:2080/hydroweb.html
E	Institute of Hydrology (IH)	http://www.nwl.ac.uk/ih/
E	Inter-American Water Resources Network	http://iwrn.ces.fau.edu/
N	International Association of Hydrogeologists	http://www.ngu.no/iah/iah.html
N	International Association of Hydraulic Engineering and Research (IAHR)	http://www.iahr.nl/
N	International Association of Hydrological Sciences	http://www.wlu.ca/~wwwiahs/index.html
N	International Association on Water Quality	http://www.iawq.org.uk/
G	International Boundary and Water Commission (IBWC): US Section	http://www.ibwc.state.gov
N	International Commission on Irrigation and Drainage	http://www.ilri.nl.icid.ciid.html
N	International Commission on Large Dams	http://genepi.louis-jean.com/cigb/anglais.html
N	International Commission on Water Quality	http://www.ex.ac.uk/~BWWebb/icwq/index.html
N	International Desalination Association	http://www.ida.bm
G	International Geosphere Biosphere Programme	http://www.igbp.kva.se/
E	International Ground Water Modeling Center	http://www.mines.edu/igwmc/
N	International Hydrological Programme	http://www.pangea.org/org/unesco/
G	International Lake Environment Committee	http://www.geic.or.jp/cop3/ngo/c0019/24.html
G	International Network of Basin Organizations	http://www.oieau.fr/riob/anglais/list_org.htm
N	International Rivers Network	http://www.irn.org/
N	International Water Academy	http://thewateracademy.no
N	International Water and Sanitation Centre	http://www.irc.nl/
E	International Water Law Project	http://home.att.net/~IntlH2OLaw/
N	International Water Management Institute (IWMI)	http://www.cgiar.org/iwmi/
N	International Water Resources Association	http://www.iwra.siu.edu
E	IRC International Water and Sanitation Centre	http://www.irc.nl/
N	Islamic Relief	http://www.irw.org/water.htm
E	Island Press Publisher, Environment	http://www.islandpress.org/
N	Middle East Desalination Research Center	http://www.medrc.org.om
E	Middle East Water Information Network (MEWIN)	http://www.ssc.upenn.edu/~mewin/
N	National Groundwater Association (US)	http://www.h2o-ngwa.org/
E	New Mexico Water Resources Research Institute	http://wrri.nmsu.edu
N	New Zealand Water and Wastes Association	http://www.nzwwa.org.nz
N	Nile Basin Initiative	http://www.nilebasin.org
N	Pacific Institute for Studies in Development, Environment, and Security	http://www.pacinst.org
E	Powell Consortium: Water Resource Research institutes	http://wrri.nmsu.edu/powell
G	Programa Hidrologico Internacional (PHI): UNESCO	http://www.unesco.org.uy/phi
G	Reseau National des Donnes sur L'eau (French Water Data Network)	http://www.rnde.tm.fr/anglais/rnde.htm

WATER-RELATED WEB SITES *Continued*

Key	Organization	URL Address
N	Rio Grande Alliance	http://www.riogrande.org
N	River Network	http://www.rivernetwork.org
G	Secretaria de Medio Ambiente, Recursos Naturales y Pesca (SEMARNAP)	http://www.semarnap.gob.mx
G	South Africa Water Research Commission	http://www.wrc.org.za/
N	Stockholm Environment Institute (SEI)	http://www.sei.se/
N	Terrene Institute	http://www.terrene.org/
E	Texas Water Resources Institute; Texas Waternet	http://twri.tamu.edu
G	Transboundary Resource Inventory Program (TRIP)	http://www.bic.state.tx.us/trip
G	U.S. Bureau of Reclamation Water	http://www.usbr.gov/water
G	U.S. Bureau of Reclamation's Water Conservation Page	http://ogee.hydlab.do.usbr.gov/rwc/rwc.html
G	U.S. Department of Agriculture: Water Quality Information Center: Water and Agriculture	http://www.nal.usda.gov/wqic/
G	U.S. Environmental Protection Agency (USEPA)	http://www.epa.gov/watrhome/national.html
G	U.S. Environmental Protection Agency Surf Your Watershed	http://www.epa.gov/surf/
G	U.S. Geological Survey: San Francisco Bay/Delta	http://sfbay.wr.usgs.gov/
G	U.S. Geological Survey: U.S. Water Data	http://h2o.usgs.gov/data.html
G	U.S. Geological Survey: U.S. Water Use Data	http://water.usgs.gov/watuse/
G	U.S. Geological Survey: Water Resources of California	http://water.wr.usgs.gov/
G	U.S. Geological Survey: Water Resources of the United States	http://h2o.er.usgs.gov/
G	U.S. National Agricultural Library: Water Quality Information Center	http://www.nal.usda.gov/wqic/
E	U.S. Water News Online	http://www.uswaternews.com/
G	UNDP-World Bank Water and Sanitation Program	http://www.wsp.org/English/index.html
G	United Nations Development Programme (UNDP)	http://www.undp.org
G	United Nations Educational, Scientific, and Cultural Organization (UNESCO)	http://www.unesco.org
G	United Nations Environment Programme (UNEP)	http://www.unep.org
G	United Nations Environment Programme (UNEP) Freshwater Unit	http://www.unep.org/unep/program/natres/water/fwu/home.htm
G	United Nations Environment Programme (UNEP) Global Environmental Monitoring System (GEMS)	http://www.cciw.ca/gems
G	United Nations Food and Agriculture Organization (UNFAO)	http://www.fao.org/
N	United States Committee on Large Dams (USCOLD)	http://www.uscold.org/~uscold/
E	Universities Council on Water Resources (UCOWR)	http://www.uwin.siu.edu/ucowr/index.html
E	Universities Water Information Network	http://www.uwin.siu.edu/
E	University of London: Water Issues Group	http://endjinn.soas.ac.uk/geography/waterissues/
E	Virtual Irrigation Library	http://www.wiz.uni-kassel.de/kww/projekte/irrig/irrig_i.html
✓ G	Vision21: Water for People	http://www.wsscc.org/vision21/index.html

(*continues*)

WATER-RELATED WEB SITES *Continued*

Key	Organization	URL Address
N	Water Aid	http://www.wateraid.org.uk/index.html
N	Water Education Foundation (US)	http://www.water-ed.org/
N	Water Environment Federation	http://www.wef.org/
N	Water For People	http://www.water4people.org/
E	Water Information Websites: Water Information Organization	http://www.waterinfo.org/interest.html
E	Water Librarian's Home Page	http://www.wco.com/~rteeter/waterlib.html
N	Water Quality Association	http://www.wqa.org/
G	Water Research Commission, South Africa	http://www.wrc.org.za/default.htm
E	Water Resources Center Archives: University of California, Berkeley	http://www.lib.berkeley.edu/WRCA/index.html
E	Water Resources Publications, LLC	http://www.wrpllc.com/
G	Water Supply and Sanitation Collaborative Council	http://www.wsscc.org/index.html
C	Water Utility Homepages: List at AWWA	http://www.awwa.org/asp/utility.asp
N	WateReuse Association of California (US)	http://www.watereuse.org/
E	WaterWiser: Water Efficiency Clearinghouse	http://www.waterwiser.org/
N	West Bengal and Bangladesh Arsenic Crisis Information Centre	http://bicn.com/acic/
G	Western Water Policy Review Advisory Commission	http://www.den.doi.gov/wwprac
G	WHO Water, Sanitation, and Health	http://www.who.int/water_sanitation_health/index.htm
G	World Bank	http://www.worldbank.org/
G	World Bank Group: Global Water Unit	http://www-esd.worldbank.org/water/
G	World Bank Institute Water Policy Reform Program	http://www.worldbank.org/wbi/edien/water.html
G	World Bank Water Policy Reform Program	http://www.worldbank.org/html/edi/edien.html
N	World Commission on Dams (WCD)	http://www.dams.org
N	World Conservation Union (IUCN)	http://www.iucn.org
G	World Health Organization (WHO)	http://www.who.org
G	World Hydrological Cycle Observing System (WHYCOS)	http://www.wmo.ch/web/homs/whycos.html
G	World Meteorological Organization (WMO)	http://www.wmo.ch
N	The World's Water	http://www.worldwater.org
N	World Water and Climate Atlas	http://www.cgiar.org/iwmi/Atlas.htm
N	World Water Council (WWC)	http://www.worldwatercouncil.org/
G	World Water Forum 2000 (The Hague)	http://www.worldwaterforum.org/
N	World Water Vision for the 21st Century	http://worldwatercouncil.org/site/vision/index.cfm

Key:

G: Government/International Sites

N: Nongovernmental Organization; Research Organization Sites

U: Universities Sites

E: Educational Resources (Publishers, Databases, References, Libraries, Information Networks)

C: Commercial Sites

Data Section

Table 1. Total Renewable Freshwater Supply, by Country

Description

This table is updated and corrected from the version found in the 1998–1999 edition of *The World's Water*. New data for Asia, parts of Oceania, and other regions are provided. Average annual renewable freshwater resources are listed by country. All quantities are in cubic kilometers per year (km³/yr). These data represent average freshwater resources in a country—actual annual renewable supply will vary from year to year. The data typically include both renewable surface water and groundwater supplies, including surface inflows from neighboring countries. The UNFAO refers to this as total natural renewable water resources. Flows to other countries are not subtracted from these numbers; they thus represent the water made available by the natural hydrologic cycle, unconstrained by political, institutional, or economic factors.

Limitations

These detailed country data should be viewed, and used, with caution. The data come from different sources and were estimated over different periods. Many countries do not directly measure or report internal water-resources data, so some of these entries were produced using indirect methods. In the past few years, new assessments have begun to standardize definitions and assumptions, particularly the work of the UNFAO.

Not all of the annual renewable water supplies are available for use by the countries to which they are credited here—some flows are committed to downstream users. For example, the Sudan is listed as having 154 cubic kilometers per year, but treaty commitments require it to pass significant flows downstream to Egypt. Other countries such as Turkey, Syria, and France, to name only a few, also pass significant amounts of water to other users. The average annual figures also hide large seasonal, interannual, and long-term variations.

Sources

a: Total natural renewable surface and groundwater. Typically includes flows from other countries. (FAO: "Natural total renewable water resources.")

b: Belyaev, 1987, Institute of Geography, National Academy of Sciences, Moscow, USSR.

c: UN Food and Agriculture Organization, 1995, *Water Resources of the African Countries: A Review*, Food and Agriculture Organization, United Nations, Rome.

d: World Resources Institute, 1994, *World Resources 1994–95*, in collaboration with the United Nations Environment Programme and the United Nations Development Programme, Oxford University Press, New York.

e: Margat, J., 1989, "The Sharing of Common Water Resources in the European Economic Community (EEC)," *Water International,* Vol. 14, pp. 59–91, as cited in P. Gleick (editor), *Water in Crisis*, Oxford University Press, New York, Table A11.

f: Shahin, M., 1989, "Review and Assessment of Water Resources in the Arab Region," *Water International,* Vol. 14, No. 4, pp. 206–219 as cited in P. Gleick (editor), *Water in Crisis*, Oxford University Press, New York, Table A17.

g: Goscomstat, USSR, 1989, *Protection of the Environment and Rational Utilization of Natural Resources in the USSR, Statistical Handbook,* Government Committee on Statistics, Moscow (in Russian), as cited in P. Gleick (editor), *Water in Crisis*, Oxford University Press, New York, Table A16.

h: World Resources Institute, 1996, *World Resources 1996–97,* A Joint Publication of the World Resources Institute, United Nations Environment Programme, United Nations Development Programme, and the World Bank, Oxford University Press, New York.

i: Economic Commission for Europe, 1992, *Environmental Statistical Database: The Environment in Europe and North America*, United Nations, New York.

j: UN Food and Agriculture Organization, 1997, *Water Resources of the Near East Region: A Review*, Food and Agriculture Organization, United Nations, Rome, Italy.

k: UN Food and Agriculture Organization, 1997, *Irrigation in the Countries of the Former Soviet Union in Figures*, Rome, Italy.

l: UN Food and Agriculture Organization, 1999, *Irrigation in Asia in Figures*. Food and Agriculture Organization, United Nations, Rome, Italy.

m: Nix, H. 1995, *Water/Land/Life: The Eternal Triangle*, Water Research Foundation of Australia, Canberra, Australia.

TABLE 1 Total Renewable Freshwater Supply, by Country

Region and Country	Annual Renewable Water Resources[a] (km³/yr)	Year of Estimate	Source of Estimate
AFRICA			
Algeria	14.3	1997	c,j
Angola	184.0	1987	b
Benin	25.8	1994	c
Botswana	14.7	1992	c
Burkina Faso	17.5	1992	c
Burundi	3.6	1987	b
Cameroon	268.0	1987	b
Cape Verde	0.3	1990	c
Central African Republic	141.0	1987	b
Chad	43.0	1987	b
Comoros	1.0	1987	b
Congo	832.0	1987	b
Congo, Democratic Republic (formerly Zaire)	1,019.0	1990	c
Cote D'Ivoire	77.7	1987	b
Djibouti	0.3	1997	j
Egypt	86.8	1997	j
Equatorial Guinea	30.0	1987	b
Eritrea	8.8	1990	c
Ethiopia	110.0	1987	b
Gabon	164.0	1987	b
Gambia	8.0	1982	c
Ghana	53.0	1970	c
Guinea	226.0	1987	b
Guinea-Bissau	27.0	1991	c
Kenya	30.2	1990	c
Lesotho	5.2	1987	b
Liberia	232.0	1987	b
Libya	0.6	1997	c,j
Madagascar	337.0	1984	c
Malawi	18.7	1994	c
Mali	67.0	1987	b
Mauritania	11.4	1997	c,j
Mauritius	2.2	1974	c
Morocco	30.0	1997	c,j
Mozambique	216.0	1992	c
Namibia	45.5	1991	c
Niger	32.5	1988	c
Nigeria	280.0	1987	b
Rwanda	6.3	1993	c
Senegal	39.4	1987	b
Sierra Leone	160.0	1987	b
Somalia	15.7	1997	j

(continues)

TABLE 1   *Continued*

Region and Country	Annual Renewable Water Resources[a] (km³/yr)	Year of Estimate	Source of Estimate
South Africa	50.0	1990	c
Sudan	154.0	1997	c,j
Swaziland	4.5	1987	b
Tanzania	89.0	1994	c
Togo	11.5	1987	b
Tunisia	4.1	1997	j
Uganda	66.0	1970	c
Zambia	116.0	1994	c
Zimbabwe	20.0	1987	b
NORTH AND CENTRAL AMERICA			
Barbados	<1.0	1962	d
Belize	16.0	1987	b
Canada	2901.0	1980	d
Costa Rica	95.0	1970	d
Cuba	34.5	1975	d
Dominican Republic	20.0	1987	b
El Salvador	19.0	1975	d
Guatemala	116.0	1970	d
Haiti	11.0	1987	b
Honduras	83.4	1992	h
Jamaica	8.3	1975	d
Mexico	357.4	1975	d
Nicaragua	175.0	1975	d
Panama	144.0	1975	d
Trinidad and Tobago	5.1	1975	d
United States of America	2478.0	1985	d
SOUTH AMERICA			
Argentina	994.0	1976	d
Bolivia	300.0	1987	b
Brazil	6950.0	1987	b
Chile	468.0	1975	d
Colombia	1070.0	1987	b
Ecuador	314.0	1987	b
Guyana	241.0	1971	d
Paraguay	314.0	1987	b
Peru	40.0	1987	b
Suriname	200.0	1987	b
Uruguay	124.0	1965	d
Venezuela	1317.0	1970	d
ASIA			
Afghanistan	65.0	1997	j
Bahrain	0.1	1997	j
Bangladesh	1210.6	1999	l
Bhutan	95.0	1987	b

T A B L E 1 *Continued*

Region and Country	Annual Renewable Water Resources[a] (km³/yr)	Year of Estimate	Source of Estimate
Brunei	8.5	1999	l
Cambodia	476.1	1999	l
China	2829.6	1999	l
Cyprus	0.9	1997	d,j
India	1907.8	1999	l
Indonesia	2838.0	1999	l
Iran	137.5	1997	j
Iraq	96.4	1997	j
Israel	2.2	1986	d
Japan	430.0	1999	l
Jordan	0.9	1997	j
Korea DPR	77.1	1999	l
Korea Rep	69.7	1999	l
Kuwait	0.0	1997	j
Laos	331.6	1999	l
Lebanon	4.8	1997	j
Malaysia	580.0	1999	l
Maldives	0.03	1999	l
Mongolia	34.8	1999	l
Myanmar	1045.6	1999	l
Nepal	210.2	1999	l
Oman	1.0	1997	j
Pakistan	429.4	1997	j
Philippines	479.0	1999	l
Qatar	0.1	1997	j
Saudi Arabia	2.4	1997	j
Singapore	0.6	1975	d
Sri Lanka	50.0	1999	l
Syria	46.1	1997	j
Thailand	409.9	1999	l
Turkey	200.7	1997	j
United Arab Emirates	0.2	1997	j
Vietnam	891.2	1999	l
Yemen	4.1	1997	j
EUROPE			
Albania	21.3	1970	d
Austria	90.3	1980	d
Belgium	12.5	1980	e
Bulgaria	205.0	1980	d
Czech Republic	58.2	1990	h
Denmark	13.0	1977	e
Finland	113.0	1980	d
France	198.0	1990	i
Germany	171.0	1991	i

(continues)

TABLE 1   *Continued*

Region and Country	Annual Renewable Water Resources[a] (km³/yr)	Year of Estimate	Source of Estimate
Greece	58.7	1980	e
Hungary	120.0	1991	i
Iceland	170.0	1987	d
Ireland	50.0	1972	e
Italy	167.0	1990	h
Luxembourg	5.0	1976	e
Malta	0.0	1997	j
Netherlands	90.0	1980	e
Norway	392.0	1991	i
Poland	56.2	1980	e
Portugal	69.6	1990	h
Romania	208.0	1980	d
Slovakia	30.8	1990	h
Spain	111.3	1985	e
Sweden	180.0	1980	d
Switzerland	50.0	1985	d
United Kingdom	120.0	1980	e
Yugoslavia	265.0	1980	d
Russia	4498.0	1997	g,k
Armenia	10.5	1997	k
Azerbaidzhan	30.3	1997	k
Belarus	58.0	1997	k
Estonia	12.8	1997	k
Georgia	63.3	1997	k
Kazakhstan	109.6	1997	k
Kyrgyzstan	20.6	1997	k
Latvia	35.4	1997	k
Lithuania	24.9	1997	k
Moldavia	11.7	1997	k
Tadjikistan	16.0	1997	k
Turkmenistan	24.7	1997	k
Ukraine	139.5	1997	k
Uzbekistan	50.4	1997	k
OCEANIA			
Australia	398.0	1995	m
Fiji	28.6	1987	b
New Zealand	397.0	1995	m
Papua New Guinea	801.0	1987	b
Solomon Islands	44.7	1987	b

Note: Compiled by P.H. Gleick, Pacific Institute.

Table 2. Freshwater Withdrawal, by Country and Sector

Description

Data on water use by regions and by different economic sectors are among the most sought after in the water resources arena. The following table presents the most up-to-date data on total freshwater withdrawals by country in cubic kilometers per year and cubic meters per person per year, using estimated water withdrawal for the year noted and UN medium variant population estimates by country for the year 2000. The table also gives the breakdown of that water use by the domestic, agricultural, and industrial sectors, in both percentage of total water use and cubic meters per person per year. The data source is identified, and the estimated year 2000 population is shown in the final column.

The use of water varies greatly from country to country and from region to region. "Withdrawal" refers to water taken from a water source for use. It does not refer to water "consumed" in that use. The domestic sector typically includes household and municipal uses as well as commercial and governmental water use. The industrial sector includes water used for power plant cooling and industrial production. The agricultural sector includes water for irrigation and livestock.

Limitations

Extreme care should be used when applying these data—they are often the least reliable and most inconsistent of all water-resources information. They come from a wide variety of sources and are collected using a wide variety of approaches, with few formal standards. As a result, this table includes data that are measured, estimated, modeled using different assumptions, or derived from other data. The data also represent many different time periods, making direct intercomparisons difficult. For example, some water use data are over 20 years old. Separate data are now provided for the independent states of the former Soviet Union. Data are no longer reported separately for former East and West Germany, but separate data for the former states of Yugoslavia are not yet available.

Another major limitation of these data is that they do not include the use of rainfall in agriculture. Many countries use a significant fraction of the rain falling on their territory for agricultural production, but this water use is neither accurately measured nor reported. In the past several years, the United Nations Food and Agriculture Organization has begun a systematic reassessment of water-use data, and there is reason to hope that over the next several years a more accurate picture of global, regional, and sectoral water use will emerge.

Sources

a: UN Food and Agriculture Organization, 1995, *Water Resources of African Countries*, Food and Agriculture Organization, United Nations, Rome, Italy.

b: UN Food and Agriculture Organization, 1997, *Water Resources of the Near East Region: A Review*, Food and Agriculture Organization, United Nations, Rome, Italy.

c: World Resources Institute, 1994, *World Resources 1994–95*, in collaboration with the United Nations Environment Programme and the United Nations Development Programme, Oxford University Press, New York.

d: UN Food and Agriculture Organization, 1997, *Irrigation in the Countries of the Former Soviet Union in Figures*, Rome, Italy.

e: U.S. Geological Survey, 1997, "Estimated Use of Water in the United States in 1995," U.S. Department of the Interior, U.S. Geological Survey.

f: Eurostat Yearbook, 1997, *Statistics of the European Union*, EC/C/6/Ser.26GT, Luxembourg.

g: World Resources Institute, 1990, *World Resources 1990–1991*, Oxford University Press, New York.

h: UN FAO 1999, *Irrigation in Asia in Figures*, Food and Agriculture Organization, United Nations, Rome, Italy.

i: Nix, H. 1995, *Water/Land/Life: The Eternal Triangle*, Water Research Foundation of Australia, Canberra.

TABLE 2 Freshwater Withdrawal, by Country and Sector

Region and Country	Year	Total Freshwater Withdrawal (km³/yr)	Estimated Year 2000 per Capita Withdrawal (m³/p/yr)	Use Domestic (%)	Industrial (%)	Agricultural (%)	Domestic m³/p/yr	Industrial m³/p/yr	Agricultural m³/p/yr	Source for Water Use Data	Estimated Population Year 2000 (millions)
AFRICA											
Algeria	1990	4.50	142	25	15	60	35	22	85	a	31.60
Angola	1987	0.48	38	14	10	76	5	4	29	a	12.80
Benin	1994	0.15	23	23	10	67	5	2	16	a	6.20
Botswana	1992	0.11	70	32	20	48	22	14	33	a	1.62
Burkina Faso	1992	0.38	31	19	0	81	6	0	25	a	12.06
Burundi	1987	0.10	14	36	0	64	5	0	9	a	6.97
Cameroon	1987	0.40	26	46	19	35	12	5	9	a	15.13
Cape Verde	1990	0.03	59	10	19	88	6	11	51	a	0.44
Central African Republic	1987	0.07	19	21	5	74	4	1	14	a	3.64
Chad	1987	0.18	25	16	2	82	4	0	20	a	7.27
Comoros	1987	0.01	14	48	5	47	7	1	7	g	0.71
Congo	1987	0.04	13	62	27	11	8	4	1	a	2.98
Congo, Democratic Republic (formerly Zaire)	1990	0.36	7	61	16	23	4	1	2	a	51.75
Cote D'Ivoire	1987	0.71	47	22	11	67	10	5	31	a	15.14
Djibouti	1973	0.01	11	28	21	51	3	2	6	a	0.69
Egypt	1993	55.10	809	6	8	86	49	65	696	b	68.12
Equatorial Guinea	1987	0.01	22	81	13	6	18	3	1	a	0.45
Ethiopia (and Eritrea)	1987	2.20	31	11	3	86	3	1	27	a	69.99
Gabon	1987	0.06	49	72	22	6	35	11	3	a	1.24
Gambia	1982	0.02	16	7	2	91	1	0	15	a	1.24
Ghana	1970	0.30	15	35	13	52	5	2	8	a	19.93

(continues)

TABLE 2 Continued

Region and Country	Year	Total Freshwater Withdrawal (km³/yr)	Estimated Year 2000 per Capita Withdrawal (m³/p/yr)	Domestic (%)	Industrial (%)	Agricultural (%)	Use Domestic m³/p/yr	Industrial m³/p/yr	Agricultural m³/p/yr	Source for Water Use Data	Estimated Population Year 2000 (millions)
Guinea	1987	0.74	94	10	3	87	9	3	82	a	7.86
Guinea-Bissau	1991	0.02	14	60	4	36	8	1	5	a	1.18
Kenya	1990	2.05	68	20	4	76	13	3	52	a	30.34
Lesotho	1987	0.05	22	22	22	56	5	5	12	a	2.29
Liberia	1987	0.13	40	27	13	60	11	5	24	a	3.26
Libya	1994	4.60	720	11	2	87	78	16	626	a	6.39
Madagascar	1984	16.30	937	1	0	99	9	0	928	a	17.40
Malawi	1994	0.94	85	10	3	86	9	3	74	a	10.98
Mali	1987	1.36	108	2	1	97	2	1	105	a	12.56
Mauritania	1985	1.63	632	6	2	92	39	11	581	a	2.58
Mauritius	1974	0.36	305	16	7	77	50	21	235	a	1.18
Morocco	1991	11.05	381	5	3	92	19	11	351	a	28.98
Mozambique	1992	0.61	31	9	2	89	3	1	28	a	19.56
Namibia	1991	0.25	144	29	3	68	41	5	98	a	1.73
Niger	1988	0.50	46	16	2	82	7	1	38	a	10.81
Nigeria	1987	3.63	28	31	15	54	9	4	15	a	128.79
Rwanda	1993	0.77	100	5	2	94	5	2	94	a	7.67
Senegal	1987	1.36	143	5	3	92	7	4	132	a	9.50
Sierra Leone	1987	0.37	76	7	4	89	5	3	68	a	4.87
Somalia	1987	0.81	70	3	0	97	2	0	68	a	11.53
South Africa	1990	13.31	288	17	11	72	49	31	207	a	46.26
Sudan	1995	17.80	597	4	1	94	27	7	563	a	29.82
Swaziland	1980	0.66	667	2	2	96	11	16	640	a	0.98

	Year										
Tanzania	1994	1.17	35	9	2	89	3	1	31	a	33.69
Togo	1987	0.09	19	62	13	26	12	3	5	a	4.68
Tunisia	1990	3.08	313	9	3	89	27	9	277	a	9.84
Uganda	1970	0.20	9	32	8	60	3	1	5	a	22.46
Zambia	1994	1.71	187	16	7	77	30	13	144	a	9.13
Zimbabwe	1987	1.22	98	14	7	79	14	7	78	a	12.42
NORTH AND CENTRAL AMERICA											427.23
Barbados	1960	0.03	115	52	41	7	60	47	8	g	0.26
Belize	1987	0.02	83	10	0	90	8	0	75	c	0.24
Canada	1990	43.89	1,431	11	80	8	157	1144	114	f	30.68
Costa Rica	1970	1.35	355	4	7	89	14	25	316	c	3.80
Cuba	1975	8.10	723	9	2	89	65	14	644	c	11.20
Dominican Republic	1987	2.97	350	5	6	89	17	21	311	c	8.50
El Salvador	1975	1.00	158	7	4	89	11	6	141	c	6.32
Guatemala	1970	0.73	60	9	17	74	5	10	44	c	12.22
Haiti	1987	0.04	5	24	8	68	1	0	3	c	7.82
Honduras	1992	1.52	234	4	5	91	9	12	213	c	6.49
Jamaica	1990	0.32	124	7	7	86	9	9	106	c	2.59
Mexico	1991	77.62	785	6	8	86	47	63	675	c	98.88
Nicaragua	1975	0.89	190	25	21	54	47	40	102	c	4.69
Panama	1975	1.30	455	12	11	77	55	50	350	c	2.86
Trinidad and Tobago	1975	0.15	112	27	38	35	30	43	39	c	1.34
United States of America	1995	469.00	1,688	12	46	42	203	777	709	e	277.83
SOUTH AMERICA											296.72
Argentina	1976	27.60	745	9	18	73	67	134	544	c	37.03
Bolivia	1987	1.24	149	10	5	85	15	7	127	c	8.33
Brazil	1990	36.47	216	43	17	40	93	37	86	c	169.20
Chile	1975	16.80	1,104	6	5	89	66	55	983	c	15.21

(continues)

TABLE 2 *Continued*

Region and Country	Year	Total Freshwater Withdrawal (km³/yr)	Estimated Year 2000 per Capita Withdrawal (m³/p/yr)	Use Domestic (%)	Industrial (%)	Agricultural (%)	Domestic m³/p/yr	Industrial m³/p/yr	Agricultural m³/p/yr	Source for Water Use Data	Estimated Population Year 2000 (millions)
Colombia	1987	5.34	137	41	16	43	56	22	59	c	38.91
Ecuador	1987	5.56	440	7	3	90	31	13	396	c	12.65
Guyana	1992	1.46	1,670	1	0	99	17	0	1654	c	0.87
Paraguay	1987	0.43	78	15	7	78	12	5	61	c	5.50
Peru	1987	6.10	238	19	9	72	45	21	171	c	25.66
Suriname	1987	0.46	1,018	6	5	89	61	51	906	c	0.45
Uruguay	1965	0.65	199	6	3	91	12	6	181	c	3.27
Venezuela	1970	4.10	170	43	11	46	73	19	78	c	24.17
ASIA											3,112.70
Afghanistan	1991	26.11	1,020	1	0	99	10	0	1010	b	25.59
Bahrain	1991	0.24	387	39	4	56	151	15	217	b	0.62
Bangladesh	1990	14.64	114	12	2	86	14	2	98	h	128.31
Bhutan	1987	0.02	10	36	10	54	4	1	5	h	2.03
Brunei	1994	0.92	2,788	nd	nd	nd	4			h	0.33
Cambodia	1987	0.52	46	5	1	94	2	0	44	h	11.21
China	1993	525.46	412	5	18	77	21	74	317	h	1,276.30
Cyprus	1993	0.21	267	7	2	91	19	5	243	b	0.79
India	1990	500.00	497	5	3	92	25	15	457	h	1,006.77
Indonesia	1990	74.35	350	6	1	93	21	3	325	h	212.57
Iran	1993	70.03	916	6	2	92	55	18	843	b	76.43
Iraq	1990	42.80	1,852	3	5	92	56	93	1704	b	23.11
Israel	1990	1.70	280	16	5	79	45	14	221	c	6.08
Japan	1992	91.40	723	19	17	64	137	123	463	h	126.43

Jordan	1993	0.98	155	22	3	75	34	5	117	b	6.33
Korea Democratic People's Republic	1987	14.16	592	11	16	73	65	95	432	h	23.91
Korea Rep	1994	23.67	505	26	11	63	131	56	318	h	46.88
Kuwait	1994	0.54	274	37	2	60	101	5	164	b	1.97
Laos	1987	0.99	174	8	10	82	14	17	143	h	5.69
Lebanon	1994	1.29	393	28	4	68	110	16	267	b	3.29
Malaysia	1995	12.73	571	10	13	77	57	74	440	h	22.30
Maldives	1987	0.003	10	98	2	0	10	0	0	h	0.29
Mongolia	1993	0.43	157	20	27	53	31	42	83	h	2.74
Myanmar	1987	3.96	80	7	3	90	6	2	72	h	49.34
Nepal	1994	28.95	1,189	1	0	99	12	0	1177	h	24.35
Oman	1991	1.22	450	5	2	94	23	9	423	b	2.72
Pakistan	1991	155.60	997	2	2	97	20	20	967	b	156.01
Philippines	1995	55.42	739	8	4	88	59	30	650	h	75.04
Qatar	1994	0.28	476	23	3	74	109	14	352	b	0.60
Saudi Arabia	1992	17.02	786	9	1	90	71	8	707	b	21.66
Singapore	1975	0.19	53	45	51	4	24	27	2	c	3.59
Sri Lanka	1990	9.77	519	2	2	96	10	10	498	h	18.82
Syria	1993	14.41	894	4	2	94	36	18	840	b	16.13
Thailand	1990	33.13	548	5	4	91	27	22	498	h	60.50
Turkey	1992	31.60	481	16	11	72	77	53	346	b	65.73
United Arab Emirates	1995	2.11	863	24	9	67	207	78	578	b	2.44
Vietnam	1990	54.33	674	4	10	86	27	67	580	h	80.55
Yemen	1990	2.93	162	7	1	92	11	2	149	c	18.12
EUROPE											498.37
Albania	1970	0.20	57	6	18	76	3	10	44	c	3.49
Austria	1991	2.52	304	19	73	8	58	222	24	f	8.29

(continues)

TABLE 2 Continued

Region and Country	Year	Total Freshwater Withdrawal (km³/yr)	Estimated Year 2000 per Capita Withdrawal (m³/p/yr)	Use Domestic (%)	Industrial (%)	Agricultural (%)	Domestic m³/p/yr	Industrial m³/p/yr	Agricultural m³/p/yr	Source for Water Use Data	Estimated Population Year 2000 (millions)
Belgium	1990	9.00	877	11	85	4	97	746	35	f	10.26
Bulgaria	1988	13.00	1,565	7	38	55	110	595	861	c	8.31
Czech Republic	1991	2.74	269	23	68	9	62	183	24	c	10.20
Denmark	1990	1.20	228	30	27	43	68	61	98	f	5.27
Finland	1994	2.43	469	12	85	3	56	399	14	f	5.18
France	1994	34.88	591	16	69	15	94	407	89	f	59.06
Germany	1990	58.85	712	14	68	18	100	484	128	f	82.69
Greece	1990	6.00	566	8	29	63	45	164	357	f	10.60
Hungary	1991	6.81	694	9	55	36	62	382	250	c	9.81
Iceland	1994	0.16	567	31	63	6	176	357	34	f	0.28
Ireland	1990	1.20	336	16	74	10	54	248	34	f	3.57
Italy	1990	56.20	983	14	27	59	138	265	580	f	57.19
Luxembourg	1994	0.06	133	42	45	13	56	60	17	f	0.43
Malta	1995	0.06	147	87	1	12	128	1	18	b	0.38
Netherlands	1991	7.80	491	5	61	34	25	300	167	f	15.87
Norway	1985	2.03	461	20	72	8	92	332	37	f	4.41
Poland	1991	12.28	317	16	60	24	51	190	76	f	38.73
Portugal	1990	7.29	745	15	37	48	112	276	357	f	9.79
Romania	1994	26.00	1,155	8	33	59	92	381	682	c	22.51
Slovakia	1991	1.78	331								5.37
Spain	1994	33.30	837	12	26	62	100	218	519	f	39.80
Sweden	1994	2.96	333	36	55	9	120	183	30	f	8.90
Switzerland	1994	2.60	351	23	73	4	81	256	14	f	7.41

United Kingdom	1994	11.75	201	20	77	3	40	155	6	f	58.34
Yugoslavia*	1980	8.77	368	16	72	12	59	265	44	c	23.81
FORMER SOVIET UNION											
Armenia	1994	2.93	800	30	4	66	240	32	528	d	3.66
Azerbaijan	1995	16.53	2,112	5	25	70	106	528	1478	d	7.83
Belarus	1990	2.73	265	22	43	35	58	114	93	d	10.28
Estonia	1995	0.16	113	56	39	5	63	44	6	d	1.42
Georgia	1990	3.47	640	21	20	59	134	128	378	d	5.42
Kazakhstan	1993	33.67	1,989	2	17	81	40	338	1611	d	16.93
Kyrgyz Republic	1994	10.09	2,221	3	3	94	67	67	2088	d	4.54
Latvia	1994	0.29	121	55	32	13	67	39	16	d	2.40
Lithuania	1995	0.25	68	81	16	3	55	11	2	d	3.69
Moldova	1992	2.96	664	9	65	26	60	432	173	d	4.46
Russian Federation	1994	77.10	527	19	62	20	100	327	105	d	146.20
Tajikistan	1994	11.87	1,855	3	4	92	56	74	1707	d	6.40
Turkmenistan	1994	23.78	5,309	1	1	98	53	53	5203	d	4.48
Ukraine	1992	25.99	512	18	52	30	92	266	153	d	50.80
Uzbekistan	1994	58.05	2,320	4	2	94	93	46	2181	d	25.02
OCEANIA											26.48
Australia	1995	17.80	945	15	10	75	142	94	709	i	18.84
Fiji	1987	0.03	35	20	20	60	7	7	21	c	0.85
New Zealand	1991	2.00	532	46	10	44	245	53	234	c	3.76
Papua New Guinea	1987	0.10	21	29	22	49	6	5	10	c	4.81
Solomon Islands	1987			40	20	40	40			c	0.44

Notes: Figures may not add to totals due to independent rounding. 2000 Population numbers: medium UN Variant.

*Includes Bosnia and Herzegovina, Macedonia, Croatia.

Table 3. World Population, Year 0 to A.D. 2050

Description

Total global population estimates from year 0 to A.D. 2000, with the medium projections of the United Nations from 2010 to the year 2050, are shown here. Units are billions of people. These data are plotted in the accompanying figure.

Limitations

No actual measurements of total world population are done. All values are based on summations of regional census records, estimations, and extrapolations. These data come from a variety of sources and should be considered approximations.

Sources

Durand, J.D., 1974, *Historical Estimates of World Population: An Evaluation*, University of Pennsylvania, Population Studies Center, Philadelphia, Pennsylvania.

United Nations, 1973, *The Determinants and Consequences of Population Trends*, Volume 1, United Nations Publications, New York.

United Nations, 1966, *World Population Prospects as Assessed in 1963*, United Nations Publications, New York.

United Nations, 1998, *World Population Prospects: The 1998 Revision*, United Nations Publications, New York.

TABLE 3   World Population,
Year 0 to A.D. 2050

Year (A.D.)	Population	Source
0	0.3	Durand
1000	0.31	Durand
1250	0.4	Durand
1500	0.5	Durand
1750	0.79	UN 1973
1800	0.98	UN 1973
1850	1.26	UN 1973
1900	1.65	UN 1973
1910	1.75	Interpolated
1920	1.86	UN 1966
1930	2.07	UN 1966
1940	2.3	UN 1966
1950	2.52	UN 1998
1960	3.02	UN 1998
1970	3.7	UN 1998
1980	4.44	UN 1998
1990	5.27	UN 1998
1998	5.9	UN 1998
2000	6.06	UN 1998
2010	6.79	UN 1998
2020	7.5	UN 1998
2030	8.11	UN 1998
2040	8.58	UN 1998
2050	8.91	UN 1998

World Population Year 0 to 2050

Table 4. Population, by Continent, 1750 to 2050

Description

Table 4 shows the total population of the world, by continents, for the years 1750, 1800, 1850, 1900, 1950, and 1998. The medium UN estimate for 2050 is also included. Units are millions of people. Note that the actual population of Europe is expected to decrease between 1998 and 2050. This table also shows the percentage distribution of these populations during each time period. Europe, for example, had an estimated 20.6 percent of the world's total population in 1750. By 1998, this had dropped to 12.4 percent, and it is expected to drop to only 7 percent by 2050. The continental data are also displayed in Chapter 4 in Figure 4.2.

Limitations

No actual measurements of continental world populations are taken. All values are based on summations of regional census records, estimations, and extrapolations. These data come from a variety of sources and should be considered approximations.

Source

United Nations, 1998, *World Population Prospects: The 1998 Revision*, United Nations Publications, New York.

TABLE 4 Population, by Continent, 1750 to 2050 (Millions and Percentage)

Area	1750	(%)	1800	(%)	1850	(%)	1900	(%)	1950	(%)	1998	(%)	2050 (est.)	(%)
Africa	106	13.4	107	10.9	111	8.8	133	8.1	221	8.8	749	12.7	1,766	19.8
Asia	502	63.5	635	64.9	809	64.1	947	57.4	1,402	55.6	3,585	60.7	5,268	59.1
Europe	163	20.6	203	20.8	276	21.9	408	24.7	547	21.7	729	12.4	628	7.0
Latin America and the Caribbean	16	2.0	24	2.5	38	3.0	74	4.5	167	6.6	504	8.5	809	9.1
Northern America	2	0.3	7	0.7	26	2.1	82	5.0	172	6.8	305	5.2	392	4.4
Oceania	2	0.3	2	0.2	2	0.2	6	0.4	13	0.5	30	0.5	46	0.5
World	791	100.0	978	100.0	1,262	100.0	1,650	100.0	2,522	100.0	5,902	100.0	8,909	100.0

Note: Percentages have been rounded.

Table 5. Renewable Water Resources and Water Availability, by Continent

Description

Statistics on the availability of renewable water resources are shown here, by continent. Average, maximum, and minimum water availability, measured over the period 1921 to 1985, are shown in cubic kilometers per year, along with the coefficient of variation (C_v), which is the standard deviation divided by the average. A list of counties included in each continent is shown below. The table also includes the area of these continents, in millions of square kilometers, and their estimated populations for 1995. Using these numbers, two relative measures of "potential" water availability by continent are shown, in thousand cubic meters per square kilometer, and in thousand cubic meters per person per year. Selected data from this table are plotted in Chapter 2, Figures 2.3 and 2.4.

Europe	Denmark, Finland, Iceland, Norway, Sweden, Belgium, Czech Republic, France, Germany, The Netherlands, Poland, Slovakia, Switzerland, United Kingdom, Albania, Bulgaria, Greece, Hungary, Italy, Portugal, Rumania, Spain, Yugoslavia (former), Estonia, Latvia, Lithuania, Byelorussia, European Russia, Moldova, Ukraine
North America	Canada, United States, Belize, Costa Rica, Cuba, Dominican Republic, El Salvador, Guatemala, Haiti, Honduras, Jamaica, Mexico, Nicaragua, Panama, Puerto Rico, Trinidad and Tobago, and other Caribbean islands
Africa	Algeria, Egypt, Libya, Morocco, Sudan, Tunisia, Angola, Botswana, Lesotho, Mozambique, Namibia, South Africa, Swaziland, Zimbabwe, Burundi, Djibouti, Ethiopia, Kenya, Madagascar, Malawi, Rwanda, Somali, Tanzania, Uganda, Zambia, Benin, Burkina-Faso, Cape Verde, Chad, Gambia, Ghana, Guinea, Guinea-Bissau, Liberia, Mali, Mauritania, Niger, Nigeria, Senegal, Sierra Leone, Togo, Cameroon, Central African Republic, Congo, Equatorial Guinea, Gabon, Sao-Tome and Principe, Zaire
Asia	China, Mongolia, North Korea, South Korea, Bangladesh, Butan, India, Nepal, Pakistan, Sri Lanka, Afghanistan, Bahrain, Cyprus, Iraq, Iran, Israel, Jordan, Kuwait, Lebanon, Oman, Qatar, Saudi Arabia, Turkey, United Arab Emirates, Yemen, Burma, Cambodia, Indonesia, Japan, Laos, Malaysia, Philippines, Thailand, Vietnam, Kazakhstan, Kyrghizstan, Tajikistan, Turkmenistan, Uzbekistan, Asian part of Russia, Armenia, Azerbaijan, Georgia

South America   Colombia, French Guyana, Guyana, Surinam, Venezuela,
Brazil, Chile, Ecuador, Peru, Argentina, Bolivia, Paraguay,
Uruguay

Australia and   Australia, New Zealand, Papua New Guinea
Oceania

Limitations

The period of record only goes back to 1921 and forward to 1985 because of limitations on data collection and availability, according to Shiklomanov, the author of this data set. More recent data are available for some regions, but were not used here. The "maximum" and "minimum" are those seen in the historical record, not the potential maximum and minimum. The sum of the continent maximum and minimum do not equal the total shown for the world, since the maximum and minimum years for each continent do not coincide.

Source

Shiklomanov, I. 1998, *World Water Resources and World Water Use*, Data archive on CD-ROM from the State Hydrological Institute, St. Petersburg, Russia.

TABLE 5   Renewable Water Resources and Water Availability, by Continent

Continent	Area (million km^2)	Population (million) 1995	Water resources (km^3/yr) Average	Water resources (km^3/yr) Max	Water resources (km^3/yr) Min	C_v	Potential water availability (1000 m^3/yr) per km^2	Potential water availability (1000 m^3/yr) per capita
Europe	10.46	685	2,900	3,410	2,254	0.08	277	4.23
North and Central America	24.3	453	7,890	8,917	6,895	0.06	324	17.4
Africa	30.1	708	4,050	5,082	3,073	0.1	134	5.72
Asia	43.5	3,445	13,510	15,008	11,800	0.06	311	3.92
South America	17.9	315	12,030	14,350	10,320	0.07	672	38.2
Australia and Oceania	8.95	29	2,360	2,843	1,850	0.1	264	82.2
The World	135.21	5,635	42,740	44,712	39,742	0.02	317	7.6

Table 6. Dynamics of Water Resources, Selected Countries, 1921 to 1985

Description

Statistics on the variations in regional water resources are shown here, for various countries. Average, maximum, and minimum water availability, measured over the period 1921 to 1985, are shown in cubic kilometers per year. The coefficient of variation (C_v), measured as the standard deviation divided by the mean, is also shown. Regions with high C_v experience greater high and low flows compared to the average. For some countries, total water availability includes both local water resources and inflows from neighboring countries.

Limitations

Only a few countries are characterized here, though included are both water "rich" and water "poor" countries. The period of record only goes back to 1921 and forward to 1985 because of limitations on data collection and availability, according to Shiklomanov, the author of this data set. More recent data are available for some regions, but were not used here. The "maximum" and "minimum" are those seen in the historical record, not the potential maximum and minimum.

Source

Shiklomanov, I. 1998, *World Water Resources and World Water Use*, Data archive on CD-ROM from the State Hydrological Institute, St. Petersburg, Russia.

TABLE 6 Dynamics of Water Resources, Selected Countries, 1921 to 1985 (km³/yr)

Country	Average	Max	Min	C_v
Brazil	6,223	7,640	5,200	0.08
Russia	4,053	4,513	3,533	0.05
Canada	3,287	3,760	2,910	0.06
USA	2,930	3,864	2,058	0.111
China	2,701	3,455	2,015	0.12
India	1,456	1,794	1,065	0.11
Zaire	989	1,328	786	0.1
Bolivia	361	487	279	0.13
Australia	352	701	228	0.24
Mexico	345	476	236	0.12
New Zealand	313	405	246	0.11
Nigeria	275	437	148	0.26
Argentina	269	610	149	0.27
France*	195	263	90	0.22
Nicaragua	176	226	134	0.13
Panama	144	196	98	0.17
Costa Rica	110	158	75	0.16
Spain	109	256	27.7	0.48
Honduras	93	128	66.2	0.14
Portugal*	53	157	15	0.55
Cuba	34.7	48.5	24.2	0.16
Albania	19.1	34.3	11.9	0.2
El Salvador	18.9	27.4	10.2	0.16
Jamaica	8.2	16	3.88	0.29

*Includes all local water resources plus inflows from neighboring countries.

Table 7. International River Basins of the World

Description

This table lists all international river basins from a new analysis published in 1999 (see source, below). A "river basin" is defined as the area that contributes hydrologically to a first-order stream that ends in the ocean or a terminal (closed) lake or inland sea. "River basin" is thus synonymous with "watershed" or "catchment." A basin is considered "international" if any perennial tributary crosses the political boundaries of two or more nations or states. The total area of each basin is provided in square kilometers, along with the countries that share each basin and the area of each country within the basin. Altogether there are 261 such international river basins, comprising just under half the land area of the earth, excluding Antarctica. See Chapter 2 for a broader description of these data, their implications, and limitations.

Limitations

Changing political boundaries constantly change the number and extent of international river basins. In the last assessment done, in 1978, 214 international river basins were identified. The increase from that number to the current estimate of 261 reflects both better mapping of the world's hydrologic boundaries and major changes in political borders in the former Soviet Union, Eastern European countries, and elsewhere. Political borders will continue to shift, and the groupings and numbers in this table will continue to change.

Major rivers that join above an outlet are grouped together here. For example, the Tigris and Euphrates rivers are distinct and separate rivers for most of their extent, but they merge upstream of the Arabian Gulf and hence are grouped in this table. Similarly, the Ganges, Brahmaputra, and Meghna rivers—each a significant river in its own right—join before they reach the ocean and are thus grouped here.

Some border areas are disputed. Where such disputes are well known, they are explicitly identified in this table. For example, portions of the Indus and the Ganges-Brahmaputra-Meghna basins are under Chinese control but claimed by India, while other portions are under Indian control but claimed by China.

Groundwater basins also often extend across international boundaries, but these are not included here. Indeed, no assessment of international groundwater basins has been published.

Source

Wolf, A.T., J.A. Natharius, J.J. Danielson, B.S. Ward, J. K. Pender, 1999, "International River Basins of the World", *International Journal of Water Resources Development,* Vol. 15, No. 4 (December).

TABLE 7 International River Basins of the World

Basin Name	Area of Basin (km²)	Countries Sharing the Basin	Area within Country (km²)	Percentage of Watershed within Country
Akpa Yafi	4,900	Cameroon	3,100	62.26
		Nigeria	1,900	37.74
Alesek	8,300	Canada	7,200	87.27
		United States	1,000	12.42
Amacuro	4,000	Venezuela	3,400	85.15
		Guyana	600	14.61
Amazon	5,866,100	Brazil	3,672,600	62.61
		Peru	974,600	16.61
		Bolivia	684,400	11.67
		Colombia	353,000	6.02
		Ecuador	137,800	2.35
		Venezuela	38,500	0.66
		Guyana	5,200	0.09
		Suriname	20	0
Amur	1,884,000	Russia	1,005,300	53.36
		China	849,900	45.11
		Mongolia	28,700	1.52
		Korea, Democratic People's Republic of	120	0.01
An Nahr Al Kabir	1,200	Syria	730	60.6
		Lebanon	470	39.4
Aral Sea	1,319,900	Kazakhstan	923,500	69.97
		Uzbekistan	236,700	17.93
		Kyrgyzstan	138,000	10.46
		Tajikistan	13,000	0.99
		Turkmenistan	1,500	0.12
		China	40	0
Artibonite	8,800	Haiti	6,600	74.37
		Dominican Republic	2,300	25.55
Asi/Orontes	18,200	Syria	10,100	55.63
		Turkey	6,600	36.13
		Lebanon	1,500	8.21
Astara Chay	560	Iran	450	81.33
		Azerbaijan	100	18.67
Atrak	34,200	Iran	23,500	68.64
		Turkmenistan	10,700	31.36
Atui	10,400	Mauritania	9,300	89.71
		Western Sahara	1,100	10.29
Aviles	260	Argentina	230	88.72
		Chile	30	11.28
Awash	155,300	Ethiopia	144,000	92.72
		Djibouti	11,100	7.15
		Somalia	240	0.15

TABLE 7 *Continued*

Basin Name	Area of Basin (km²)	Countries Sharing the Basin	Area within Country (km²)	Percentage of Watershed within Country
Aysen	13,300	Chile	11,300	85.06
		Argentina	2,000	14.94
Baker	30,800	Chile	21,000	68.29
		Argentina	9,800	31.69
Bangau	430	Malaysia	230	53.04
		Brunei	200	46.26
Bann	5,600	United Kingdom	5,400	97.15
		Ireland	160	2.85
Baraka	66,600	Eritrea	41,800	62.84
		Sudan	24,700	37.16
Barima	8,700	Guyana	7,700	87.86
		Venezuela	1,000	11.79
Barta	1,800	Latvia	1,100	60.86
		Lithuania	670	37.73
Beilun	960	China	700	73.61
		Vietnam	250	26.39
Belize	11,500	Belize	7,000	60.86
		Guatemala	4,500	39.14
Benito	12,600	Equatorial Guinea	11,100	88.57
		Gabon	1,400	11.16
Bia	11,900	Ghana	6,900	57.83
		Ivory Coast	4,800	40.01
Bidasoa	530	Spain	470	89.52
		France	60	10.48
Buzi	27,900	Mozambique	24,700	88.81
		Zimbabwe	3,100	11.18
Ca/Song-Koi	33,800	Vietnam	19,600	57.83
		Laos, People's Democratic Republic of	14,300	42.12
Cancoso/Lauca	32,100	Bolivia	26,200	81.57
		Chile	5,900	18.43
Candelaria	12,800	Mexico	11,300	88.24
		Guatemala	1,500	11.74
Castletown	380	United Kingdom	290	76.12
		Ireland	90	23.88
Catatumbo	26,100	Colombia	16,700	64.02
		Venezuela	9,400	35.97
Cavally	30,600	Ivory Coast	16,600	54.31
		Liberia	12,700	41.48
		Guinea	1,300	4.21

(continues)

TABLE 7 *Continued*

Basin Name	Area of Basin (km²)	Countries Sharing the Basin	Area within Country (km²)	Percentage of Watershed within Country
Cestos	15,000	Liberia	12,700	84.54
		Ivory Coast	2,300	15.32
		Guinea	20	0.14
Changuinola	3,200	Panama	2,900	91.29
		Costa Rica	270	8.34
Chico/Carmen Silva	1,700	Argentina	1,000	59.70
		Chile	680	40.3
Chilkat	4,100	United States	2,400	58.01
		Canada	1,700	41.99
Chiloango	11,700	Congo, Democratic Republic of the (Kinshasa)	7,700	65.91
		Angola	3,700	32.11
		Congo, Republic of the (Brazzaville)	230	1.97
Chira	16,700	Peru	9,200	55.33
		Ecuador	7,500	44.67
Chiriqui	1,700	Panama	1,500	86.17
		Costa Rica	240	13.83
Choluteca	7,400	Honduras	7,200	97.68
		Nicaragua	170	2.32
Chuy	180	Brazil	110	64.57
		Uruguay	60	32.57
Coatan Achute	2,000	Mexico	1,700	86.27
		Guatemala	270	13.73
Coco/Segovia	25,400	Nicaragua	17,900	70.52
		Honduras	7,500	29.48
Colorado	651,100	United States	640,700	98.4
		Mexico	10,400	1.6
Columbia	668,400	United States	566,500	84.75
		Canada	101,900	15.24
Comau	920	Chile	840	90.91
		Argentina	80	9.09
Congo/Zaire	3,699,100	Congo, Democratic Republic of the (Kinshasa)	2,307,800	62.39
		Central African Republic	402,000	10.87
		Angola	291,500	7.88
		Congo, Republic of the (Brazzaville)	248,400	6.72
		Zambia	176,600	4.77
		Tanzania, United Republic of	166,800	4.51
		Cameroon	85,300	2.31

T A B L E 7 *Continued*

Basin Name	Area of Basin (km²)	Countries Sharing the Basin	Area within Country (km²)	Percentage of Watershed within Country
Congo/Zaire (*continued*)		Burundi	14,300	0.39
		Rwanda	4,500	0.12
		Gabon	460	0.01
		Malawi	90	0
Corubal	24,100	Guinea	17,600	72.89
		Guinea-Bissau	6,500	26.82
Coruh	20,700	Turkey	18,800	90.93
		Georgia	1,900	9.07
Courantyne/ Corantijn	67,700	Suriname	36,900	54.46
		Guyana	30,800	45.45
Cross	52,800	Nigeria	40,300	76.39
		Cameroon	12,400	23.56
Cullen	590	Chile	490	83
		Argentina	100	17
Danube	779,500	Romania	228,800	29.35
		Hungary	92,800	11.9
		Yugoslavia (Serbia and Montenegro)	81,000	10.4
		Austria	80,300	10.3
		Germany	52,100	6.68
		Bulgaria	47,300	6.06
		Slovakia	46,800	6.01
		Bosnia and Herzegovina	37,800	4.85
		Croatia	34,000	4.37
		Ukraine	25,600	3.29
		Czech Republic	21,300	2.74
		Slovenia	16,400	2.1
		Moldova	12,100	1.55
		Switzerland	1,700	0.21
		Italy	740	0.09
		Poland	550	0.07
		Albania	140	0.02
Daoura	34,600	Morocco	18,300	52.82
		Algeria	16,300	47.18
Dasht	31,800	Pakistan	25,100	78.87
		Iran	6,700	21.13
Daugava	79,600	Byelarus	28,300	35.55
		Russia	27,100	34.02
		Latvia	23,200	29.14
		Lithuania	1,000	1.29
Dnieper	495,500	Ukraine	296,800	59.9
		Byelarus	116,700	23.55
		Russia	81,900	16.53

(*continues*)

TABLE 7 *Continued*

Basin Name	Area of Basin (km²)	Countries Sharing the Basin	Area within Country (km²)	Percentage of Watershed within Country
Dniester	72,200	Ukraine	52,900	73.37
		Moldova	19,200	26.6
		Poland	20	0.03
Don	425,600	Russia	371,200	87.21
		Ukraine	54,400	12.78
Douro/Duero	96,200	Spain	77,900	81.01
		Portugal	18,300	18.99
Dra	54,900	Morocco	40,600	73.97
		Algeria	14,300	26.03
Drin	18,500	Yugoslavia (Serbia and Montenegro)	9,000	48.55
		Albania	7,200	39.23
		Macedonia	2,300	12.21
Ebro	85,100	Spain	84,200	98.96
		France	470	0.55
		Andorra	410	0.49
Elancik	1,400	Russia	940	68.19
		Ukraine	440	31.81
Elbe	139,500	Germany	88,600	63.54
		Czech Republic	49,600	35.6
		Austria	1,100	0.77
		Poland	140	0.1
Erne	3,500	Ireland	2,000	56.39
		United Kingdom	1,500	43.59
Essequibo	154,300	Guyana	115,400	74.79
		Venezuela	38,800	25.12
		Brazil	140	0.09
Etosha-Cuvelai	167,600	Namibia	114,300	68.24
		Angola	53,200	31.76
Fane	200	Ireland	190	96.46
		United Kingdom	10	3.54
Fenney	2,800	India	1,800	65.88
		Bangladesh	950	34.12
Firth	6,000	Canada	3,800	63.6
		United States	2,200	36.40
Flurry	70	United Kingdom	50	73.77
		Ireland	20	26.23
Fly	64,600	Papua New Guinea	60,400	93.42
		Indonesia	4,300	6.58
Foyle	2,900	United Kingdom	2,000	67.23
		Ireland	960	32.77

TABLE 7 *Continued*

Basin Name	Area of Basin (km²)	Countries Sharing the Basin	Area within Country (km²)	Percentage of Watershed within Country
Fraser	239,700	Canada	239,100	99.74
		United States	620	0.26
Gallegos-Chico	11,600	Argentina	7,000	60.15
		Chile	4,600	39.85
Gambia	70,000	Senegal	50,800	72.55
		Guinea	13,300	18.95
		Gambia, The	5,900	8.41
Ganges-Brahmaputra-Meghna	1,675,700	India	974,300	58.14
		China	320,400	19.12
		Nepal	147,300	8.79
		Bangladesh	123,400	7.36
		India, Claimed by China	67,100	4.00
		Bhutan	39,900	2.38
		Myanmar	2,100	0.13
		Indian Control, Claimed by China	1,200	0.07
Garonne	55,800	France	55,100	98.8
		Spain	620	1.11
		Andorra	40	0.07
Gash	31,700	Eritrea	17,700	55.75
		Sudan	8,500	26.85
		Ethiopia	5,500	17.41
Gauja	8,100	Latvia	6,900	85.87
		Estonia	1,100	14.13
Geba	12,800	Guinea-Bissau	8,700	67.71
		Senegal	4,100	31.86
		Guinea	50	0.41
Goascoran	2,800	Honduras	1,500	53.36
		El Salvador	1,300	46.64
Golok	1,800	Thailand	1,100	58.09
		Malaysia	780	41.91
Great Scarcies	11,400	Guinea	9,000	79.3
		Sierra Leone	2,300	20.53
Grijalva	126,800	Mexico	78,900	62.26
		Guatemala	47,800	37.73
Guadiana	65,700	Spain	55,300	84.13
		Portugal	10,400	15.87
Guir	79,100	Algeria	61,400	77.6
		Morocco	17,700	22.4
Han	35,300	Korea, Republic of	25,100	71.18
		Korea, Democratic People's Republic of	10,100	28.63

(*continues*)

T A B L E 7   *Continued*

Basin Name	Area of Basin (km²)	Countries Sharing the Basin	Area within Country (km²)	Percentage of Watershed within Country
Har Us Nur	197,800	Mongolia	195,400	98.77
		Russia	2,300	1.17
		China	120	0.06
Hari/Harirud	92,600	Afghanistan	41,100	44.33
		Iran	35,400	38.17
		Turkmenistan	16,200	17.5
Helmand	345,200	Afghanistan	283,800	82.23
		Iran	48,300	13.99
		Pakistan	13,100	3.78
Hondo	14,600	Mexico	8,900	61.14
		Guatemala	4,200	28.5
		Belize	1,500	10.36
Hsi/Bei Jiang	361,500	China	351,700	97.28
		Vietnam	9,800	2.72
Ili/Kunes He	161,200	Kazakhstan	97,200	60.28
		China	55,300	34.31
		Kyrgyzstan	8,700	5.4
Incomati	46,200	South Africa	29,200	63.19
		Mozambique	14,300	30.97
		Swaziland	2,700	5.84
Indus	1,086,000	Pakistan	609,100	56.09
		India	282,200	25.98
		China	111,000	10.22
		Afghanistan	72,500	6.68
		Chinese Control, Claimed by India	9,600	0.89
		Indian Control, Claimed by China	1,600	0.15
Irrawaddy	404,100	Myanmar	368,400	91.15
		China	18,600	4.6
		India	14,200	3.52
		India, Claimed by China	1,200	0.29
Isonzo	3,000	Slovenia	1,800	59.55
		Italy	1,200	40.05
Jacobs	440	Norway	300	68.55
		Russia	140	31.45
Jordan (Dead Sea)	42,800	Jordan	20,600	48.44
		Israel	9,100	21.35
		Syria	4,900	11.54
		West Bank	3,200	7.4
		Egypt	2,700	6.39
		Golan Heights	1,500	3.54
		Lebanon	570	1.34

TABLE 7 *Continued*

Basin Name	Area of Basin (km²)	Countries Sharing the Basin	Area within Country (km²)	Percentage of Watershed within Country
Juba-Shibeli	805,100	Ethiopia	367,700	45.67
		Somalia	221,500	27.52
		Kenya	215,900	26.81
Jurado	820	Colombia	580	70.52
		Panama	240	28.75
Kaladan	30,500	Myanmar	22,700	74.39
		India	7,400	24.24
Karnafauli	15,000	Bangladesh	11,200	74.78
		India	3,700	24.98
Kemi	55,800	Finland	52,700	94.38
		Russia	3,100	5.56
		Norway	10	0.01
Klaralven	51,500	Sweden	43,400	84.15
		Norway	8,200	15.84
Kogilnik	6,100	Moldova	3,600	57.85
		Ukraine	2,600	42.15
Komoe	78,500	Ivory Coast	58,500	74.52
		Burkina Faso	17,100	21.74
		Ghana	2,300	2.93
		Mali	630	0.81
Kowl-E-Namaksar	40,100	Iran	26,600	66.5
		Afghanistan	13,400	33.5
Krka	1,300	Croatia	1,100	89.84
		Bosnia and Herzegovina	110	8.96
Kunene	110,300	Angola	95,500	86.57
		Namibia	14,800	13.43
Kura-Araks	193,800	Azerbaijan	59,800	30.86
		Georgia	34,500	17.78
		Iran	33,500	17.28
		Armenia	29,900	15.42
		Turkey	28,500	14.7
		Russia	110	0.06
La Plata	2,966,900	Brazil	1,366,700	46.06
		Argentina	817,900	27.57
		Paraguay	400,100	13.49
		Bolivia	270,200	9.11
		Uruguay	111,600	3.76
Lagoon Mirim	54,900	Uruguay	31,200	56.75
		Brazil	23,700	43.18
Lake Chad	2,394,200	Chad	1,082,000	45.19
		Niger	675,700	28.22
		Central African Republic	218,900	9.14

(*continues*)

TABLE 7 *Continued*

Basin Name	Area of Basin (km²)	Countries Sharing the Basin	Area within Country (km²)	Percentage of Watershed within Country
Lake Chad (*continued*)		Nigeria	180,800	7.55
		Algeria	90,000	3.76
		Sudan	83,100	3.47
		Cameroon	46,900	1.96
		Chad, Claimed by Libya	12,300	0.51
		Libya	4,600	0.19
Lake Fagnano	3,800	Argentina	2,800	74.95
		Chile	950	25.05
Lake Natron	55,600	Tanzania, United Republic of	37,300	67.06
		Kenya	18,300	32.94
Lake Prespa	1,400	Macedonia	610	42.76
		Albania	420	29.54
		Greece	390	27.71
Lake Titicaca- Poopo	116,500	Bolivia	61,700	52.99
		Peru	53,600	45.96
		Chile	1,200	1.05
Lake Turkana	207,600	Ethiopia	113,600	54.75
		Kenya	89,900	43.28
		Uganda	2,600	1.23
		Sudan	1,500	0.7
		Sudan, Administered by Kenya	70	0.03
Lake Ubsa-Nur	74,800	Mongolia	52,700	70.48
		Russia	22,100	29.52
Lava-Pregel	8,800	Russia	6,400	72.54
		Poland	2,200	25.36
Lempa	18,000	El Salvador	9,500	52.45
		Honduras	5,800	32.01
		Guatemala	2,800	15.54
Lielupe	27,200	Lithuania	19,000	70.03
		Latvia	8,100	29.71
Lima	2,300	Spain	1,200	50.88
		Portugal	1,100	49.04
Limpopo	415,500	South Africa	184,100	44.31
		Mozambique	87,300	21.01
		Botswana	81,500	19.61
		Zimbabwe	62,600	15.06
Little Scarcies	19,300	Sierra Leone	13,300	68.86
		Guinea	6,000	31.11
Loffa	11,400	Liberia	10,000	87.49
		Guinea	1,400	12.43

TABLE 7   *Continued*

Basin Name	Area of Basin (km²)	Countries Sharing the Basin	Area within Country (km²)	Percentage of Watershed within Country
Lotagipi Swamp	38,900	Kenya	20,500	52.52
		Sudan	10,000	25.58
		Sudan, Administered by Kenya	3,300	8.44
		Ethiopia	3,200	8.30
		Uganda	2,000	5.16
Ma	24,600	Vietnam	16,800	68.20
		Laos, People's Democratic Republic of	7,800	31.80
Mana-Morro	6,900	Liberia	5,800	83.67
		Sierra Leone	1,100	16.31
Maputo	31,300	South Africa	18,600	59.43
		Swaziland	11,000	35.02
		Mozambique	1,700	5.55
Maritsa	52,800	Bulgaria	35,000	66.38
		Turkey	14,300	27.16
		Greece	3,400	6.46
Maroni	65,900	Suriname	37,000	56.20
		French Guiana	28,100	42.66
		Brazil	640	0.96
Massacre	800	Haiti	500	62.03
		Dominican Republic	290	35.96
Mataje	730	Ecuador	540	73.98
		Colombia	190	26.02
Mbe	7,000	Gabon	6,500	92.65
		Equatorial Guinea	500	7.18
Medjerda	23,100	Tunisia	15,600	67.28
		Algeria	7,600	32.72
Mekong	780,300	Laos, People's Democratic Republic of	198,400	25.42
		Thailand	194,100	24.87
		China	168,400	21.58
		Cambodia	157,000	20.11
		Vietnam	35,000	4.49
		Myanmar	27,500	3.53
Merauke	6,500	Indonesia	4,000	61.35
		Papua New Guinea	2,500	38.65
Mino/Minho	16,600	Spain	16,000	96.42
		Portugal	590	3.56
Mira	11,700	Colombia	7,100	61.01
		Ecuador	4,600	38.99

(continues)

TABLE 7 *Continued*

Basin Name	Area of Basin (km²)	Countries Sharing the Basin	Area within Country (km²)	Percentage of Watershed within Country
Mississippi	3,226,300	United States	3,176,500	98.46
		Canada	49,800	1.54
Mius	7,100	Ukraine	4,800	67.83
		Russia	2,300	31.53
Moa	22,600	Sierra Leone	10,900	48.16
		Guinea	8,700	38.58
		Liberia	3,000	13.27
Mono	23,400	Togo	22,400	95.43
		Benin	1,100	4.57
Motaqua	16,100	Guatemala	14,600	90.85
		Honduras	1,500	9.11
Murgab	60,900	Afghanistan	36,400	59.78
		Turkmenistan	24,500	40.22
Naatamo	710	Norway	530	74.08
		Finland	180	24.37
Nahr El Kebir	2,000	Syria	1,700	81.77
		Turkey	370	18.18
Narva	58,200	Russia	29,300	50.26
		Estonia	16,800	28.84
		Latvia	12,200	20.89
		Byelarus	10	0.02
Negro	2,500	Honduras	1,300	52.34
		Nicaragua	1,200	47.66
Nelson-Saskatchewan	1,109,400	Canada	952,000	85.81
		United States	157,400	14.19
Neman	93,000	Byelarus	41,500	44.64
		Lithuania	39,700	42.73
		Poland	6,600	7.10
		Russia	4,800	5.15
		Latvia	330	0.35
Neretva	10,800	Bosnia and Herzegovina	9,900	91.99
		Croatia	500	4.68
		Yugoslavia (Serbia and Montenegro)	360	3.32
Nestos	12,000	Greece	8,500	70.88
		Bulgaria	3,500	28.75
Niger	2,117,700	Nigeria	563,000	26.59
		Mali	541,600	25.58
		Niger	499,200	23.57
		Algeria	161,500	7.63
		Guinea	96,300	4.55

Table 7 *Continued*

Basin Name	Area of Basin (km²)	Countries Sharing the Basin	Area within Country (km²)	Percentage of Watershed within Country
Niger (*continued*)		Cameroon	88,200	4.17
		Burkina Faso	83,100	3.93
		Benin	45,200	2.14
		Ivory Coast	22,800	1.08
		Chad	16,600	0.78
		Sierra Leone	30	0
Nile	3,038,100	Sudan	1,931,300	63.57
		Ethiopia	356,900	11.75
		Egypt	273,100	8.99
		Uganda	238,900	7.86
		Tanzania, United Republic of	120,300	3.96
		Kenya	50,900	1.68
		Congo, Democratic Republic of the (Kinshasa)	21,700	0.71
		Rwanda	20,800	0.69
		Burundi	13,000	0.43
		Egypt, Administered by Sudan	4,400	0.14
		Eritrea	3,500	0.12
		Sudan, Administered by Egypt	2,000	0.07
Ntem	35,000	Cameroon	20,400	58.26
		Gabon	9,400	26.99
		Equatorial Guinea	5,200	14.75
Nyanga	12,400	Gabon	11,500	93.30
		Congo, Republic of the (Brazzaville)	830	6.70
Ob	2,734,800	Russia	2,109,600	77.14
		Kazakhstan	573,400	20.97
		China	50,400	1.84
Oder/Odra	116,500	Poland	103,000	88.43
		Czech Republic	7,400	6.35
		Germany	6,100	5.22
		Slovakia	10	0
Ogooue	223,400	Gabon	189,800	84.93
		Congo, Republic of the (Brazzaville)	26,600	11.91
		Cameroon	5,100	2.30
		Equatorial Guinea	1,900	0.87
Okavango	708,600	Botswana	359,000	50.67
		Namibia	176,800	24.95
		Angola	150,100	21.18
		Zimbabwe	22,700	3.20

(continues)

TABLE 7 *Continued*

Basin Name	Area of Basin (km²)	Countries Sharing the Basin	Area within Country (km²)	Percentage of Watershed within Country
Olanga	18,800	Russia	16,800	89.38
		Finland	2,000	10.62
Oral (Ural)	260,400	Kazakhstan	142,400	54.69
		Russia	118,000	45.31
Orange	947,700	South Africa	565,600	59.68
		Namibia	240,600	25.39
		Botswana	121,600	12.84
		Lesotho	19,900	2.09
Orinoco	958,500	Venezuela	607,400	63.37
		Colombia	351,100	36.63
Oued Bon Naima	510	Morocco	350	68.87
		Algeria	160	31.13
Oueme	59,500	Benin	49,500	83.24
		Nigeria	9,500	16.04
		Togo	430	0.72
Oulu	24,800	Finland	23,600	95.25
		Russia	1,200	4.75
Oyupock/ Oiapoque	27,100	Brazil	14,200	52.38
		French Guiana	12,800	47.28
Pakchan	2,700	Thailand	1,600	59.64
		Myanmar	1,100	38.86
Palena	13,300	Chile	7,300	54.58
		Argentina	6,100	45.42
Pandaruan	810	Brunei	410	50.25
		Malaysia	400	49.75
Parnu	5,900	Estonia	5,900	99.78
		Latvia	10	0.22
Pascua	13,700	Chile	7,400	53.72
		Argentina	6,300	46.22
Pasvik	16,900	Finland	16,200	95.71
		Norway	700	4.13
		Russia	30	0.15
Patia	21,300	Colombia	20,900	97.97
		Ecuador	430	2.03
Paz	2,200	Guatemala	1,400	64.47
		El Salvador	770	35.53
Pedernales	360	Haiti	240	67.32
		Dominican Republic	120	32.68
Po	87,100	Italy	82,600	94.83
		Switzerland	4,100	4.71
		France	380	0.44
		Austria	30	0.03

TABLE 7 *Continued*

Basin Name	Area of Basin (km²)	Countries Sharing the Basin	Area within Country (km²)	Percentage of Watershed within Country
Prohladnaja	620	Russia	480	77.06
		Poland	140	22.94
Puelo	8,200	Argentina	5,100	62.33
		Chile	3,100	37.63
Pu-Lun-To	88,400	China	76,300	86.27
		Mongolia	12,100	13.65
		Kazakhstan	40	0.04
		Russia	40	0.04
Red/Song Hong	164,600	China	84,400	51.28
		Vietnam	79,000	47.98
		Laos, People's Democratic Republic of	1,200	0.75
Rezvaya	670	Turkey	500	74.66
		Bulgaria	170	25.34
Rhine	195,000	Germany	106,800	54.74
		Switzerland	34,700	17.78
		France	15,400	13
		Netherlands	11,900	6.11
		Belgium	11,200	5.74
		Luxembourg	2,500	1.29
		Austria	2,300	1.19
		Liechtenstein	160	0.08
		Italy	140	0.07
Rhone	84,700	France	84,000	99.08
		Switzerland	730	0.86
		Italy	50	0.06
Rio Grande	548,800	United States	325,100	59.25
		Mexico	223,600	40.75
Rio Grande	7,900	Chile	4,000	50.86
		Argentina	3,900	49.14
Roia	660	France	450	67.78
		Italy	200	30.09
Rudkhaneh-ye/ BahuKalat	20,600	Iran	20,600	99.68
		Pakistan	50	0.25
Ruvuma	152,200	Mozambique	99,500	65.39
		Tanzania, United Republic of	52,200	34.30
		Malawi	470	0.31
Sabi	116,100	Zimbabwe	85,700	73.88
		Mozambique	30,300	26.12

(continues)

TABLE 7   *Continued*

Basin Name	Area of Basin (km²)	Countries Sharing the Basin	Area within Country (km²)	Percentage of Watershed within Country
Saigon/ Song Nha Be	29,400	Vietnam	25,000	98.93
		Cambodia	230	0.92
Salaca	4,000	Latvia	2,700	66.27
		Estonia	1,400	33.71
Salween	244,100	China	128,000	52.43
		Myanmar	107,000	43.85
		Thailand	9,100	3.71
Samur	6,800	Russia	6,300	92.74
		Azerbaijan	430	6.38
San Juan	42,200	Nicaragua	30,400	72.02
		Costa Rica	11,800	27.93
San Martin	640	Chile	580	90.22
		Argentina	60	9.78
Sarata	1,800	Ukraine	1,100	63.78
		Moldova	640	36.16
Sarstun	2,100	Guatemala	1,800	87.63
		Belize	260	12.37
Sassandra	68,200	Ivory Coast	59,700	87.51
		Guinea	8,500	12.49
Schelde	17,500	France	8,900	51.11
		Belgium	8,400	48.22
		Netherlands	80	0.46
Seine	86,100	France	84,200	97.78
		Belgium	1,800	2.13
		Luxembourg	80	0.09
Sembakung	15,200	Indonesia	8,200	53.53
		Malaysia	7,100	46.46
Senegal	437,000	Mauritania	219,100	50.14
		Mali	151,300	34.61
		Senegal	35,700	8.16
		Guinea	31,000	7.08
Seno Union/ Serrano	6,500	Chile	5,700	87.93
		Argentina	670	10.34
Sepik	73,400	Papua New Guinea	71,100	96.94
		Indonesia	2,200	3.06
Sixaola	2,900	Costa Rica	2,500	88.68
		Panama	290	9.96
Song Vam Co Dong	15,300	Vietnam	7,700	50.22
		Cambodia	7,600	49.72

TABLE 7   *Continued*

Basin Name	Area of Basin (km²)	Countries Sharing the Basin	Area within Country (km²)	Percentage of Watershed within Country
St. Croix	4,600	United States	3,300	70.86
		Canada	1,400	29.14
St. John	55,100	Canada	35,600	64.60
		United States	19,400	35.25
St. John	15,600	Liberia	13,000	83.55
		Guinea	2,600	16.44
		Ivory Coast	2	0.01
St. Lawrence	1,055,200	Canada	559,000	52.98
		United States	496,100	47.02
St. Paul	21,200	Liberia	11,800	55.47
		Guinea	9,500	44.52
Stikine	50,900	Canada	50,000	98.32
		United States	850	1.68
Struma	16,800	Bulgaria	8,400	49.84
		Greece	6,000	35.45
		Macedonia	1,800	10.63
		Yugoslavia (Serbia and Montenegro)	690	4.08
Suchiate	1,600	Guatemala	1,100	68.79
		Mexico	490	31.21
Sujfun	16,800	China	9,800	58.39
		Russia	7,000	41.61
Sulak	14,800	Russia	13,800	92.02
		Georgia	1,000	6.70
		Azerbaijan	20	0.11
Tafna	9,500	Algeria	7,000	74.16
		Morocco	2,400	25.84
Tagus/Tejo	69,900	Spain	55,500	79.29
		Portugal	14,500	20.71
Taku	18,000	Canada	17,600	98.20
		United States	320	1.80
Tami	89,900	Indonesia	87,600	97.54
		Papua New Guinea	2,200	2.46
Tana	16,100	Norway	9,400	58.34
		Finland	6,700	41.60
Tano	14,300	Ghana	13,900	97.07
		Ivory Coast	420	2.93
Tarim	950,200	China	901,700	94.90
		Kyrgyzstan	23,900	2.51
		Chinese Control, Claimed by India	21,500	2.27

(continues)

TABLE 7 *Continued*

Basin Name	Area of Basin (km²)	Countries Sharing the Basin	Area within Country (km²)	Percentage of Watershed within Country
Tarim (*continued*)		Pakistan	1,900	0.20
		Tajikistan	1,000	0.11
		Kazakhstan	110	0.01
		Afghanistan	20	0
Terek	43,800	Russia	41,800	95.43
		Georgia	2,000	4.57
Tigris-Euphrates/ Shatt al Arab	793,600	Iraq	318,900	40.19
		Turkey	197,100	24.84
		Iran	155,600	19.60
		Syria	119,400	15.04
		Jordan	2,200	0.28
		Saudi Arabia	240	0.03
Tijuana	4,400	Mexico	3,100	70.57
		United States	1,300	29.43
Torne/Tornealven	37,300	Sweden	25,300	67.86
		Finland	10,600	28.50
		Norway	1,400	3.64
Tuloma	26,100	Russia	23,400	89.91
		Finland	2,600	10.04
Tumbes-Poyango	5,000	Ecuador	3,500	71.04
		Peru	1,400	28.96
Tumen	33,000	China	22,600	68.56
		Korea, Democratic People's Republic of	10,200	30.90
		Russia	180	0.54
Umba	8,200	Tanzania, United Republic of	6,900	83.83
		Kenya	1,300	16.17
Umbeluzi	5,400	Swaziland	3,100	57.17
		Mozambique	2,300	41.63
		South Africa	70	1.19
Utamboni	7,700	Gabon	4,600	59.81
		Equatorial Guinea	3,000	39.31
Valdivia	11,400	Chile	11,300	99.09
		Argentina	100	0.89
Vardar	33,200	Macedonia	20,400	61.29
		Yugoslavia (Serbia and Montenegro)	8,900	26.79
		Greece	4,000	11.92
Velaka	1,100	Bulgaria	780	72.47
		Turkey	300	27.53

TABLE 7 *Continued*

Basin Name	Area of Basin (km²)	Countries Sharing the Basin	Area within Country (km²)	Percentage of Watershed within Country
Venta	7,700	Latvia	5,400	70.20
		Lithuania	2,200	28.33
Vijose	9,000	Albania	6,500	72.40
		Greece	2,500	27.60
Vistula/Wista	193,900	Poland	169,200	87.28
		Ukraine	13,000	6.70
		Byelarus	9,700	5.03
		Slovakia	1,900	0.99
		Czech Republic	10	0
Volga	1,553,900	Russia	1,849,800	99.74
		Kazakhstan	2,400	0.15
		Byelarus	1,600	0.10
Volta	414,000	Burkina Faso	174,200	42.07
		Ghana	166,500	40.21
		Togo	25,900	6.25
		Mali	18,900	4.57
		Benin	15,000	3.62
		Ivory Coast	13,400	3.24
Vuoksa	62,700	Finland	54,100	86.26
		Russia	8,600	13.74
Wadi Al Izziyah	580	Lebanon	380	65.91
		Israel	190	33.74
Whiting	2,600	Canada	2,000	80.06
		United States	510	19.94
Yalu	63,000	Korea, Democratic People's Republic of	31,700	50.38
		China	31,200	49.59
Yaqui	74,700	Mexico	70,100	93.87
		United States	4,600	6.13
Yelcho	10,600	Argentina	6,900	65.06
		Chile	3,700	34.88
Yenisey/Jenisej	2,497,600	Russia	2,169,800	86.88
		Mongolia	327,600	13.12
Yser	920	France	500	53.63
		Belgium	430	46.37
Yukon	829,700	United States	496,400	59.83
		Canada	333,300	40.17

(continues)

T A B L E 7   *Continued*

Basin Name	Area of Basin (km²)	Countries Sharing the Basin	Area within Country (km²)	Percentage of Watershed within Country
Zambezi	1,388,200	Zambia	577,900	41.63
		Angola	255,500	18.40
		Zimbabwe	215,800	15.55
		Mozambique	163,800	11.80
		Malawi	110,700	7.97
		Tanzania, United Republic of	27,300	1.97
		Botswana	19,100	1.38
		Namibia	17,100	1.23
		Congo, Democratic Republic of the (Kinshasa)	1,000	0.07
Zapaleri	3,600	Chile	2,400	68.42
		Bolivia	610	16.99
		Argentina	520	14.59
Zarumilla	670	Ecuador	580	87.29
		Peru	90	12.71

Table 8. Fraction of a Country's Area in International River Basins

Description

The data from Table 7 are sorted here by country and ordered by the fraction of a country's total area lying within international river basins. The total area of each country and the area in international basins are listed in square kilometers. The names of the international river basins in each country are also shown.

Many countries lie entirely or largely within international river basins: 39 countries have more than 90 percent of their total area in international river basins, including several large countries, such as the Democratic Republic of the Congo, Niger, the Central African Republic Zambia, Bolivia, Chad, Paraguay, the Czech Republic and others. Many more have more than half of their total area in international basins.

Limitations

These data do not provide information on the fraction of actual water flows provided from each country or region. Thus a country can have a significant fraction of a watershed but generate only a small or negligible fraction of total river flow, or conversely, may have a small fraction of the watershed, but be responsible for generating a large amount of flow. Some country watershed area totals may not add up to 100 percent due to rounding.

Some border areas are disputed. Where such disputes are well known, they are explicitly identified in Table 7. For example, portions of the Indus and the Ganges-Brahmaputra-Meghna basins are under Chinese control but claimed by India, while other portions are under Indian control but claimed by China. Portions of the Nile are administered by Egypt but claimed by the Sudan and portions are administered by the Sudan and claimed by Egypt. See Table 7 for details.

Source

Wolf, A.T., J.A. Natharius, J.J. Danielson, B.S. Ward, J.K. Pender, 1999, "International River Basins of the World," *International Journal of Water Resources Development,* Vol. 15, No. 4 (December).

TABLE 8   Fraction of a Country's Area in International River Basins

Country	Total Country Area (km²)	Area in International Basins (km²)	Fraction of Country in International River Basin (%)	International River Basins
Armenia	29,900	29,900	100	Kura-Araks
Austria	83,730	83,730	100	Rhine, Po, Elbe, Danube
Botswana	581,200	581,200	100	Zambezi, Orange, Okavango, Limpopo
Burkina Faso	274,400	274,400	100	Volta, Niger, Komoe
Burundi	27,300	27,300	100	Nile, Congo/Zaire
Congo, Democratic Republic of the	2,338,200	2,338,200	100	Zambezi, Nile, Congo/Zaire, Chiloango
Hungary	92,800	92,800	100	Danube
Moldova	35,540	35,540	100	Sarata, Kogilnik, Dniester, Danube
Nepal	147,300	147,300	100	Ganges-Brahmaputra-Meghna
Paraguay	400,100	400,100	100	La Plata
Rwanda	25,300	25,300	100	Nile, Congo/Zaire
Slovakia	48,710	48,710	100	Vistula/Wista, Oder/Odra, Danube
Switzerland	41,230	41,230	100	Rhone, Rhine, Po, Danube
Uganda	243,500	243,500	100	Nile, Lotagipi Swamp, Lake Turkana
Zambia	754,500	754,500	100	Zambezi, Congo/Zaire
Bhutan	39,927	39,900	99.9	Ganges-Brahmaputra-Meghna
Central African Republic	621,499	620,900	99.9	Lake Chad, Congo/Zaire
Zimbabwe	390,804	389,900	99.8	Zambezi, Sabi, Okavango, Limpopo, Buzi
Czech Republic	78,495	78,310	99.8	Vistula/Wista, Oder/Odra, Elbe, Danube
Luxembourg	2,594	2,580	99.5	Seine, Rhine
Andorra	452	450	99.5	Garonne, Ebro
Macedonia	25,321	25,110	99.2	Vardar, Struma, Lake Prespa, Drin
Niger	1,186,021	1,174,900	99.1	Niger, Lake Chad

Country				Rivers
Yugoslavia (Serbia and Montenegro)	101,945	99,950	98.0	Vardar, Struma, Neretva, Drin, Danube
Bangladesh	138,507	135,550	97.9	Karnafauli, Ganges-Brahmaputra-Meghna, Fenney
Swaziland	17,164	16,800	97.9	Umbeluzi, Maputo, Incomati
Liechtenstein	165	160	97.1	Rhine
Romania	236,654	228,800	96.7	Danube
Lithuania	64,849	62,570	96.5	Venta, Neman, Lielupe, Daugava, Barta
Laos, People's Democratic Republic of	230,566	221,700	96.2	Red/Song Hong, Mekong, Ma, Ca/Song-Koi
Byelarus	206,681	197,810	95.7	Volga, Vistula/Wista, Neman, Narva, Dnieper, Daugava
Bolivia	1,090,353	1,043,110	95.7	Zapaleri, Lake Titicaca-Poopo, La Plata, Cancoso/Lauca, Amazon
Chad	1,168,002	1,110,900	95.1	Niger, Lake Chad
Benin	116,515	110,800	95.1	Volta, Oueme, Niger, Mono
Malawi	119,028	111,260	93.5	Zambezi, Ruvuma, Congo/Zaire
Latvia	64,299	59,940	93.2	Venta, Salaca, Parnu, Neman, Narva, Lielupe, Gauja, Daugava, Barta
Bosnia and Herzegovina	51,403	47,810	93.0	Neretva, Krka, Danube
Poland	310,715	281,850	90.7	Vistula/Wista, Prohladnaja, Oder/Odra, Neman, Lava-Pregel, Elbe, Dniester, Danube
Cambodia	182,612	164,830	90.3	Song Vam Co Dong, Saigon/Song Nha Be, Mekong
Slovenia	20,246	18,200	89.9	Isonzo, Danube
Nigeria	912,039	795,500	87.2	Oueme, Niger, Lake Chad, Cross, Akpa Yafi
Bulgaria	110,802	95,150	85.9	Velaka, Struma, Rezvaya, Nestos, Maritsa, Danube
Kyrgyzstan	199,340	170,600	85.6	Tarim, Ili/Kunes He, Aral Sea
Gabon	261,689	223,660	85.5	Utamboni, Ogooue, Nyanga, Ntem, Mbe, Congo/Zaire, Benito
Togo	57,300	48,730	85.0	Volta, Oueme, Mono
Ethiopia	1,132,328	960,900	84.9	Nile, Lotagipi Swamp, Lake Turkana, Juba-Shibeli, Gash, Awash
Guinea	246,077	205,270	83.4	St. Paul, St. John, Senegal, Sassandra, Niger, Moa, Loffa, Little Scarcies, Great Scarcies, Geba, Gambia, Corubal, Cestos, Cavally
Sudan	2,490,409	2,064,470	82.9	Nile, Lotagipi Swamp, Lake Turkana, Lake Chad, Gash, Baraka
Uruguay	178,141	142,860	80.2	Lagoon Mirim, La Plata, Chuy

(continues)

241

TABLE 8 *Continued*

Country	Total Country Area (km²)	Area in International Basins (km²)	Fraction of Country in International River Basin (%)	International River Basins
Equatorial Guinea	27,085	21,700	80.1	Utamboni, Ogooue, Ntem, Mbe, Benito
Peru	1,296,912	1,038,890	80.1	Zarumilla, Tumbes-Poyango, Lake Titicaca-Poopo, Chira, Amazon
Congo, Republic of the	345,430	276,060	79.9	Ogooue, Nyanga, Congo/Zaire, Chiloango
Ghana	239,981	189,600	79.0	Volta, Tano, Komoe, Bia
Myanmar	669,821	528,800	78.9	Salween, Pakchan, Mekong, Kaladan, Irrawaddy, Ganges-Brahmaputra-Meghna
Venezuela	916,561	698,500	76.2	Orinoco, Essequibo, Catatumbo, Barima, Amazon, Amacuro
Ukraine	596,041	451,640	75.8	Vistula/Wista, Sarata, Mius, Kogilnik, Elancik, Don, Dniester, Dnieper, Danube
Guyana	211,241	159,700	75.6	Essequibo, Courantyne/Corantijn, Barima, Amazon, Amacuro
Pakistan	877,753	649,250	74.0	Tarim, Rudkhaneh-ye/BahuKalat, Indus, Helmand, Dasht
Iraq	436,422	318,900	73.1	Tigris-Euphrates/Shatt al Arab
Guatemala	109,502	79,970	73.0	Suchiate, Sarstun, Paz, Motaqua, Lempa, Hondo, Grijalva, Coatan Achute, Candelaria, Belize
Syria	187,937	136,830	72.8	Tigris-Euphrates/Shatt al Arab, Nahr El Kebir, Jordan (Dead Sea), Asi/Orontes, An Nahr Al Kabir
Liberia	96,296	69,000	71.7	St. Paul, St. John, Moa, Mana-Morro, Loffa, Cestos, Cavally
Belgium	30,480	21,830	71.6	Yser, Seine, Schelde, Rhine
Germany	356,109	253,600	71.2	Rhine, Oder/Odra, Elbe, Danube
Azerbaijan	85,808	60,350	70.3	Sulak, Samur, Kura-Araks, Astara Chay
Afghanistan	641,869	447,220	69.7	Tarim, Murgab, Kowl-E-Namaksar, Indus, Helmand, Hari/Harirud
Namibia	825,632	563,600	68.3	Zambezi, Orange, Okavango, Kunene, Etosha-Cuvelai
Kenya	584,429	396,800	67.9	Umba, Nile, Lotagipi Swamp, Lake Turkana, Lake Natron, Juba-Shibeli

Country			
Angola	1,252,421	67.8	Zambezi, Okavango, Kunene, Etosha-Cuvelai, Congo/Zaire, Chiloango
Colombia	1,141,962	65.6	Patia, Orinoco, Mira, Mataje, Jurado, Catatumbo, Amazon
Lesotho	30,352	65.6	Orange
South Africa	1,223,111	65.2	Umbeluzi, Orange, Maputo, Limpopo, Incomati
Kazakhstan	2,715,976	64.0	Volga, Tarim, Pu-Lun-To, Oral (Ural), Ob, Ili/Kunes He, Aral Sea
Croatia	56,288	63.2	Neretva, Krka, Danube
United States	9,450,720	62.4	Yukon, Yaqui, Whiting, Tijuana, Taku, Stikine, St. Lawrence, St. John, St. Croix, Rio Grande, Nelson-Saskatchewan, Mississippi, Fraser, Firth, Columbia, Colorado, Chilkat, Alesek
Ecuador	256,932	60.3	Zarumilla, Tumbes-Poyango, Patia, Mira, Mataje, Chira, Amazon
Brazil	8,507,128	59.7	Oyupock/Oiapoque, Maroni, Lagoon Mirim, La Plata, Essequibo, Chuy, Amazon
Vietnam	327,123	59.0	Song Vam Co Dong, Saigon/Song Nha Be, Red/Song Hong, Mekong, Ma, Hsi/Bei Jiang, Ca/Song-Koi, Beilun
Spain	505,674	57.6	Tagus/Tejo, Mino/Minho, Lima, Guadiana, Garonne, Ebro, Douro/Duero, Bidasoa
Mali	1,256,747	56.7	Volta, Senegal, Niger, Komoe
Georgia	69,943	56.3	Terek, Sulak, Kura-Araks, Coruh
Cameroon	466,307	56.1	Ogooue, Ntem, Niger, Lake Chad, Cross, Congo/Zaire, Akpa Yafi
El Salvador	20,697	55.9	Paz, Lempa, Goascoran
Ivory Coast	322,216	55.4	Volta, Tano, St. John, Sassandra, Niger, Komoe, Cestos, Cavally, Bia
Estonia	45,545	55.3	Salaca, Parnu, Narva, Gauja
Gambia, The	10,678	55.3	Gambia
West Bank	5,816	55.0	Jordan (Dead Sea)
Mozambique	788,629	53.8	Zambezi, Umbeluzi, Sabi, Ruvuma, Maputo, Limpopo, Incomati, Buzi
Uzbekistan	445,711	53.1	Aral Sea

(continues)

TABLE 8 *Continued*

Country	Total Country Area (km²)	Area in International Basins (km²)	Fraction of Country in International River Basin (%)	International River Basins
Eritrea	121,941	63,000	51.7	Nile, Gash, Baraka
Djibouti	21,638	11,100	51.3	Awash
Suriname	145,498	73,920	50.8	Maroni, Courantyne/Corantijn, Amazon
Finland	333,797	168,680	50.5	Vuoksa, Tuloma, Torne/Tornealven, Tana, Pasvik, Oulu, Olanga, Naatamo, Kemi
Albania	28,755	14,260	49.6	Vijose, Lake Prespa, Drin, Danube
French Guiana	83,811	40,900	48.8	Oyupock/Oiapoque, Maroni
Portugal	92,098	44,890	48.7	Tagus/Tejo, Mino/Minho, Lima, Guadiana, Douro/Duero
Russia	16,851,940	7,923,810	47.0	Yenisey/Jenisej, Vuoksa, Volga, Tumen, Tuloma, Terek, Sulak, Sujfun, Samur, Pu-Lun-To, Prohladnaja, Pasvik, Oulu, Oral (Ural), Olanga, Ob, Neman, Narva, Mius, Lava-Pregel, Lake Ubsa-Nur, Kura-Araks, Kemi, Jacobs, Har Us Nur, Elancik, Don, Dnieper, Daugava, Amur
Senegal	196,911	90,600	46.0	Senegal, Geba, Gambia
France	546,729	249,460	45.6	Yser, Seine, Schelde, Roia, Rhone, Rhine, Po, Garonne, Ebro, Bidasoa
Guinea-Bissau	33,635	15,200	45.2	Geba, Corubal
Israel	20,774	9,290	44.7	Wadi Al Izziyah, Jordan (Dead Sea)
India	3,089,282	1,354,400	43.8	Karnafauli, Kaladan, Irrawaddy, Indus, Ganges-Brahmaputra-Meghna, Fenney
Tanzania, United Republic of	944,977	410,800	43.5	Zambezi, Umba, Ruvuma, Nile, Lake Natron, Congo/Zaire
Korea, Democratic People's Republic of	122,473	52,120	42.6	Yalu, Tumen, Han, Amur
Thailand	515,144	205,900	40.0	Salween, Pakchan, Mekong, Golok
Mongolia	1,559,176	616,500	39.5	Yenisey/Jenisej, Pu-Lun-To, Lake Ubsa-Nur, Har Us Nur, Amur
Belize	22,175	8,760	39.5	Sarstun, Hondo, Belize
Nicaragua	129,047	49,670	38.5	San Juan, Negro, Coco/Segovia, Choluteca

Country			River basins
Sierra Leone	72,531	38.1	Niger, Moa, Mana-Morro, Little Scarcies, Great Scarcies
Somalia	639,065	34.7	Juba-Shibeli, Awash
China	9,338,902	34.4	Yalu, Tumen, Tarim, Sujfun, Salween, Red/Song Hong, Pu-Lun-To, Ob, Mekong, Irrawaddy, Ili/Kunes He, Hsi/Bei Jiang, Har Us Nur, Ganges-Brahmaputra-Meghna, Beilun, Aral Sea, Amur, Indus
Turkey	779,986	34.2	Velaka, Tigris-Euphrates/Shatt al Arab, Rezvaya, Nahr El Kebir, Maritsa, Kura-Araks, Coruh, Asi/Orontes
Netherlands	35,493	33.8	Schelde, Rhine
Argentina	2,781,013	31.3	Zapaleri, Yelcho, Valdivia, Seno Union/Serrano, San Martin, Rio Grande, Puelo, Pascua, Palena, Lake Fagnano, La Plata, Gallegos-Chico, Cullen, Comau, Chico/Carmen Silva, Baker, Aysen, Aviles
Papua New Guinea	466,161	29.2	Tami, Sepik, Merauke, Fly
Costa Rica	51,608	28.7	Sixaola, San Juan, Chiriqui, Changuinola
Lebanon	10,240	28.5	Wadi Al Izziyah, Jordan (Dead Sea), Asi/Orontes, An Nahr Al Kabir
Egypt	982,910	28.5	Nile, Jordan (Dead Sea)
Italy	300,980	28.2	Roia, Rhone, Rhine, Po, Isonzo, Danube
Haiti	27,157	27.0	Pedernales, Massacre, Artibonite
Jordan	89,275	25.5	Tigris-Euphrates/Shatt al Arab, Jordan (Dead Sea)
Korea, Republic of	98,339	25.5	Han
Canada	9,904,700	23.8	Yukon, Whiting, Taku, Stikine, St. Lawrence, St. John, St. Croix, Nelson-Saskatchewan, Mississippi, Fraser, Firth, Columbia, Chilkat, Alesek
Honduras	112,852	22.0	Negro, Motaqua, Lempa, Goascoran, Coco/Segovia, Choluteca
Mauritania	1,041,570	21.9	Senegal, Atui
Iran	1,624,760	21.6	Tigris-Euphrates/Shatt al Arab, Rudkhaneh-ye/BahuKalat, Kura-Araks, Kowl-E-Namaksar, Helmand, Hari/Harirud, Dasht, Atrak, Astara Chay
Mexico	1,962,939	20.8	Yaqui, Tijuana, Suchiate, Rio Grande, Hondo, Grijalva, Colorado, Coatan Achute, Candelaria

(continues)

TABLE 8 *Continued*

Country	Total Country Area (km²)	Area in International Basins (km²)	Fraction of Country in International River Basin (%)	International River Basins
Morocco	403,860	79,350	19.6	Tafna, Oued Bon Naima, Guir, Dra, Daoura
Greece	131,852	24,790	18.8	Vijose, Vardar, Struma, Nestos, Maritsa, Lake Prespa
Sweden	443,800	68,700	15.5	Torne/Tornealven, Klaralven
Algeria	2,320,972	358,260	15.4	Tafna, Oued Bon Naima, Niger, Medjerda, Lake Chad, Guir, Dra, Daoura
Chile	742,298	92,470	12.5	Zapaleri, Yelcho, Valdivia, Seno Union/Serrano, San Martin, Rio Grande, Puelo, Pascua, Palena, Lake Titicaca-Poopo, Lake Fagnano, Gallegos-Chico, Cullen, Comau, Chico/Carmen Silva, Cancoso/Lauca, Baker, Aysen, Aviles
Turkmenistan	471,429	52,900	11.2	Murgab, Hari/Harirud, Atrak, Aral Sea
Brunei	5,770	610	10.6	Pandaruan, Bangau
Tunisia	155,402	15,600	10.0	Medjerda
Tajikistan	142,410	14,000	9.8	Tarim, Aral Sea
Panama	74,697	4,930	6.6	Sixaola, Jurado, Chiriqui, Changuinola
Norway	316,962	20,540	6.5	Torne/Tornealven, Tana, Pasvik, Naatamo, Klaralven, Kemi, Jacobs
Dominican Republic	48,445	2,710	5.6	Pedernales, Massacre, Artibonite
Indonesia	1,910,842	106,300	5.6	Tami, Sepik, Sembakung, Merauke, Fly
Ireland	69,384	3,420	4.9	Foyle, Flurry, Fane, Erne, Castletown, Bann
United Kingdom	243,137	9,250	3.8	Foyle, Flurry, Fane, Erne, Castletown, Bann
Malaysia	330,270	8,500	2.6	Sembakung, Pandaruan, Golok, Bangau
Western Sahara	269,602	1,100	0.41	Atui
Libya	1,620,515	4,600	0.28	Lake Chad
Saudi Arabia	1,960,175	240	0.01	Tigris-Euphrates/Shatt al Arab

Table 9. International River Basins, by Country

Description

The data from Tables 7 and 8 are sorted here by country and ordered alphabetically. The total area of each country and the area in international basins are listed in square kilometers. The names of the international river basins in each country are also shown.

Limitations

These data do not provide information on the fraction of actual water flows provided from each country or region. Thus a country can have a significant fraction of a watershed but generate only a small or negligible fraction of total river flow, or conversely, may have a small fraction of the watershed, but be responsible for generating a large amount of flow. Some country watershed area totals may not add up to 100 percent due to rounding.

Some border areas are disputed. Where such disputes are well known, they are explicitly identified in Table 7. For example, portions of the Indus and the Ganges-Brahmaputra-Meghna basins are under Chinese control but claimed by India, while other portions are under Indian control but claimed by China. Portions of the Nile are administered by Egypt but claimed by the Sudan and portions are administered by the Sudan and claimed by Egypt. See Table 7 for details.

Source

Wolf, A.T., J.A. Natharius, J.J. Danielson, B.S. Ward, J. K. Pender, 1999, "International River Basins of the World," *International Journal of Water Resources Development,* Vol. 15, No. 4 (December).

TABLE 9 International River Basins, by Country

Country	Total Area (km²)	Area in International Basins (km²)	Fraction of Country in International Basins (%)	International River Basins
Afghanistan	641,869	447,220	69.7	Tarim, Murgab, Kowl-E-Namaksar, Indus, Helmand, Hari/Harirud
Albania	28,755	14,260	49.6	Vijose, Lake Prespa, Drin, Danube
Algeria	2,320,972	358,260	15.4	Tafna, Oued Bon Naima, Niger, Medjerda, Lake Chad, Guir, Dra, Daoura
Andorra	452	450	99.5	Garonne, Ebro
Angola	1,252,421	849,500	67.8	Zambezi, Okavango, Kunene, Etosha-Cuvelai, Congo/Zaire, Chiloango
Argentina	2,781,013	870,560	31.3	Zapaleri, Yelcho, Valdivia, Seno Union/Serrano, San Martin, Rio Grande, Puelo, Pascua, Palena, Lake Fagnano, La Plata, Gallegos-Chico, Cullen, Comau, Chico/Carmen Silva, Baker, Aysen, Aviles
Armenia	29,900	29,900	100	Kura-Araks
Austria	83,730	83,730	100	Rhine, Po, Elbe, Danube
Azerbaijan	85,808	60,350	70.3	Sulak, Samur, Kura-Araks, Astara Chay
Bangladesh	138,507	135,550	97.9	Karnafauli, Ganges-Brahmaputra-Meghna, Fenney
Belgium	30,480	21,830	71.6	Yser, Seine, Schelde, Rhine
Belize	22,175	8,760	39.5	Sarstun, Hondo, Belize
Benin	116,515	110,800	95.1	Volta, Oueme, Niger, Mono
Bhutan	39,927	39,900	99.9	Ganges-Brahmaputra-Meghna
Bolivia	1,090,353	1,043,110	95.7	Zapaleri, Lake Titicaca-Poopo, La Plata, Cancoso/Lauca, Amazon
Bosnia and Herzegovina	51,403	47,810	93.0	Neretva, Krka, Danube
Botswana	581,200	581,200	100	Zambezi, Orange, Okavango, Limpopo
Brazil	8,507,128	5,078,090	59.7	Oyupock/Oiapoque, Maroni, Lagoon Mirim, La Plata, Essequibo, Chuy, Amazon
Brunei	5,770	610	10.6	Pandaruan, Bangau
Bulgaria	110,802	95,150	85.9	Velaka, Struma, Rezvaya, Nestos, Maritsa, Danube

248

Country				
Burkina Faso	274,400	274,400	100	Volta, Niger, Komoe
Burundi	27,300	27,300	100	Nile, Congo/Zaire
Byelarus	206,681	197,810	95.7	Volga, Vistula/Wista, Neman, Narva, Dnieper, Daugava
Cambodia	182,612	164,830	90.3	Song Vam Co Dong, Saigon/Song Nha Be, Mekong
Cameroon	466,307	261,400	56.1	Ogooue, Ntem, Niger, Lake Chad, Cross, Congo/Zaire, Akpa Yafi
Canada	9,904,700	2,354,400	23.8	Yukon, Whiting, Taku, Stikine, St. Lawrence, St. John, St. Croix, Nelson-Saskatchewan, Mississippi, Fraser, Firth, Columbia, Chilkat, Alesek
Central African Republic	621,499	620,900	99.9	Lake Chad, Congo/Zaire
Chad	1,168,002	1,110,900	95.1	Niger, Lake Chad
Chile	742,298	92,470	12.5	Zapaleri, Yelcho, Valdivia, Seno Union/Serrano, San Martin, Rio Grande, Puelo, Pascua, Palena, Lake Titicaca-Poopo, Lake Fagnano, Gallegos-Chico, Cullen, Comau, Chico/Carmen Silva, Cancoso/Lauca, Baker, Aysen, Aviles
China	9,338,902	3,211,660	34.4	Yalu, Tumen, Tarim, Sujfun, Salween, Red/Song Hong, Pu-Lun-To, Ob, Mekong, Irrawaddy, Ili/Kunes He, Hsi/Bei Jiang, Har Us Nur, Ganges-Brahmaputra-Meghna, Beilun, Aral Sea, Amur, Indus
Colombia	1,141,962	749,570	65.6	Patia, Orinoco, Mira, Mataje, Jurado, Catatumbo, Amazon
Congo, Democratic Republic of the	2,338,200	2,338,200	100	Zambezi, Nile, Congo/Zaire, Chiloango
Congo, Republic of the	345,430	276,060	79.9	Ogooue, Nyanga, Congo/Zaire, Chiloango
Costa Rica	51,608	14,810	28.7	Sixaola, San Juan, Chiriqui, Changuinola
Croatia	56,288	35,600	63.2	Neretva, Krka, Danube
Czech Republic	78,495	78,310	99.8	Vistula/Wista, Oder/Odra, Elbe, Danube
Djibouti	21,638	11,100	51.3	Awash
Dominican Republic	48,445	2,710	5.6	Pedernales, Massacre, Artibonite
Ecuador	256,932	154,950	60.3	Zarumilla, Tumbes-Poyango, Patia, Mira, Mataje, Chira, Amazon

(continues)

TABLE 9 Continued

Country	Total Area (km²)	Area in International Basins (km²)	Fraction of Country in International Basins (%)	International River Basins
Egypt	982,910	280,200	28.5	Nile, Jordan (Dead Sea)
El Salvador	20,697	11,570	55.9	Paz, Lempa, Goascoran
Equatorial Guinea	27,085	21,700	80.1	Utamboni, Ogooue, Ntem, Mbe, Benito
Eritrea	121,941	63,000	51.7	Nile, Gash, Baraka
Estonia	45,545	25,200	55.3	Salaca, Parnu, Narva, Gauja
Ethiopia	1,132,328	960,900	84.9	Nile, Lotagipi Swamp, Lake Turkana, Juba-Shibeli, Gash, Awash
Finland	333,797	168,680	50.5	Vuoksa, Tuloma, Torne/Tornealven, Tana, Pasvik, Oulu, Olanga, Naatamo, Kemi
France	546,729	249,460	45.6	Yser, Seine, Schelde, Roia, Rhone, Rhine, Po, Garonne, Ebro, Bidasoa
French Guiana	83,811	40,900	48.8	Oyupock/Oiapoque, Maroni
Gabon	261,689	223,660	85.5	Utamboni, Ogooue, Nyanga, Ntem, Mbe, Congo/Zaire, Benito
Gambia, The	10,678	5,900	55.3	Gambia
Georgia	69,943	39,400	56.3	Terek, Sulak, Kura-Araks, Coruh
Germany	356,109	253,600	71.2	Rhine, Oder/Odra, Elbe, Danube
Ghana	239,981	189,600	79.0	Volta, Tano, Komoe, Bia
Greece	131,852	24,790	18.8	Vijose, Vardar, Struma, Nestos, Maritsa, Lake Prespa
Guatemala	109,502	79,970	73.0	Suchiate, Sarstun, Paz, Motaqua, Lempa, Hondo, Grijalva, Coatan Achute, Candelaria, Belize
Guinea	246,077	205,270	83.4	St. Paul, St. John, Senegal, Sassandra, Niger, Moa, Loffa, Little Scarcies, Great Scarcies, Geba, Gambia, Corubal, Cestos, Cavally
Guinea-Bissau	33,635	15,200	45.2	Geba, Corubal
Guyana	211,241	159,700	75.6	Essequibo, Courantyne/Corantijn, Barima, Amazon, Amacuro
Haiti	27,157	7,340	27.0	Pedernales, Massacre, Artibonite
Honduras	112,852	24,800	22.0	Negro, Motaqua, Lempa, Goascoran, Coco/Segovia, Choluteca

Hungary	92,800	100	Danube
India	3,089,282	43.8	Karnafauli, Kaladan, Irrawaddy, Indus, Ganges-Brahmaputra-Meghna, Fenney
Indonesia	1,910,842	5.6	Tami, Sepik, Sembakung, Merauke, Fly
Iran	1,624,760	21.6	Tigris-Euphrates/Shatt al Arab, Rudkhaneh-ye/BahuKalat, Kura-Araks, Kowl-E-Namaksar, Helmand, Hari/Harirud, Dasht, Atrak, Astara Chay
Iraq	436,422	73.1	Tigris-Euphrates/Shatt al Arab
Ireland	69,384	4.9	Foyle, Flurry, Fane, Erne, Castletown, Bann
Israel	20,774	44.7	Wadi Al Izziyah, Jordan (Dead Sea)
Italy	300,980	28.2	Roia, Rhone, Rhine, Po, Isonzo, Danube
Ivory Coast	322,216	55.4	Volta, Tano, St. John, Sassandra, Niger, Komoe, Cestos, Cavally, Bia
Jordan	89,275	25.5	Tigris-Euphrates/Shatt al Arab, Jordan (Dead Sea)
Kazakhstan	2,715,976	64.0	Volga, Tarim, Pu-Lun-To, Oral (Ural), Ob, Ili/Kunes He, Aral Sea
Kenya	584,429	67.9	Umba, Nile, Lotagipi Swamp, Lake Turkana, Lake Natron, Juba-Shibeli
Korea, Democratic People's Republic of	122,473	42.6	Yalu, Tumen, Han, Amur
Korea, Republic of	98,339	25.5	Han
Kyrgyzstan	199,340	85.6	Tarim, Ili/Kunes He, Aral Sea
Laos, People's Democratic Republic of	230,566	96.2	Red/Song Hong, Mekong, Ma, Ca/Song-Koi
Latvia	64,299	93.2	Venta, Salaca, Parnu, Neman, Narva, Lielupe, Gauja, Daugava, Barta,
Lebanon	10,240	28.5	Wadi Al Izziyah, Jordan (Dead Sea), Asi/Orontes, An Nahr Al Kabir
Lesotho	30,352	65.6	Orange
Liberia	96,296	71.7	St. Paul, St. John, Moa, Mana-Morro, Loffa, Cestos, Cavally
Libya	1,620,515	0.3	Lake Chad
Liechtenstein	165	97.1	Rhine
Lithuania	64,849	96.5	Venta, Neman, Lielupe, Daugava, Barta
Luxembourg	2,594	99.5	Seine, Rhine

(continues)

TABLE 9 *Continued*

Country	Total Area (km²)	Area in International Basins (km²)	Fraction of Country in International Basins (%)	International River Basins
Macedonia	25,321	25,110	99.2	Vardar, Struma, Lake Prespa, Drin
Malawi	119,028	111,260	93.5	Zambezi, Ruvuma, Congo/Zaire
Malaysia	330,270	8,500	2.6	Sembakung, Pandaruan, Golok, Bangau
Mali	1,256,747	712,430	56.7	Volta, Senegal, Niger, Komoe
Mauritania	1,041,570	228,400	21.9	Senegal, Atui
Mexico	1,962,939	408,490	20.8	Yaqui, Tijuana, Suchiate, Rio Grande, Hondo, Grijalva, Colorado, Coatan Achute, Candelaria
Moldova	35,540	35,540	100	Sarata, Kogilnik, Dniester, Danube
Mongolia	1,559,176	616,500	39.5	Yenisey/Jenisej, Pu-Lun-To, Lake Ubsa-Nur, Har Us Nur, Amur
Morocco	403,860	79,350	19.6	Tafna, Oued Bon Naima, Guir, Dra, Daoura
Mozambique	788,629	423,900	53.8	Zambezi, Umbeluzi, Sabi, Ruvuma, Maputo, Limpopo, Incomati, Buzi
Myanmar	669,821	528,800	78.9	Salween, Pakchan, Mekong, Kaladan, Irrawaddy, Ganges-Brahmaputra-Meghna
Namibia	825,632	563,600	68.3	Zambezi, Orange, Okavango, Kunene, Etosha-Cuvelai
Nepal	147,300	147,300	100	Ganges-Brahmaputra-Meghna
Netherlands	35,493	11,980	33.8	Schelde, Rhine
Nicaragua	129,047	49,670	38.5	San Juan, Negro, Coco/Segovia, Choluteca
Niger	1,186,021	1,174,900	99.1	Niger, Lake Chad
Nigeria	912,039	795,500	87.2	Oueme, Niger, Lake Chad, Cross, Akpa Yafi
Norway	316,962	20,540	6.5	Torne/Tornealven, Tana, Pasvik, Naatamo, Klaralven, Kemi, Jacobs
Pakistan	877,753	649,250	74.0	Tarim, Rudkhaneh-ye/BahuKalat, Indus, Helmand, Dasht
Panama	74,697	4,930	6.6	Sixaola, Jurado, Chiriqui, Changuinola
Papua New Guinea	466,161	136,200	29.2	Tami, Sepik, Merauke, Fly
Paraguay	400,100	400,100	100	La Plata

Country				River basins
Peru	1,296,912	1,038,890	80.1	Zarumilla, Tumbes-Poyango, Lake Titicaca-Poopo, Chira, Amazon
Poland	310,715	281,850	90.7	Vistula/Wista, Prohladnaja, Oder/Odra, Neman, Lava-Pregel, Elbe, Dniester, Danube
Portugal	92,098	44,890	48.7	Tagus/Tejo, Mino/Minho, Lima, Guadiana, Douro/Duero
Romania	236,654	228,800	96.7	Danube
Russia	16,851,940	7,923,810	47.0	Yenisey/Jenisej, Vuoksa, Volga, Tumen, Tuloma, Terek, Sulak, Sujfun, Samur, Pu-Lun-To, Prohladnaja, Pasvik, Oulu, Oral (Ural), Olanga, Ob, Neman, Narva, Mius, Lava-Pregel, Lake Ubsa-Nur, Kura-Araks, Kemi, Jacobs, Har Us Nur, Elancik, Don, Dnieper, Daugava, Amur
Rwanda	25,300	25,300	100	Nile, Congo/Zaire
Saudi Arabia	1,960,175	240	0.0	Tigris-Euphrates/Shatt al Arab
Senegal	196,911	90,600	46.0	Senegal, Geba, Gambia
Sierra Leone	72,531	27,630	38.1	Niger, Moa, Mana-Morro, Little Scarcies, Great Scarcies
Slovakia	48,710	48,710	100	Vistula/Wista, Oder/Odra, Danube
Slovenia	20,246	18,200	89.9	Isonzo, Danube
Somalia	639,065	221,740	34.7	Juba-Shibeli, Awash
South Africa	1,223,111	797,570	65.2	Umbeluzi, Orange, Maputo, Limpopo, Incomati
Spain	505,674	291,190	57.6	Tagus/Tejo, Mino/Minho, Lima, Guadiana, Garonne, Ebro, Douro/Duero, Bidasoa
Sudan	2,490,409	2,064,470	82.9	Nile, Lotagipi Swamp, Lake Turkana, Lake Chad, Gash, Baraka
Suriname	145,498	73,920	50.8	Maroni, Courantyne/Corantijn, Amazon
Swaziland	17,164	16,800	97.9	Umbeluzi, Maputo, Incomati
Sweden	443,800	68,700	15.5	Torne/Tornealven, Klaralven
Switzerland	41,230	41,230	100	Rhone, Rhine, Po, Danube
Syria	187,937	136,830	72.8	Tigris-Euphrates/Shatt al Arab, Nahr El Kebir, Jordan (Dead Sea), Asi/Orontes, An Nahr Al Kabir
Tajikistan	142,410	14,000	9.8	Tarim, Aral Sea
Tanzania, United Republic of	944,977	410,800	43.5	Zambezi, Umba, Ruvuma, Nile, Lake Natron, Congo/Zaire
Thailand	515,144	205,900	40.0	Salween, Pakchan, Mekong, Golok

(continues)

253

TABLE 9 *Continued*

Country	Total Area (km²)	Area in International Basins (km²)	Fraction of Country in International Basins (%)	International River Basins
Togo	57,300	48,730	85.0	Volta, Oueme, Mono
Tunisia	155,402	15,600	10.0	Medjerda
Turkey	779,986	266,470	34.2	Velaka, Tigris-Euphrates/Shatt al Arab, Rezvaya, Nahr El Kebir, Maritsa, Kura-Araks, Coruh, Asi/Orontes
Turkmenistan	471,429	52,900	11.2	Murgab, Hari/Harirud, Atrak, Aral Sea
Uganda	243,500	243,500	100	Nile, Lotagipi Swamp, Lake Turkana
Ukraine	596,041	451,640	75.8	Vistula/Wista, Sarata, Mius, Kogilnik, Elancik, Don, Dniester, Dnieper, Danube
United Kingdom	243,137	9,250	3.8	Foyle, Flurry, Fane, Erne, Castletown, Bann
United States	9,450,720	5,895,200	62.4	Yukon, Yaqui, Whiting, Tijuana, Taku, Stikine, St. Lawrence, St. John, St. Croix, Rio Grande, Nelson-Saskatchewan, Mississippi, Fraser, Firth, Columbia, Colorado, Chilkat, Alesek
Uruguay	178,141	142,860	80.2	Lagoon Mirim, La Plata, Chuy
Uzbekistan	445,711	236,700	53.1	Aral Sea
Venezuela	916,561	698,500	76.2	Orinoco, Essequibo, Catatumbo, Barima, Amazon, Amacuro
Vietnam	327,123	193,150	59.0	Song Vam Co Dong, Saigon/Song Nha Be, Red/Song Hong, Mekong, Ma, Hsi/Bei Jiang, Ca/Song-Koi, Beilun
West Bank	5,816	3,200	55.0	Jordan (Dead Sea)
Western Sahara	269,602	1,100	0.4	Atui
Yugoslavia (Serbia and Montenegro)	101,945	99,950	98.0	Vardar, Struma, Neretva, Drin, Danube
Zambia	754,500	754,500	100	Zambezi, Congo/Zaire
Zimbabwe	390,804	389,900	99.8	Zambezi, Sabi, Okavango, Limpopo, Buzi

Table 10. Irrigated Area, by Country and Region, 1961 to 1997

Description

Total irrigated areas by country and continental region are listed here for 1961, 1965, 1970, 1975, 1980, 1985, 1990, 1995, and 1997—the latest year for which reliable data are available. Units are thousands of hectares. At the bottom of the table, the rates of change over each time period, and an average annual rate of change, are also given, in percentage and percentage per year, respectively. While total irrigated area worldwide continues to increase, the average annual rate of change has been dropping.

Limitations

These data depend on in-country surveys, national reports, and estimates by the Food and Agriculture Organization. In some regions, multiple cropping may increase the apparent area in production. These data are not reported here. No differentiation is made about the quality of the land in production. Recent changes in political borders and the independence of several countries make certain time-series comparisons very difficult. Data for the Soviet Union, Yugoslavia, and Czechoslovakia are provided here through 1990; thereafter the irrigated areas of the newly independent states are reported separately. See the description for Table 11 for a discussion of how to evaluate overall continental irrigated area data.

Source

Food and Agriculture Organization, 1999, Web site at www.fao.org

TABLE 10 Irrigated Area, by Country, 1961 to 1997 (thousand hectares)

Country and Region	1961	1965	1970	1975	1980	1985	1990	1995	1997
Africa									
Algeria	229	233	238	244	253	338	384	555	560
Angola	75	75	75	75	75	75	75	75	75
Benin	0	2	2	4	5	6	6	10	20
Botswana	1	2	1	1	2	2	2	1	1
Burkina Faso	2	2	4	8	10	12	20	25	25
Burundi	3	5	5	5	10	14	14	14	14
Cameroon	2	4	7	10	14	21	21	21	21
Cape Verde	2	2	2	2	2	2	3	3	3
Chad	5	5	5	6	6	10	14	17	20
Congo	0	0	1	2	1	1	1	1	1
Congo, Dem. Rep. (formerly Zaire)				0	7	9	10	11	11
Cote d'Ivoire	4	6	20	34	44	54	66	73	73
Djibouti	1	1	1	1	1	1	1	1	1
Egypt	2,568	2,672	2,843	2,825	2,445	2,497	2,648	3,283	3,300
Eritrea								28	28
Ethiopia	150	150	155	158	160	162	162	190	190
Gabon	4	4	4	4	4	4	4	7	7
Gambia	1	1	1	1	1	1	1	2	2
Ghana	0	0	7	7	7	7	6	11	11
Guinea	20	20	50	50	90	90	90	93	95
Guinea Bissau	17	17	17	17	17	17	17	17	17
Kenya	14	14	29	40	40	42	54	67	67
Lesotho	3	3	3	3	3	3	3	3	3

Liberia	0	0	2	2	2	2	2	2	2
Libya	121	130	175	200	225	300	470	470	470
Madagascar	300	330	330	465	645	826	1,000	1,087	1,090
Malawi	1	1	4	13	18	18	20	28	28
Mali	60	60	61	60	60	60	78	85	86
Mauritania	20	20	30	30	49	49	49	49	49
Mauritius	8	12	15	15	16	17	17	18	18
Morocco	875	895	920	1,060	1,217	1,245	1,258	1,258	1,251
Mozambique	8	16	26	40	65	93	105	107	107
Namibia	4	4	4	4	4	4	4	7	7
Niger	16	16	18	18	23	30	66	66	66
Nigeria	200	200	200	200	200	200	230	235	233
Reunion	3	5	5	5	5	8	11	12	12
Rwanda	4	4	4	4	4	4	4	4	4
Sao Tome and Principe	10	10	10	10	10	10	10	10	10
Senegal	70	85	78	78	62	90	94	71	71
Sierra Leone	1	2	6	13	20	28	28	29	29
Somalia	90	90	95	100	125	180	180	200	200
South Africa	808	890	1,000	1,017	1,128	1,128	1,290	1,270	1,270
Sudan	1,480	1,550	1,625	1,700	1,800	1,946	1,946	1,946	1,950
Swaziland	36	40	47	56	58	62	67	69	69
Tanzania	20	28	38	52	120	127	144	150	155
Togo	2	2	4	6	6	7	7	7	7
Tunisia	100	100	200	200	243	300	300	361	380
Uganda	2	3	4	4	6	9	9	9	9
Zambia	2	2	9	18	19	28	30	46	46
Zimbabwe	22	34	46	70	80	90	100	150	150

(continues)

TABLE 10 *Continued*

Irrigated Area (thousand hectares)

Country and Region	1961	1965	1970	1975	1980	1985	1990	1995	1997
North and Central America									
Barbados	1	1	1	1	1	1	1	1	1
Belize	0	0	1	1	1	2	2	3	3
Canada	350	380	421	500	596	748	718	720	720
Costa Rica	26	26	26	36	61	110	118	126	126
Cuba	230	330	450	580	762	861	900	910	910
Dominican Republic	110	115	125	140	165	198	225	259	259
El Salvador	18	20	20	33	110	110	120	120	120
Guadaloupe	1	1	2	1	2	2	2	2	2
Guatamala	32	43	56	72	87	102	117	125	125
Haiti	35	40	60	70	70	70	75	90	90
Honduras	50	66	66	70	72	72	74	74	74
Jamaica	22	24	24	32	33	33	33	33	33
Martinique	1	1	1	2	5	4	4	3	3
Mexico	3,000	3,200	3,583	4,479	4,980	5,285	5,600	6,100	6,500
Nicaragua	18	18	40	67	80	83	85	88	88
Panama	14	18	20	23	28	30	31	32	32
Puerto Rico	39	39	39	39	39	39	39	40	40
Saint Lucia	1	1	1	1	1	1	2	3	3
St. Vincent	0	1	1	1	1	1	1	1	1
Trindad and Tobago	11	11	15	18	21	22	22	22	22
United States	14,000	15,200	16,000	16,690	20,582	19,831	20,900	21,400	21,400

South America

Country									
Argentina	980	1,110	1,280	1,440	1,580	1,620	1,680	1,700	1,700
Bolivia	72	75	80	120	140	125	110	78	88
Brazil	490	610	796	1,100	1,600	2,100	2,700	3,169	3,169
Chile	1,075	1,100	1,180	1,242	1,255	1,257	1,265	1,265	1,270
Colombia	226	235	250	300	400	465	680	1,037	1,061
Ecuador	440	450	470	506	500	300	290	240	250
French Guiana	1	1	1	1	1	2	2	2	2
Guyana	90	109	115	120	125	127	130	130	130
Paraguay	30	30	40	55	60	65	67	67	67
Peru	1,016	1,060	1,106	1,130	1,160	1,210	1,450	1,753	1,760
Suriname	14	15	28	33	42	55	59	60	60
Uruguay	27	35	52	57	79	97	120	140	140
Venezuela	60	62	70	90	137	171	180	200	205
Asia									
Afghanistan	2,160	2,260	2,340	2,430	2,505	2,586	3,000	2,800	2,800
Armenia	Formerly included in Soviet Union							290	290
Azerbaijan	Formerly included in Soviet Union						1,455	1,453	1,455
Bahrain	1	1	1	1	1	1	2	4	5
Banglaesh	426	572	1,058	1,441	1,569	2,073	2,936	3,429	3,693
Bhutan	8	10	18	22	26	30	39	39	40
Brunei Darsm			1	0	1	1	1		1
Cambodia	62	100	89	89	100	130	160	270	270
China	30,402	33,579	38,113	42,776	45,467	44,581	47,965	49,857	51,819
Cyprus	30	30	30	30	30	30	36	40	40
Gaza Strip	8	8	9	10	10	11	11	12	12
Georgia	Formerly included in Soviet Union							469	470

(*continues*)

TABLE 10 Continued

Irrigated Area (thousand hectares)

Country and Region	1961	1965	1970	1975	1980	1985	1990	1995	1997
Hong Kong	9	8	8	6	3	3	2	2	2
India	24,685	26,510	30,440	33,730	38,478	41,779	45,144	53,000	57,000
Indonesia	3,900	3,900	3,900	3,900	4,301	4,300	4,410	4,687	4,815
Iran	4,700	4,900	5,200	5,900	4,948	6,800	7,000	7,264	7,265
Iraq	1,250	1,350	1,480	1,567	1,750	1,750	3,525	3,525	3,525
Israel	136	151	172	180	203	233	206	199	199
Japan	2,940	2,943	3,415	3,171	3,055	2,952	2,846	2,745	2,701
Jordan	31	32	34	36	37	48	63	75	75
Kazakhstan	Formerly included in Soviet Union							2,380	2,149
Korea, DPR	500	500	500	900	1,120	1,270	1,420	1,460	1,460
Korea, Rep.	1,150	1,199	1,184	1,277	1,307	1,325	1,345	1,206	1,163
Kuwait	0	0	1	1	1	2	3	5	5
Kyrgyzstan	Formerly included in Soviet Union							1,077	1,074
Laos	12	13	17	40	115	119	130	155	164
Lebanon	41	61	68	86	86	86	86	105	117
Malaysia	228	236	262	308	320	334	335	340	340
Mongolia	5	5	10	23	35	60	77	84	84
Myanmar (Burma)	536	753	839	976	999	1,085	1,005	1,555	1,556
Nepal	70	86	117	230	520	760	900	1,134	1,135
Oman	20	23	29	34	38	41	58	62	62
Pakistan	10,751	11,472	12,950	13,630	14,680	15,760	16,940	17,200	17,580
Philippines	690	730	826	1,040	1,219	1,440	1,560	1,550	1,550
Qatar	1	1	1	1	3	5	6	13	13

Saudi Arabia	343	353	365	375	600	800	900	1,620	1,620
Sri Lanka	335	341	465	480	525	583	520	570	600
Syria	558	522	451	516	539	652	693	1,089	1,168
Tajikistan	Formerly included in Soviet Union							719	720
Thailand	1,621	1,768	1,960	2,419	3,015	3,822	4,238	4,642	5,010
Turkey	1,310	1,400	1,800	2,200	2,700	3,200	3,800	4,186	4,200
Turkmenistan	Formerly included in Soviet Union							1,750	1,800
United Arab Emirates	30	35	45	50	53	58	63	68	72
Uzbekistan	Formerly included in Soviet Union							4,281	4,281
Vietnam	1,000	980	980	1,000	1,542	1,770	1,840	2,000	2,300
West Bank	10	10	9	8	9	9	10	9	9
Yemen	207	231	260	282	289	302	348	485	485
Europe									
Albania	156	205	284	331	371	399	423	340	340
Austria	4	4	4	4	4	4	4	4	4
Bel-Lux	1	1	1	1	1	1	1	24	35
Belarus	Formerly included in Soviet Union							115	115
Bulgaria	720	945	1,001	1,128	1,197	1,229	1,263	800	800
Bosnia Herzegovinia	Formerly included in Yugoslavia							2	2
Croatia	Formerly included in Yugoslavia							3	3
Czechoslovakia	108	116	126	136	123	187	282		
Czech Republic	Formerly included in Czechoslovakia							24	24
Denmark	40	65	90	180	391	410	430	481	476
Estonia	Formerly included in Soviet Union							4	4
Finland	2	7	16	40	60	62	64	64	64

(continues)

Irrigated Area (thousand hectares)

TABLE 10 *Continued*

Country and Region	1961	1965	1970	1975	1980	1985	1990	1995	1997
France	360	440	539	680	870	1,050	1,300	1,630	1,670
Germany	321	390	419	448	460	470	482	475	475
Greece	430	576	730	875	961	1,099	1,195	1,328	1,385
Hungary	133	100	109	156	134	138	204	210	210
Italy	2,400	2,400	2,400	2,400	2,400	2,425	2,711	2,698	2,698
Latvia	Formerly included in Soviet Union							20	20
Lithuania	Formerly included in Soviet Union							9	9
Macedonia	Formerly included in Yugoslavia							61	55
Malta	1		1	1	1	1	1	1	2
Moldova Rep.	Formerly included in Soviet Union							309	309
Netherlands	290	330	380	430	480	530	555	565	565
Norway	18	25	30	40	74	90	97	127	127
Poland	295	275	213	231	100	100	100	100	100
Portugal	620	621	622	625	630	630	630	632	632
Romania	206	230	731	1,474	2,301	2,956	3,109	3,110	3,089
Russia	Formerly included in Soviet Union							5,362	4,990
Slovakia	Formerly included in Czechoslovakia							217	190
Slovenia	Formerly included in Yugoslavia							2	2
Spain	1,950	2,226	2,379	2,818	3,029	3,217	3,402	3,527	3,603
Sweden	20	22	33	45	70	99	114	115	115
Switzerland	20	23	25	25	25	25	25	25	25
Ukraine	Formerly included in Soviet Union							2,585	2,466
United Kingdom	108	105	88	86	140	152	164	108	108
Yugoslavia	121	118	130	133	145	164	170		
Yugoslav SFR	Formerly included in Yugoslavia							65	65

Former Soviet Union	9,400	9,900	11,100	14,500	17,200	19,689	20,800		
Oceania									
Australia	1,001	1,274	1,476	1,469	1,500	1,700	1,832	2,500	2,700
Fiji	1	1	1	1	1	1	1	3	3
New Zealand	77	93	111	150	183	256	280	285	285
Total Irrigated Area (000 ha)	**138,813**	**149,740**	**167,331**	**187,559**	**209,233**	**223,304**	**242,185**	**260,083**	**267,727**
Rate of Change over Period		0.08	0.12	0.12	0.12	0.07	0.08	0.07	0.03
Average Annual Change (%)		1.97	2.35	2.42	2.31	1.35	1.69	1.48	1.47

Note: Data for the former Soviet Union after 1990 are split among the separate independent states, now included in Asia and Europe. Data from Yugoslavia and Czechoslovakia after 1990 are now split among several independent states.

Table 11. Irrigated Area, by Continent, 1961 to 1997

Description

Total irrigated areas by continental area are listed here for 1961, 1965, 1970, 1975, 1980, 1985, 1990, 1995, and 1997—the latest year for which reliable data are available. These data sum the individual country data in Table 10. Units are thousands of hectares.

Limitations

These data depend on in-country surveys, national reports, and estimates by the Food and Agriculture Organization. In some regions, multiple cropping may increase the apparent area in production. These data are not reported here. No differentiation is made about the quality of the land in production. Recent changes in political borders and the independence of several countries make certain continental time-series comparisons misleading. Data for the Soviet Union, Yugoslavia, and Czechoslovakia are provided in Table 10 through 1990; thereafter the irrigated areas of the newly independent states are reported in their own continental regions. When summing by continental area, therefore, trends will appear misleading because some of the newly independent states are now included in Asia, while others are in Europe. No meaningful time-series trends by continent can thus be seen for these areas. The time-series for Africa, North and Central America, South America, and Oceania do not suffer from this problem.

Source

Food and Agriculture Organization, 1999, Web site at www.fao.org.

TABLE 11 Irrigated Area, by Continent, 1961 to 1997 (thousand hectares)

Country and Region	1961	1965	1970	1975	1980	1985	1990	1995	1997
Africa	7,364	7,747	8,426	8,937	9,407	10,229	11,121	12,254	12,314
North and Central America	17,959	19,535	20,952	22,856	27,697	27,605	29,069	30,152	30,552
South America	4,521	4,892	5,468	6,194	7,079	7,594	8,733	9,841	9,902
Asia	90,166	97,073	109,446	121,165	132,199	140,792	153,623	179,906	187,194
Europe	8,324	9,225	10,351	12,287	13,967	15,438	16,726	25,142	24,777
Former Soviet Union	9,400	9,900	11,100	14,500	17,200	19,689	20,800		
Oceania	1,079	1,368	1,588	1,620	1,684	1,957	2,113	2,788	2,988
Totals	138,813	149,740	167,331	187,559	209,233	223,304	242,185	260,083	267,727

Note: Data for the former Soviet Union after 1990 are split among various newly separate independent states.

Table 12. Human-Induced Soil Degradation, by Type and Cause, Late 1980s

Description

Estimates of the extent of human-induced soil degradation by type and cause for the late 1980s are shown here. Units are millions of hectares. Four categories of degradation are included: light soil degradation implies somewhat reduced productivity (with no quantification of that reduction provided by the authors) "manageable by local farming systems"; moderate soil degradation implies greatly reduced productivity, requiring improvements "often beyond the means of local farmers"; strongly degraded soils are no longer reclaimable at the farm level; extremely degraded soils are considered "unreclaimable and beyond restoration." No data are available for certain categories and regions. Types of soil degradation included here are those caused by water use, wind, application of chemicals, and physical effects. The vast majority of soil degradation is caused by wind and water erosion from bad farming practices, leading to loss of topsoil.

Limitations

Not all regions have been adequately studied, so these data are rough estimates of the overall extent of degradation. The categories are somewhat subjective. No estimates of the economic costs of reclaiming lightly or moderately degraded lands are included, nor are estimates available for the economic losses associated with land degradation. There are significant differences in whether or not these types of degradation can be reversed or prevented. Soil erosion can rarely be reversed, though the rate of loss can be slowed. Salinization can be reversed with careful actions.

Source

Ghassemi, F., A.J. Jakeman, and H.A. Nix, 1995, *Salinisation of Land and Water Resources: Human Causes, Extent, Management and Case Studies*, Center for Resource and Environmental Studies, University of New South Wales Press, Ltd. Sydney, Australia.

TABLE 12 Human-Induced Soil Degradation, by Type and Cause, Late 1980s (million hectares)

Type	Light	Moderate	Strong	Extreme	Total
Water-Induced					
Loss of topsoil	301.2	454.5	161.2	3.8	920.7
Terrain deformation	42.0	72.2	56.0	2.8	173.0
Wind-Induced					
Loss of topsoil	230.5	213.5	9.4	0.9	454.3
Terrain deformation	38.1	30.0	14.4		82.5
Overblowing		10.1	0.5	1.0	11.6
Chemically Induced					
Loss of nutrients	52.4	63.1	19.8		135.3
Salinization	34.8	20.4	20.3	0.8	76.3
Pollution	4.1	17.1	0.5		21.7
Acidification	1.7	2.7	1.3		5.7
Physically Induced					
Compaction	34.8	22.1	11.3		68.2
Waterlogging	6.0	3.7	0.8		10.5
Subsidence of organic soils	3.4	1.0	0.2		4.6
Total (Million hectares)	749.0	910.4	295.7	9.3	1,964.4
Total (Percent)	38.1	46.3	15.1	0.5	100.0

Table 13. Continental Distribution of Human-Induced Salinization

Description

Estimates of the extent of human-induced soil salinization by continent for the late 1980s are shown here. Units are millions of hectares. Four categories of salinization are included: light soil degradation implies somewhat reduced productivity (with no quantification of that reduction provided by the authors) "manageable by local farming systems"; moderate soil degradation implies greatly reduced productivity, requiring improvements "often beyond the means of local farmers"; strongly degraded soils are no longer reclaimable at the farm level; extremely degraded soils are considered "unreclaimable and beyond restoration." No data ("nd") are available for certain categories and regions.

Limitations

Not all regions have been adequately studied, so these data are rough estimates of overall salinization extent. The categories are somewhat subjective. No estimates of the economic costs of reclaiming lightly or moderately degraded lands are included, nor are estimates available for the economic losses associated with land degradation. Differences with Table 12 are associated with rounding errors.

Source

Ghassemi, F., A.J. Jakeman, and H.A. Nix, 1995, *Salinisation of Land and Water Resources: Human Causes, Extent, Management and Case Studies*, Center for Resource and Environmental Studies, University of New South Wales Press, Ltd. Sydney, Australia.

TABLE 13   Continental Distribution of Human-Induced Salinization (million hectares)

Continent	Light	Moderate	Strong	Extreme	Total
Africa	4.7	7.7	2.4	nd	14.8
Asia	26.8	8.5	17.0	0.4	52.7
South America	1.8	0.3	nd	nd	2.1
North and Central America	0.3	1.5	0.5	nd	2.3
Europe	1.0	2.3	0.5	nd	3.8
Oceania	nd	0.5	nd	0.4	0.9
Total	34.6	20.8	20.4	0.8	76.6

Note: nd = no data.

Table 14. Salinization, by Country, Late 1980s

Description

Estimates of the extent of human-induced soil salinization for countries thought to be among the worst affected are shown here for the late 1980s. Total area of cultivated land and the area irrigated are shown in units of millions of hectares. The area affected by salt is shown in the same units. The percentage of irrigated land is also provided. The bottom row shows total global areas, including regions and countries not included in the list above: thus a global total of more than 45 million hectares of irrigated land are thought to be "affected" by salinization. This total includes salt-affected lands only in the world's irrigated areas. Another 31.2 million hectares are estimated to be salinized in nonirrigated areas.

Limitations

Not all regions have been adequately studied, so these data are rough estimates of overall salinization extent. Different lands are affected by different degrees of salinization. The totals here do not discriminate between lands that are lightly affected and those for which little possibility exists of reclamation.

Source

Ghassemi, F., A.J. Jakeman, and H.A. Nix, 1995, *Salinisation of Land and Water Resources: Human Causes, Extent, Management and Case Studies*, Center for Resource and Environmental Studies, University of New South Wales Press, Ltd. Sydney, Australia.

TABLE 14   Salinization, by Country, Late 1980s

Country	Million hectares			Percentage of Irrigated Land Affected
	Cultivated Land Area: Total	Irrigated Area	Area Affected by Salt	
Argentina	35.8	1.5	0.6	38.7
Australia	47.1	1.8	0.2	8.9
China	100.0	48.0	6.7	14.0
Former Soviet Union	232.6	20.5	3.7	18.0
Egypt	2.7	2.7	0.9	32.6
India	169.0	42.1	7.0	16.6
Iran	14.8	5.7	1.7	30.2
Pakistan	20.8	16.1	4.2	26.2
South Africa	13.2	1.1	0.1	9.1
Thailand	20.0	4.0	0.4	10.0
USA	189.9	18.7	4.2	22.2
World Total[a]	1,473.7	227.1	45.4[b]	20.0

[a] The totals includes countries not on this list.

[b] This total represents salt-affected lands only in the world's irrigated areas. Another 31.2 million hectares are salinized in nonirrigated areas.

Table 15. Total Number of Reservoirs, by Continent and Volume

Description

The total number of reservoirs with a storage capacity greater than 0.1 cubic kilometers is listed here by continent, as of the mid-1990s. More than 2,800 such reservoirs are included. The total potential storage capacity of these reservoirs is also included in cubic kilometers.

Limitations

Many reservoirs have been built with volumes under 0.1 cubic kilometers, but these are not included here. No estimate is available of the total volume of water actually stored behind these reservoirs. The actual volume of water stored behind reservoirs is different than the available capacity, depending on the purpose for which a reservoir is operated and on the inflow of water to the system. The data only go up to 1996.

Source

Avakyan, A.B., and V.B. Iakovleva, 1998, "Status of global reservoirs: The position in the late twentieth century," *Lakes and Reservoirs: Research and Management*, Vol. 3, pp. 45–52

TABLE 15   Total Number of Reservoirs, by Continent and Volume

Continent	Number of Reservoirs	Volume of Reservoirs (km^3)
North America	915	1,692.1
Central and South America	265	971.5
Europe	576	645.0
Asia	815	1,980.4
Africa	176	1,000.7
Australia/New Zealand	89	94.8
Total	2,836	6,384.5

Note: Only reservoirs with a total volume of more than 0.1 cubic kilometers are included.

Table 16. Number of Reservoirs Larger Than 0.1 km³, by Continent, Time Series

Description

The total number of reservoirs with a storage capacity greater than 0.1 cubic kilometers is listed here by continent, for different time periods. Data are presented for reservoirs constructed before 1900, from 1901 to 1950, and then by decade to the mid-1990s. More than 2,800 reservoirs are included. Figure 6.3 in Chapter 6 plots the average yearly number of reservoirs built, by region, for each of the time periods listed here.

Limitations

Many reservoirs have been built with volumes under 0.1 cubic kilometers, but these are not included here. The data only go up to 1996.

Source

Avakyan, A.B. and V.B. Iakovleva, 1998, "Status of global reservoirs: The position in the late twentieth century," *Lakes and Reservoirs: Research and Management,* Vol. 3, pp. 45–52

TABLE 16 Number of Reservoirs Larger than 0.1 km³, by Continent, Time Series

	Up to 1900	1901–1950	1951–1960	1961–1970	1971–1980	1981–1990	1990–present	Total
North America	25	342	178	216	113	34	7	915
Central and South America	1	22	30	54	88	51	19	265
Europe	9	104	113	172	94	76	8	576
Asia	5	47	161	215	222	138	27	815
Africa	1	15	21	24	57	52	6	176
Australia/New Zealand		10	21	18	27	12	1	89
Totals	41	540	524	699	601	363	68	2,836
Cumulative Totals	41	581	1,105	1,804	2,405	2,768	2,836	

Table 17. Volume of Reservoirs Larger Than 0.1 km³, by Continent, Time Series

Description

The total volume of reservoirs larger than 0.1 cubic kilometers is shown here, by continent, for different time periods. Data are presented for reservoirs constructed before 1900, from 1901 to 1950, and then by decade to the mid-1990s. Total volume of large reservoirs is estimated to be 6,385 cubic kilometers. Also shown are the average annual additions to volume during these periods. The greatest volumes were added during the 1960s, with large additions in North America and Asia. South America dam construction peaked in the 1970s and 1980s.

Limitations

Many reservoirs have been built with volumes under 0.1 cubic kilometers, but these are not included here. No estimate is available of the total volume of water actually stored behind these reservoirs. The actual volume of water stored behind reservoirs is different than the available capacity, depending on the purpose for which a reservoir is operated and on the inflow of water to the system. The data only go up to 1996.

Source

Avakyan, A.B. and V.B. Iakovleva, 1998, "Status of global reservoirs: The position in the late twentieth century," *Lakes and Reservoirs: Research and Management,* Vol. 3, pp. 45–52

TABLE 17 Volume of Reservoirs Larger than 0.1 km³, by Continent, Time Series (cubic kilometers)

	Up to 1900	1901–1950	1951–1960	1961–1970	1971–1980	1981–1990	1991–1996	Total
North America	8.4	344.7	254.4	534.0	339.0	176.9	34.7	1,692.1
Central and South America	0.3	8.8	28.8	96.9	251.5	349.1	236.1	971.5
Europe	3.3	121.7	175.0	189.4	103.6	49.3	2.7	645.0
Asia	1.7	17.9	293.6	640.0	484.1	321.5	221.6	1,980.4
Africa	0.1	15.0	381.1	364.4	173.7	56.6	9.8	1,000.7
Australia/New Zealand		10.6	20.1	15.5	42.4	5.9	0.3	94.8
Totals	13.8	518.7	1,153.0	1,840.2	1,394.3	959.3	505.2	6,384.5
Average Annual Additions (km³/yr)	10.4		115.3	184.0	139.4	95.9	84.2	

Table 18. Dams Removed or Decommissioned in the United States, 1912 to Present

Description

This list offers the first effort to compile information on dams removed or decommissioned from rivers in the United States since 1912. More than 460 dams are listed here, from all areas of the country. The name of the dam, the river, the state, information on dam height and length (in meters), and reasons for removal are provided, when available. Blank spaces mean no information was available for that item. The data in this table were collected by three nongovernmental environmental organizations (American Rivers [www.amriver.org], Friends of the Earth [www.foe.org], and Trout Unlimited [www.tu.org]) from state dam safety offices, federal agencies, river conservation and fishing organizations, dam owners, media reports, and academic institutions.

When information was available about the reason for a dam's removal, it was included here in one of six broad categories. Many of these categories overlap and few dams are removed for a single reason. The categories are described as follows:

Ecology: dam was removed to restore fish and wildlife habitat; to provide fish passage; to improve water quality.

Economics: maintenance of dam was too costly; removal was cheaper than repair; dam was no longer used; dam was in deteriorating condition.

Failure: dam failed; dam was damaged in flooding.

Recreation: dam was removed to increase recreational opportunities.

Safety: dam was deemed unsafe; owner no longer wanted liability associated with the dam.

Unauthorized dam: dam was built without a needed permit; dam was built improperly.

"NPS" sometimes appears in the "Reasons for Removal" column of the list. The information for dams with this designation was provided by the National Park Service, which documents deactivation of dams on or having an impact on National Park Service lands.

Limitations

Until recently, no effort was made to record dam removals. Little information is available on the history of taking down dams and no comprehensive review of dam removal experiences exists. This list, compiled by three nongovernmental environmental organizations (American Rivers, Friends of the Earth, and Trout Unlimited) is the first effort to pull together records from widely divergent sources. As a result, this list should only be considered preliminary, not comprehensive, and the actual number of dams removed is likely to be much higher. Many smaller dam removals (e.g., less than 2 meters high) are often not documented at all. Many states and federal agencies have not kept thorough records of dams under their jurisdiction that have been removed. No information on dams removed outside of the United States is included, though Chapter 6 discusses some specific international examples.

Readers with information on other dams removed, or with corrections or additions to this list, are encouraged to contact one of the organizations listed above. Updates will regularly be made available at www.worldwater.org and www.amrivers.org.

Source

American Rivers, Friends of the Earth, and Trout Unlimited. 1999. *Dam Removal Success Stories: Restoring Rivers through Selective Removal of Dams that Don't Make Sense.* December. Special thanks to Margaret Bowman and Elizabeth Maclin.

TABLE 18   Dams Removed or Decommissioned in the United States, 1912 to Present

State	River	Project Name	Removed	H (m)	L (m)	Reason
AK	Switzer Creek (trib.)	Switzer One Dam	1988	5		S
AK	Switzer Creek (trib.)	Switzer Two Dam	1988	5		S
AR		Hot Springs Park Ricks Lower #1 Dam	1986	3		NPS
AR		Winton Spring Dam		1		NPS
AR	Coop Creek	Mansfield Dam		6		
AR	Crow Creek	Lake St. Francis Dam	1989	14		U
AZ	Canada del Oro	Golder Dam	1980			S
AZ	Walsh Canyon	Concrete Dam	1982	12		S
AZ	Walsh Canyon	Perrin Dam	1980	10		S
CA		Arco Pond Dam		3		NPS
CA		Bear Valley Dam	1982	5		NPS
CA		C-Line Dam #1	1993	17		NPS
CA		Hagmmaier North Dam		9		NPS
CA		Happy Isles Dam	1987	2		NPS
CA		John Muir #1 Dam				NPS
CA		Lower Murphy Dam		2		NPS
CA		Rogers Dam	1983	12		NPS
CA		Upper Murphy Dam		8		NPS
CA	Beaver Creek	Three C. Picket Dam	1949			
CA	Big Creek	Big Creek Mfg. Dam		4		
CA	Butte Creek	McGowan Dam	1998	2		E
CA	Butte Creek	McPherrin Dam	1998	4		E
CA	Butte Creek	Point Four Dam	1993	2		
CA	Butte Creek	Western Canal East Channel Dam	1998	3		E
CA	Butte Creek	Western Canal Main Dam	1998	3		E
CA	Canyon Creek	Henry Danninbrink Dam	1927			
CA	Canyon Creek	Red Hill Mining Do. Dam	1951	9		
CA	Cold Creek	Lake Christopher Dam	1994	3	122	
CA	Guadalupe River	unnamed small dam #1	1998			
CA	Guadalupe River	unnamed small dam #2	1998			
CA	Hayfork Creek	Hessellwood Dam	1925	3		E
CA	Hayfork Creek	Russell (Hinkley) Dam	1922	3		E
CA	Horse Creek	Big Nugget Mine Dam	1949	4	12	
CA	Indian Creek	D.B. Fields Dam	1947	2		
CA	Indian Creek	D.B. Fields/Johnson Dam	1946			
CA	Indian Creek	Minnie Reeves Dam		6		
CA	Kidder Creek	Altoona Dam	1947	4	18	
CA	Lost Man Creek	Upper Dam	1989	2	17	
CA	Mad River	Sweasey Dam	1970	17		
CA	Monkey Creek	Trout Haven Dam				E
CA	Redding Creek	Clarissa V. Mining Dam	1950	6		
CA	Rock Creek	Rock Creek Dam	1985	4	19	
CA	Rush Creek	Anderline Dam	1936	6		
CA	Salmon River	Bennett-Smith Dam	1950	3		
CA	Salmon River	Bonally Mining Co. Dam	1946	3	54	
CA	Salt Creek	Salt Creek Dam		3		
CA	Scott River	Barton Dam	1950	4	8	
CA	Swillup Creek	Moser Dam	1949			
CA	Trinity River	Lone Jack Dam		7		
CA	Trinity River	North Fork Placers Dam	1950	5		
CA	Trinity River	Quinn Dam	1951	4		

(continues)

TABLE 18 *Continued*

State	River	Project Name	Removed	H (m)	L (m)	Reason
CA	Trinity River	Todd Dam	1949	4		
CA	Trinity River	Trinity Cty. Water & Power Co. Dam	1946	3		
CA	White's Gulch	Smith Dam	1949	2	8	
CA	Wildcat Creek	unnamed dam #1	1992	2		E
CA	Wildcat Creek	unnamed dam #2	1992	2		E
CO		Glacier #1 Dam	1985	3		NPS
CO		No Name #15 Dam		5		NPS
CO		No Name #17 Dam		5		NPS
CO		No Name #21 Dam	1990			NPS
CO		No Name #22 Dam		5		NPS
CO		No Name #8 Dam	1990	4		NPS
CO	Cony Creek	Pear Lake Dam	1988	9		NPS
CO	Ouzel Creek	Blue Bird Dam	1990	17	61	NPS;S
CO	Sand Beach Creek	Sand Beach Dam	1988	8		NPS
CT	Bigelow Creek (trib.)	Little Pond Dam	1994	3		U
CT	Blackwelll Brook (trib.)	Paradise Lake Dam	1991	2		$
CT	Bradley Brook	unnamed dam	1993	3		S
CT	Cedar Swamp Brook	Lower Pond Dam	1991	4		$
CT	Indian River	Indian Lake Dam	1994	4		$
CT	Mad River	John Dee's Dam		5	14	
CT	Mad River (trib.)	Frost Road Pond Dam	1983	2		S
CT	Mill Brook	Sprucedale Water Dam	1980	3		
CT	Muddy Brook	Muddy Pond Dam	1992	2		S
CT	Naugatuck River	Anaconda Dam	1999	3	101	E
CT	Naugatuck River	Freight Street Dam	1999	1	48	E
CT	Naugatuck River	Platts Mill Dam	1999	3	70	E
CT	Naugatuck River	Union City Dam	1999	5	61	E
CT	Qunnipiac River (trib.)	Woodings Pond Dam	1971	5		
CT	Shetucket River	Baltic Mills Dam	1938	8		F
CT	Wharton Brook	Simpson's Pond Dam	1995	2		$
DC	Rock Creek	Ford Dam #3	1991			
DC	Rock Creek	Millrace Dam		5		NPS
FL		Pace's Dike Dam	1991	2		NPS
FL	Chipola River	Dead Lakes Dam	1987	5	240	E
FL	Withlacoochee River	Wysong Dam	1988	1		$
GA	Wahoo Creek	Hamilton Mill Lake Dam				
ID	Clearwater River	Grangeville Dam	1963	17	134	E
ID	Clearwater River	Lewiston Dam	1973	14	323	E
ID	Colburn Creek	Colburn Mill Pond Dam	1999	4	11	E
ID	Dip Creek	Dip Creek Dam				
ID	Elkhorn Gulch	Lane Dam				
ID	Garden Creek	Buster Lake Dam				
ID	John Day Creek (trib.)	Kshmitter Dam	1988			U
ID	Lake Fork Creek	Malony Lake Dam	1986			E
ID	Little Timber Creek	Timber Creek Dam	1970			
ID	Packsaddle Creek	Packsaddle Dam				
ID	Salmon River	Sunbeam Dam	1931			E

TABLE 18 *Continued*

State	River	Project Name	Removed	H (m)	L (m)	Reason
ID	Skein Lake	Skein Lake Dam	1980			
ID	Soldier Creek	Kunkel Dam	1994			
IL		Woodhaven North Impoundment Dam		4		
IL		Woodhaven South Impoundment Dam		3		
IL	Brush Creek (trib.)	Amax Delta Basin 31 Dam		3		
IL	Cypress Ditch (trib.)	Peabody #1A Dam		7		
IL	Cypress Ditch (trib.)	Peabody #5 Dam		13		
IL	Delta Creek	Lake Marion Dam				
IL	Ewing Creek (trib.)	Old Ben Dam		9		
IL	Little Muddy River (trib.)	Consol/Burning Star 5/20 Dam		5		
IL	Mississippi River	Mississippi River Lock & Dam #26		30		
IL	Mississippi River (trib.)	Turkey Bluff Dam		13		
IL	Negro Creek (trib.)	Lake Adelpha Dam		5		
IL	Sangamon River (trib.)	Faries Park Dredge Disposal Dam		9		
IL	Sevenmile Branch	Olsens Lake Dam		5		
IL	Tributary to Sugar Creek	Springfield Dam		8		
IL	Waubonsie Creek	Stone Gate Dam	1999	1	30	E;$;F
IL	Wolf Branch (trib.)	Garden Forest Pond Dam				
IL	Wood River (trib.)	Paradise Lake Dam		6		
IN		Pinhook Dam		5		NPS
KS		Chapman Lake Dam		12		
KS		City of Wellington Dam		11		
KS		Edwin K. Simpson Dam		8		
KS		Kansas Gas & Electric Dam				
KS		Lake Bluestem Dam		21		
KS		Moline Middle City Lake Dam		6		
KS		Mott Dam		6		
KS		Robert Yonally Dam				
KS		Soldier Lake Dam		4		
KS		Wyandotte County Dam		42		
KY	Great Onyx Pond	No Name #1 Dam	1982	2		NPS
KY	Great Onyx Pond	No Name #2 Dam	1982	2		NPS
KY	Little Flat Creek	Sharpsburgh Reservoir Dam	1985	11		
KY	Pond Creek (trib.)	Ebenezer Lake Dam	1985	5		
KY	Pond River	West Fork Pond River #2 Dam		5		
LA	Bayou Dorcheat	Shirley Willis Pond Dam		3		
LA	Bayou Dupont (trib.)	Bayou Dupont #13 Dam		7		
LA	Dry Pong Creek	Kisathie Lake Dam		8		
LA	Pond Branch	Castor Lake Dam		3		
MD	Bacon Ridge Branch	Bacon Ridge Branch Weir	1991			
MD	Deep Run	Deep Run Dam	1989			
MD	Dorsey Run	Railroad Trestle Dam	1994			
MD	Horsepen Branch	Horsepen Branch Dam	1995			
MD	Little Elk Creek	Railroad Bridge at Elkton Dam	1992			
MD	Stony Run	Stony Run Dam	1990			
MD	Western Branch	Route 214 Dam	1998			

(continues)

TABLE 18 *Continued*

State	River	Project Name	Removed	H (m)	L (m)	Reason
ME	Kennebec River	Edwards Dam	1999	7	279	E
ME	Machias River	Canaan Lake Outlet Dam	1999			
ME	Penobscot River	Bangor Dam	1995			
ME	Pleasant River	Brownville Dam	1999	4	91	
ME	Pleasant River	Columbia Falls Dam	1998	3	107	$
ME	Souadabscook Stream	Grist Mill Dam	1998	4	23	E
ME	Souadabscook Stream	Hampden Recreation Area Dam	1999	1		
ME	Souadabscook Stream	Souadabscook Falls Dam	1999		46	
ME	Stetson Stream	Archer's Mill Dam	1999	4	15	
MI		Foster Trout Pond Dam	1983	1		NPS
MI		Three River City Dam	1992	4		
MI	Au Sable River	Salling Dam	1991	5	76	E;R
MI	Dead River	Marquette Dam	1912			
MI	Grand River	Wager Dam	1985	3		E
MI	Looking Glass River	Wacousta Dam	1966	1		
MI	Muskegon River	Newago Dam	1969	5		
MI	Pine River	Stronach Dam	*	5	107	
MI	Silver Lead Creek	Air Force Dam	1998			
MN		Stockton Dam	1994	9		F
MN	Cannon River	Welch Dam	1994	3	37	E;S
MN	Cottonwood River	Flandrau Dam	1995	4		E
MN	Crow River	Berning Mill Dam	1986	3		F
MN	Crow River	Hanover Dam	1984	4		F
MN	Garvin Brook	Stockton Dam				
MN	Kettle River	Sandstone Dam	1995	6	46	E;R
MN	Pomme de Terre River	Pomme de Terre River Dam				
MN	Root River	Lake Florence Dam		4		
MO		Alkire Lake Dam	1990	9		
MO		Goose Creek Lake Dam	1987	16		
MO		Indian Rock Lake Dam	1986	17		S
MT		Three Bears Lake-East Dam		3		NPS
MT	Bear Creek	Three Bears Lake-West Dam		6		NPS
MT	Lone Tree Creek	Vaux #1 Dam	1995	10		S
MT	Lone Tree Creek	Vaux #2 Dam	1995	17		S
MT	Peet Creek	Peet Creek Dam	1994	13		S
MT	Rock Creek	small dam				
MT	Wallace Creek	Wallace Creek Dam	1997	9	219	
NC		Ash Bear Pen Dam	1990	3		NPS;$
NC		Forny Ridge Dam	1988	1		NPS
NC	Little River	Cherry Hospital Dam	1998	2	41	E
NC	Neuse River	Quaker Neck Dam	1998	2	79	E
ND	Knife River	Antelope Creek Dam	1979	7		S
ND	Little Missouri River	Kunick Dam		7		U
ND	Stony Creek	Epping Dam	1979	14		
NE	Bozle Creek	Lake Crawford Dam	1987	8		S
NE	Camp Creek	Diehl Dam	1981	10		$
NE	Cedar River	Fullerton Power Plant Dam		5		F

TABLE 18 *Continued*

State	River	Project Name	Removed	H (m)	L (m)	Reason
NE	Lodgepole Creek	Bennet Dam	1982	6		$
NE	Timber Creek (trib.)	Helen Fehrs Trust Dam	1995	11		U
NE		Golf Course Dam		8		NPS
NJ		Pool Colony Dam		2		NPS
NJ	Cold Brook	Pottersville Dam	1985	6	55	S
NJ	Crooked Brook	Patex Pond Dam	1990	6	104	S
NJ	Delaware River (trib.)	Lake Success Dam	1995	6	91	S
NJ	Raritan River	Fieldsville Dam	1990	3	122	E
NJ	S.B. Timber Creek	Glenside Dam	1997	4	40	S
NJ	Van Camptens Brook	Upper Blue Mountain Dam	1995	8	64	NPS;S
NJ	Whippany River (trib.)	Knox Hill Dam	1996	5	46	S
NM	Pecos River	McMillan Dam	1989	20		S
NM	Sante Fe River	Two Mile Dam	1994	26	219	S
NV		Katherine Borrow Pit Embankment	1992	5		NPS
NY		Curry Pond Dam		1		NPS
NY		Luxton Lake Dam		5		NPS
NY	Hudson River	Fort Edward Dam	1973	9	179	S
OH		Armington Dam #2	1991	5		NPS
OH		Foxtail Dam		9		NPS
OH		Slippery Run (Stahl) Dam	1990	4		NPS
OH	Black Fork (trib.)	Altier Pond Dam	1989	10		
OH	Brannon Fork	Ohio Power Company Pond Dam	1987	5		
OH	Brush Creek (trib.)	Williams Dam		12		
OH	Collins Fork	Ohio Power Company Pond Dam		4		
OH	East Reservoir (trib.)	Wonder Lake Dam	1986	5		
OH	Hamley Run (trib.)	Poston Fresh Water Pond Dam	1988	13		
OH	Hocking River (trib.)	Cottingham Lake Dam	1991	5		
OH	Ice Creek (trib.)	Fair Haven Lake Dam	1980	9		
OH	Jackson Run (trib.)	Howard's Lake Dam				
OH	Johnny Woods River (trib.)	Carr Lake Dam	1985	3		
OH	Licking River (trib.)	Dutiel Pond Dam	1986	4		
OH	Little Auglaize River (trib.)	Burt Lake Dam	1992	5		
OH	Little Darby Creek	Little Darby Dam	1989	6		
OH	Little Darby Creek	Okie Rice Dam	1990	4		S
OH	Little Miami River	Foster Dam	1984			
OH	Little Miami River	Jacoby Road Dam	1997	2	30	E
OH	Little Pine Creek (trib.)	Mastrine Pond Dam	1978	5		
OH	Little Yellow Creek (trib.)	Old Jenkins Lake Dam		7		
OH	McLuney Creek (trib.)	Strip Mine Pond Dam		8		
OH	Modoc Run	Modoc Reservoir Dam	1981	7		
OH	Ogg Creek	Jones Lake Dam		6		
OH	Porter Creek	Marshfield Lake Dam	1973	5		
OH	Robinson Run (trib.)	Lake Hill #2 Dam		9		
OH	Robinson Run (trib.)	Lake Hill Dam #1		9		
OH	Rocky Fork (trib.)	Village at Rocky Fork Lake Dam		2		rebuilt

(continues)

TABLE 18   *Continued*

State	River	Project Name	Removed	H (m)	L (m)	Reason
OH	Seven Mile Creek (trib.)	Ashworth Lake Dam		8		
OH	Silver Creek	Silver Creek Dam				
OH	Silver Creek (trib.)	Chapel Church Lake Dam	1989			
OH	South Fork (trib.)	Georgetoen Freshwater Dam	1988	4		
OH	Spencer Creek (trib.)	State Route 800 Dam	1989	8		
OH	Stillwater Creek	Consol Pond Dam		4		
OH	Sugartree Creek (trib.)	Brashear Lake Dam	1991	5		
OH	Timber Run (trib.)	Derby Petroleum Lake Dam	1984	9		
OH	Town Fork	Toronto Band Father's Lake Dam	1991	5		
OH	Wills Creek (trib.)	Killiany Lake Dam		2		rebuilt
OH	Yankee Run	Yankee Lake Dam	1980	8		
OR	Bear Creek	Jackson Street Dam	1998	3	37	E
OR	Evans Creek	Alphonso Dam	1999	3	17	E
OR	Walla Walla River	Marie Dorian Dam	1997	2	30	E
OR	Willamette River	Catching Dam	1994	9	69	
OR	Yamhill Basin	Lafayette Locks Dam	1963			
PA		Butterfield Pond Dam	1992	4		NPS
PA		Carpenters Pond Dam		5		NPS
PA		Fire Pond Dam at Incline #10		5		NPS
PA		Lake Lettini Dam		2		NPS
PA		Lemon House Pond Dam	1984	5		NPS
PA		Lower Friendship Dam	1982	9		NPS
PA		unnamed dam, Peace Light Inn	1991	2		NPS
PA		Upper Friendship Dam	1982	4		NPS
PA		Van Horn #2 Dam		3		NPS
PA		Van Horn Dam #1	1991	2		NPS
PA		Van Horn Dam #5	1991	4		NPS
PA	Clear Shade Creek	Clear Shade Creek Reservoir Dam	1998	4	58	
PA	Coal Creek	Coal Creek Dam #2	1995	7	35	
PA	Coal Creek	Coal Creek Dam #3	1995	7		
PA	Coal Creek	Coal Creek Dam #4	1995	4	109	
PA	Coal Creek	Diverting Dam		2	17	
PA	Codorus River (trib.)	Yorkane Dam	1997			
PA	Conestoga River	American Paper Products Dam	1998	1	40	E
PA	Conestoga River	Rock Hill Dam	1997	4	91	E
PA	Fishing Creek	Snavely's Mill Dam	1997	1	32	
PA	Gillians Run	Maple Hollow Reservoir Dam	1995	7	59	
PA	Juniata River	Williamsburg Station Dam	1996	4	79	E
PA	Kettle Creek	Rose Hill Intake Dam	1998	4	46	
PA	Kishacoquillas Creek	unnamed dam	1998	3	53	
PA	Laural Run	unnamed dam	1998	2	15	
PA	Lititz Run	Mill Port Conservancy Dam	1998	3	3	E
PA	Lititz Run	unnamed dam	1998	1	3	E
PA	Little Conestoga River	East Petersburg Authority Dam	1998	1	6	E
PA	Little Conestoga River	Maple Grove Dam	1997	2	18	E
PA	Middle Creek	Mussers Dam	1992	9	117	
PA	Mill Creek	Niederriter Farm Pond Dam	1995	6	107	
PA	Mill Creek	Yorktowne Paper Dam	1997	2	18	

TABLE 18 *Continued*

State	River	Project Name	Removed	H (m)	L (m)	Reason
PA	Muddy Creek	Amish Dam		1	12	E
PA	Muddy Creek	Castle Fin Dam	1997	2	117	
PA	Red Run	Red Run Dam	1996	2	12	
PA	Spring Creek	Cabin Hill Dam	1998			
PA	Sugar Creek	Pomeroy Memorial Dam	1996	7	135	
PA	Tinicum Creek (trib.)	unnamed dam	1998	2	12	
RI	Pawtuxet River	Jackson Pond Dam	1979	6		$
SC	Burgess Creek	Gallagher Pond Dam	1989	13		S
SC	Cowpens National Battlefield	unnamed dam, State Road 11-58	1979	2		NPS
SC	Pole Branch River	Pole Branch Dam	1990	8		F
SC	Tools Fork (trib.)	Miller Trust Pond Dam	1993	12		S
SC	Turkey Quarter Creek	Old City Reservoir Dam	1988	8		S
SD		Arikara Dam	1978	12		
SD		Farmingdale Dam	1986	7		
SD		Lake Farley Dam	1980	8		rebuilt
SD		Menno Lake Dam	1984	12		
SD		Mission Dam	1987	8		
SD		Norbeck Dam & SD Highway 87		12		NPS
SD		P6L-Lower Bigger Dam		3		NPS
SD		unnamed dam #26	1987	3		NPS
SD		unnamed dam #30	1987	3		NPS
SD		unnamed dam #32		3		NPS
SD		unnamed dam #35	1987	3		NPS
TN		L. Thompson Dam #1	1990	3		NPS;$
TN		L.C. Hancock #1	1990	2		NPS
TN		L.C. Hancock #3		2		NPS
TN	Adkinson Creek	Gin House Lake Dam	1994	10		$
TN	Burra-Burra Creek	Cities Service Company Dam	1995	9		
TN	Decant Pipes	Monsanto Dam #3	1988	12		
TN	Duck Creek	Occidental Chem Pond Dam A	1995	37		
TN	Duck Creek	Occidental Chem Pond Dam D	1995	49		
TN	Duck River	Monsanto Dam #7	1990	24		
TN	Flat Creek	Sandy Stand Dam	1987	12		$
TN	Flat Creek	Shangri-la Lake Dam	1985	10		S
TN	Fork Creek (trib.)	Ballard Mill Mine Dam	1992	9		
TN	Greenlick Creek	Monsanto Dam #4	1990	16		
TN	Greenlick Creek	Monsanto Dam #5A	1990	16		
TN	Helms Branch	Monsanto Dam #9	1990	10		
TN	Hurricane Creek	Cumberland Springs Dam	1989	9		$
TN	Johnson Creek	Lake Deforest Dam	1991	11		$
TN	Ollis Creek	Eblen-Powell Dam #1		10		$
TN	Quality Creek	Rhone Poulenc Dam #17	1995	10		
TN	Quality Creek	Rhone Poulenc Dam #19	1995	18		
TN	Quality Creek	Rhone Poulenc Dam #20	1995	10		
TN	Rocky Branch	Monsanto Dam #12	1990	38		
TN	Rutherford Creek (trib.)	Occidental Chem Dam #6	1991	16		
TN	Snake Creek (trib.)	Spence Farm Pond Dam #5	1983	11		S
TN	Tipton Branch	Laurel Lake Dam	1990	13		S
TN	Walker Stream	Walkers Dam	1992	10		

(continues)

TABLE 18 *Continued*

State	River	Project Name	Removed	H (m)	L (m)	Reason
TX		Alamo Arroyo Dam	1979	15		
TX		Boot Spring Dam		5		NPS
TX		Duke Dam		2		NPS
TX		H and H Feedlot Dam	1980	11		
TX		Harris Back Lake Dam		5		
TX	Big Sandy Creek (trib.)	Lake Downs Dam		8		
TX	Daves White Branch	Millsap Reservoir Dam		8		
TX	Mill Creek	Barefoot Lake Dam		8		
TX	Mustang Creek (trib.)	Bland Lake Dam	1989	6		
TX	Pecan River (trib.)	Hilsboro Lake Park Dam		6		
TX	Tributary to Willis Creek	Railroad Reservoir Dam		3		
TX	Wasson Branch	Nix Lake Dam		7		
UT		Atlas Mineral Dam	1994	28		
UT		Bell Canyon Dam	1979	9		$
UT	Box Elder Creek	Box Elder Creek Dam	1995	15		S
UT	Muddy Creek	Brush Dam	1983	15		$
VA		Adney Gap Pond Dam	1984	4		NPS
VA		Berryville Reservoir		5		NPS
VA		Fredricksburgh & Spotsylvania Dam #2		2		NPS
VA		Fredricksburgh & Spotsylvania Dam #3		2		NPS
VA		Fredricksburgh & Spotsylvania Dam #5		2		NPS
VA		Fredricksburgh & Spotsylvania Dam #6		1		NPS
VA		Osborne Dam		4		NPS
VA		Sykes Dam	1992	7		NPS
VA	Manassas NP Battlefield	Picnic Area Dam	1984	2		NPS
VT	Batten Kill River	Red Mill Dam				
VT	Charles Brown Brook	Norwich Reservoir Dam		6		S
VT	Clyde River	Newport No. 11 Dam	1996	6	27	E
VT	Mussey Brook	Lower Eddy Pond Dam	1981	6		S
VT	Passumpsic River (trib.)	Lyndon State College Lower Dam				
VT	Wells River	Groton Dam	1998	2		
VT	Winooski River (trib.)	Winooski Water Supply Upper Dam	1983	6		S
VT	Youngs Brook	Youngs Brook Dam	1995	14		S
WA		Black Mud Waste Pond A Dam		5		
WA		Black Mud Waste Pond B Dam		5		
WA		Black Mud Waste Pond C Dam		5		
WA		Bow Lake Reservoir		2		
WA		City Lakes Dam		5		
WA		North End Reservoir		9		
WA		Pomeroy Gulch Dam		12		
WA	Boise Creek	White River Mill Pond Dam		1		
WA	Coffee Creek	Coffee Creek Dam		3		
WA	Columbia River (trib.)	Stromer Lake Dam		2		
WA	Hanford Creek (trib.)	PEO Dam #32A		4		
WA	Hanford Creek (trib.)	PEO Dam #48		1		
WA	Hunters Creek	Hunters Dam		20		
WA	Mill Creek	Mill Creek Settling Basin Dam		5		
WA	Sauk River (trib.)	Darrington Water Works Dam		6		

Table 18 *Continued*

State	River	Project Name	Removed	H (m)	L (m)	Reason
WA	Touchet River	Maiden Dam	1998			
WA	Wagleys Creek	Sultan Mill Pond Dam		5		
WA	Whitestone Creek	Rat Lake Dam	1989	10	73	S
WA	Wind River	Wind River Dam		6		
WI		McNally Trout Pond Dam	1983	2		NPS
WI		Poppe Dam	1982	1		NPS
WI		Rassussen #1 Dam				NPS
WI		Rassussen #2 Dam	1982	1		NPS
WI		Rassussen #3 Dam	1982	1		NPS
WI		Schaaf #1 Dam	1982	1		NPS
WI		Schaaf #2 Dam	1982	1		NPS
WI		unnamed dam #1 (Larrabee Tract)	1990			NPS
WI		Weingarten Dam	1982	1		NPS
WI	Apple River	Huntington Dam	1968			
WI	Apple River	McClure Dam	1968			E
WI	Apple River	Somerset Dam	1965			
WI	Bad River	Mellen Dam	1967			E
WI	Baraboo River	Island Woolen Co. Dam	1972			
WI	Baraboo River	Oak Street Dam	*	4	63	
WI	Baraboo River	Reedsburg Dam	1973	3		
WI	Baraboo River	Waterworks Dam	1998	3	67	$
WI	Baraboo River	Wonewoc Dam	1996	9		
WI	Bark River	Hebron Dam	1996	5	52	
WI	Bark River	Slabtown Dam	1992	3	18	
WI	Beaver Creek	Ettrick Dam	1976			
WI	Black Earth Creek	Black Earth Dam	1957	3		
WI	Black Earth Creek	Cross Plains Dam	1955	3		
WI	Black River	Greenwood Dam	1994	5		
WI	Carpenter Creek	Carpenter Creek Dam	1995			
WI	Cedar Creek	Hamilton Mill Dam	1996	2	30	
WI	Centerville Creek	Centerville Dam	1996	4		
WI	City Creek	Mellen Waterworks Dam	1995	4		
WI	Dunlop Creek	Dunlop Creek Dam	1955			
WI	Eau Galle River	Spring Valley Dam	1997	1		
WI	Eighteen Mile Creek	Colfax Dam	1998	6	107	
WI	Embarrass River	Hayman Falls Dam	1995	5	61	
WI	Embarrass River	Upper Tigerton Dam	1997	3		
WI	Flambeau River	Port Arthur Dam	1968			E
WI	Flume Creek	Northland Dam	1992	3		
WI	Fox River	Wilmot Dam	1992	2	61	
WI	Handsaw Creek	Huigen Dam	1970	2		
WI	Handsaw Creek	Schiek Dam	1970	2		
WI	Iron River	Orienta Falls Dam	*	13		
WI	Kickapoo River	Ontario Dam	1992			E
WI	Kickapoo River	Readstown Dam	1985			
WI	Lemonweir River	Lemonweir Dam	1992	4		
WI	Lowe Creek	Lowe Creek 1 Dam				
WI	Lowe Creek	Lowe Creek 2 Dam				
WI	Madden Branch (trib.)	Beardsley Dam	1990	4		
WI	Manitowoc River	Manitowoc Rapids Dam	1984	5	122	E
WI	Manitowoc River	Oslo Dam	1991	2		E
WI	Marengo River	Marengo Dam	1993	5		E

(continues)

TABLE 18 *Continued*

State	River	Project Name	Removed	H (m)	L (m)	Reason
WI	Maunesha River	Upper Waterloo Dam	1995	5	35	
WI	Milwaukee River	North Avenue Dam	1997	6	132	
WI	Milwaukee River	Woolen Mills Dam	1988	5		$
WI	Milwaukee River	Young America Dam	1994	3		
WI	Oconomowoc River	Funks Dam	1993	2		
WI	Oconto River	Pulcifer Dam	1994	2		
WI	Otter Creek	Klondike Dam	1978	9		
WI	Peshtigo River	Crivitz Dam	1993			
WI	Pine River	Bowen Mill Dam	1996	4		
WI	Pine River	Parfrey Dam	1996	6	137	
WI	Prairie River	Prairie Dells Dam	1991	18		
WI	Prairie River	Prairie Dells Dam	1991	18		E
WI	Prairie River	Ward Paper Mill Dam	1999	5	24	
WI	Rathbone Creek	Evans Pond Dam	1998	3		
WI	Red Cedar River	Colfax Light Power Dam	1969	6		
WI	Sheboygan River	FranklinDam	*	4	41	
WI	Shell Creek	Cartwright Dam	1995	2		
WI	Sugar River	Mount Vernon Dam	1950	3		
WI	Token Creek	Token Creek Dam	*	4		
WI	Tomorrow/Waupaca River	Nelsonville Dam	1988			
WI	Trempealeau River	Whitehall Dam	1988			
WI	Turtle Creek	Shopiere Dam	*	4	42	
WI	Willow River	Mounds Dam	1998	18	131	$
WI	Willow River	Willow Falls Dam	1992	18	49	$
WI	Yahara River	Fulton Dam	1993	5		E
WV		Ladoucer Pond Dam	1993			NPS
WY		East Dam		2		NPS
WY		No Name Dam #1		2		NPS
WY		North Dam		5		NPS
WY		South Dam		2		NPS
WY		West Dam		2		NPS
WY		White Grass Dude Ranch Dam	1988			NPS
WY	City of Sheridan (trib.)	Sheridan Heights Reservoir				$
WY	Laramie River	unnamed dam	1997			

* = Removal in progress $ = Economics   E = Ecology   F = Failure   NPS = National Park Service   R = Recreation
S = Safety   U = Unauthorized dam

Table 19. Desalination Capacity, by Country, January 1999

Description

Desalination provides fresh water from brackish water and seawater for wealthy coastal regions or regions with few other sources of supply. This table presents total installed capacity of desalination plants, by country, for land-based desalting plants rated at more than 500 cubic meters per day and delivered or contracted as of January 1999. Almost 75 percent of total desalination capacity is in the 10 countries leading the list; 6 of these countries are in the Middle East and North Africa.

Limitations

These data were collected from a wide range of sources, from desalting plant suppliers to plant operators, and therefore depend on the accuracy of the information supplied. Plants with capacities less than 500 cubic meters per day are not included. The comprehensive list is known to include plants that are no longer operating, but no estimate of this capacity exists.

Source

Wangnick, K, 1998, *1998 IDA Worldwide Desalting Plants Inventory, Report No. 15*, Wangnick Consulting, Gnarrenburg, Germany. With permission. Also, personal communication, K. Wangnick, 1999.

TABLE 19 Desalination Capacity, by Country, January 1999 (Plants rated at 500 cubic meters per day or more)

Country	Total Capacity (m³/day)	Country	Total Capacity (m³/day)
Saudi Arabia	5,106,742	Philippines	20,854
United States	3,234,042	Cayman Islands	20,621
United Arab Emirates	2,184,968	Austria	20,620
Kuwait	1,285,527	Brazil	20,399
Spain	797,511	Gibralter	20,079
Japan	777,838	Morocco	19,700
Libya	703,027	Cuba	18,926
Qatar	567,414	Argentina	18,889
Italy	521,298	Lebanon	17,083
Iran	437,771	Maldives	16,940
Bahrain	419,155	Malaysia	13,699
India	342,219	Bermudas	13,171
Korea	341,769	Azerbaijan	12,680
Iraq	324,476	Belarus	12,640
Netherlands Antilles	230,273	Cape Verde	10,500
Germany	223,719	French Antigua	10,400
Algeria	190,837	Ireland	10,312
Hong Kong	183,079	Colombia	8,765
China	181,983	Portugal	8,320
Oman	181,621	Jordan	8,231
Kazakhstan	167,379	Turkey	7,320
Virgin Islands (Combined)	156,437	Switzerland	7,306
Great Britain	150,887	Sahara	7,002
Malta	146,331	Jamaica	6,094
Egypt	139,611	Nigeria	6,000
Singapore	137,123	Denmark	5,960
Indonesia	136,463	Syria	5,488
Taiwan	134,945	Pakistan	5,310
Mexico	131,470	Mauritania	4,440
Holland	128,319	Ecuador	4,433
Russia	116,920	Belgium	3,900
Israel	90,478	Marshall Island	2,650
South Africa	85,591	Yugoslavia	2,204
Australia	84,749	Sweden	2,020
Tunisia	72,002	Belize	1,617
Cyprus	68,075	Turks and Caicos	1,540
Bahamas	53,800	Sudan	1,450
Canada	46,629	Ascension	1,362
Turmenistan	43,707	Bulgaria	1,320
Chile	40,132	Norway	1,200
France	39,664	Nauru Pacific	1,136
Greece	39,220	Dominican Republic	1,135
Yemen	36,996	Namibia	1,090
Uzbekistan	31,200	Paraguay	1,000
Czech Republic	30,445	Trinidad and Tobago	900
Antigua	28,533	Honduras	651
Peru	24,538	Finland	600
Thailand	24,075	Hungary	500
Poland	23,594	No country specified	28,643
Venezuela	22,129		
Ukraine	21,000	Total Capacity (m³/day)	21,104,811

Table 20. Total Desalination Capacity, by Process, June 1999

Description

Several processes are used for desalinating water. These processes, and their installed capacity, are listed here. Both multistage-flash distillation and reverse osmosis continue to dominate the desalination field. Together these two processes make up 84 percent of total desalination capacity. The data include land-based desalting plants rated at more than 500 cubic meters per day and in operation or contracted as of June 1999.

Limitations

These data were collected from a wide range of sources, from desalting plant suppliers to plant operators, and therefore depend on the accuracy of the information supplied. Plants with capacities less than 500 cubic meters per day are not included. The comprehensive list is known to include plants that are no longer operating, but no estimate of this capacity exists. For example, no freeze desalination plants remain in operation. Slight differences in the total capacity between this table and Table 21 are the result of slight differences in the data included in each.

Source

Wangnick, K, 1999. Personal communication. Gnarrenburg, Germany. With permission.

TABLE 20   Total Desalination Capacity, by Process, June 1999

Process	Capacity (m³/d)	Percentage
Multistage flash distillation (MSF)	10,020,672	44.05
Reverse osmosis (RO)	9,000,939	39.57
Electrodialysis (ED)	1,262,929	5.55
Vapor compression (VC)	972,792	4.28
Multi-effect distillation (ME)	917,727	4.03
Membrane softening (MS)	449,454	1.98
Other	70,480	0.31
Hybrid	50,714	0.22
Freezing	210	0.00
Total	22,745,917	

Table 21. Desalination Capacity, by Source of Water, June 1999

Description

Desalination technology can be used on water of different qualities, depending on the methods and water needs. This table shows the principal source of water for all desalination capacity in operation or contracted as of June 1999. More than 85 percent of all water desalinated is either brackish water or seawater, with nearly 60 percent of total desalination capacity used to desalt seawater. "River" water is water with small amounts of salts. "Pure" water refers to high-quality waters that need additional purification for specialty purposes. "Waste" and "brine" refer to waters with high concentrations of impurities or salts.

Limitations

These data were collected from a wide range of sources, from desalting plant suppliers to plant operators, and therefore depend on the accuracy of the information supplied. Plants with capacities less than 500 cubic meters per day are not included. The comprehensive list is known to include plants that are no longer operating, but no estimate of this capacity exists. A small number of plants can operate on multiple water sources. Slight differences in the total capacity between this table and Table 20 are the result of slight differences in the data included in each.

Source

Wangnick, K, 1999. Personal communication. Gnarrenburg, Germany. With permission.

TABLE 21 Desalination Capacity, by Source of Water, June 1999

Source Water	Capacity (m³/d)	Percentage
Seawater	13,349,246	58.82
Brackish	5,827,072	25.67
River	1,360,716	6.00
Pure	1,112,214	4.90
Waste	994,492	4.38
Brine	52,328	0.23
Total	22,696,068	

Table 22. Number of Threatened Species, by Country/Area, by Group, 1997

Description

Human manipulation of aquatic systems can have adverse effects on aquatic species. Dams, irrigation systems, and other major engineering projects disrupt habitat and impinge on feeding and breeding habits. Changes in water quality, including pH and temperature, also impact species viability, as do fishing and the introduction of exotic species. Mammals, reptiles, amphibians, birds, invertebrates, and fish species known to be threatened are shown here by country or contiguous area. Species included are those that are "critically endangered," "endangered," or "vulnerable" according to the World Conservation Union's definitions. Critically endangered species face an "extremely high risk of extinction in the wild in the immediate future," endangered species face a "very high risk of extinction in the near future," and vulnerable species face a "high risk of extinction in the wild in the medium-term future."

Limitations

Listings of threatened species are not consistent because of problems with definitions, lack of adequate research, and inadequate reporting. While some countries may have no endangered species, no data are available for many countries. Both of these situations are marked with a zero ("0"). Differences in the number of species listed from country to country often reflect differences in funding and research, rather than the true state of species well-being.

Source

World Conservation Monitoring Centre, 1999, World Conservation Monitoring Centre searchable databases at http://www.wcmc.org.uk/species/animals/table3.html.

TABLE 22   Number of Threatened Species, by Country/Area, by Group, 1997

Country/Area	Mammals	Birds	Reptiles	Amphibians	Fishes	Invertebrates
Afghanistan	11	13	1	1	0	1
Albania	2	7	1	0	7	3
Algeria	15	8	1	0	1	11
American Samoa	2	2	2	0	0	4
Andorra	0	0	0	0	0	2
Angola	17	13	5	0	0	6
Anguilla	0	0	5	0	0	0
Antigua and Barbuda	0	1	5	0	0	0
Argentina	27	41	5	5	1	11
Armenia	4	5	3	0	0	6
Aruba	1	1	2	0	0	1
Australia	58	45	37	25	37	281
Austria	7	5	1	0	7	41
Azerbaijan	11	8	3	0	5	6
Bahamas	4	4	7	0	1	1
Bahrain	1	1	0	0	0	0
Bangladesh	18	30	13	0	0	0
Barbados	0	1	2	0	0	0
Belarus	4	4	0	0	0	6
Belgium	6	3	0	0	1	13
Belize	5	1	5	0	4	1
Benin	9	1	2	0	0	0
Bermuda	0	2	1	0	0	25
Bhutan	20	14	1	0	0	1
Bolivia	24	27	3	0	0	1
Bosnia and Herzegovina	10	2	0	1	6	6
Botswana	5	7	0	0	0	0
Brazil	71	103	15	5	12	34
British Indian Ocean Territory	0	0	2	0	0	0
Brunei	9	14	4	0	2	0
Bulgaria	13	12	1	0	8	7
Burkina Faso	6	1	1	0	0	0
Burundi	5	6	0	0	0	3
Cambodia	23	18	9	0	5	0
Cameroon	32	14	3	1	26	4
Canada	7	5	3	1	13	11
Cape Verde	1	3	3	0	1	0
Cayman Islands	0	1	2	0	0	1
Central African Republic	11	2	1	0	0	0
Chad	14	3	1	0	0	1
Chile	16	18	1	3	4	0

TABLE 22 *Continued*

Country/Area	Mammals	Birds	Reptiles	Amphibians	Fishes	Invertebrates
China	75	90	15	1	28	4
Christmas Island	0	3	2	0	0	0
Cocos (Keeling) Islands	0	3	0	0	0	0
Colombia	35	64	15	0	5	0
Comoros	3	6	2	0	1	4
Congo	10	3	2	0	0	1
Cook Islands	0	6	2	0	0	0
Costa Rica	14	13	7	1	0	9
Cote d'Ivoire	16	12	4	1	0	1
Croatia	10	4	0	1	20	8
Cuba	9	13	7	0	4	3
Cyprus	3	4	3	0	0	1
Czech Republic	7	6	0	0	6	17
Denmark	3	2	0	0	0	10
Djibouti	3	3	2	0	0	0
Dominica	1	2	4	0	0	0
Dominican Republic	4	11	10	1	0	2
Ecuador	28	53	12	0	1	23
Egypt	15	11	6	0	0	1
El Salvador	2	0	6	0	0	1
Equatorial Guinea	12	4	2	1	0	2
Eritrea	6	3	3	0	0	0
Estonia	4	2	0	0	1	3
Ethiopia	35	20	1	0	0	4
Faeroe Islands	1	0	0	0	0	0
Falkland Islands	0	1	0	0	0	0
Federated States of Micronesia	6	6	2	0	0	4
Fiji	4	9	6	1	0	2
Finland	4	4	0	0	1	8
France	13	7	3	2	3	61
French Guiana	9	1	8	0	0	0
French Polynesia	0	22	2	0	3	29
French Southern Territories	0	3	0	0	0	0
Gabon	12	4	3	0	0	1
Gambia	4	1	1	0	0	0
Georgia	10	5	7	0	3	9
Germany	8	5	0	0	7	29
Ghana	13	10	4	0	0	0
Gibraltar	1	1	0	0	0	2
Greece	13	10	6	1	16	9
Greenland	2	0	0	0	0	0
Grenada	0	1	4	0	0	0

(continues)

TABLE 22   *Continued*

Country/Area	Mammals	Birds	Reptiles	Amphibians	Fishes	Invertebrates
Guadeloupe	5	0	6	0	0	1
Guam	2	3	2	0	1	6
Guatemala	8	4	9	0	0	8
Guinea	11	12	3	1	0	3
Guinea-Bissau	4	1	3	0	0	1
Guyana	10	3	8	0	0	1
Haiti	4	11	6	1	0	2
Honduras	7	4	7	0	0	2
Hong Kong	0	14	1	0	0	1
Hungary	8	10	1	0	11	26
Iceland	1	0	0	0	0	0
India	75	73	16	3	4	22
Indonesia	128	104	19	0	60	29
Iran	20	14	8	2	7	3
Iraq	7	12	2	0	2	2
Ireland	2	1	0	0	1	2
Israel	13	8	5	0	0	10
Italy	10	7	4	4	9	41
Jamaica	4	7	8	4	0	5
Japan	29	33	8	10	7	45
Jordan	7	4	1	0	0	3
Kazakhstan	15	15	1	1	5	4
Kenya	43	24	5	0	20	15
Kiribati	0	4	2	0	0	1
Kuwait	1	3	2	0	0	0
Kyrgyzstan	6	5	1	0	0	3
Laos	30	27	7	0	4	0
Latvia	4	6	0	0	1	6
Lebanon	5	5	2	0	0	1
Lesotho	2	5	0	0	1	1
Liberia	11	13	3	1	0	2
Libya	11	2	3	0	0	0
Liechtenstein	0	1	0	0	0	4
Lithuania	5	4	0	0	1	5
Luxembourg	3	1	0	0	0	4
Macedonia	10	3	1	0	4	2
Madagascar	46	28	17	2	13	14
Malawi	7	9	0	0	0	8
Malaysia	42	34	14	0	14	3
Maldives	0	1	2	0	0	0
Mali	13	6	1	0	0	0
Malta	0	2	0	0	0	3
Marshall Islands	0	1	2	0	0	1
Martinique	0	2	5	0	0	1
Mauritania	14	3	3	0	0	0

TABLE 22 *Continued*

Country/Area	Mammals	Birds	Reptiles	Amphibians	Fishes	Invertebrates
Mauritius	4	10	6	0	0	32
Mayotte	0	2	2	0	0	1
Mexico	64	36	18	3	86	40
Moldova	2	7	1	0	9	5
Monaco	0	0	0	0	0	0
Mongolia	12	14	0	0	0	3
Montserrat	1	0	5	0	0	0
Morocco	18	11	2	0	1	7
Mozambique	13	14	5	0	2	7
Myanmar	31	44	20	0	1	2
Namibia	11	8	3	1	3	1
Nauru	0	2	0	0	0	0
Nepal	28	27	5	0	0	1
Netherlands	6	3	0	0	1	9
Netherlands Antilles	1	1	6	0	0	0
New Caledonia	5	10	3	0	0	11
New Zealand	3	44	11	1	8	15
Nicaragua	4	3	7	0	0	2
Niger	11	2	1	0	0	1
Nigeria	26	9	4	0	0	1
Niue	0	1	1	0	0	0
Norfolk Island	0	8	2	0	0	12
North Korea	7	19	0	0	0	1
Northern Marianas	1	7	2	0	0	3
Norway	4	3	0	0	1	8
Oman	9	5	4	0	3	1
Pakistan	13	25	6	0	1	0
Palau	3	2	2	0	0	4
Panama	17	10	7	0	1	2
Papua New Guinea	57	31	10	0	13	11
Paraguay	10	26	3	0	0	0
Peru	46	64	9	1	0	2
Philippines	49	86	7	2	26	18
Pitcairn Islands	0	5	0	0	0	5
Poland	10	6	0	0	2	13
Portugal	13	7	0	1	9	67
Puerto Rico	3	11	9	3	0	1
Qatar	0	1	2	0	0	0
Reunion	2	3	4	0	0	17
Romania	16	11	2	0	11	21
Russia	31	38	5	0	13	26
Rwanda	9	6	0	0	0	2
Sao Tome and Principe	3	9	2	0	0	2
Saudi Arabia	9	11	2	0	0	1
Senegal	13	6	7	0	0	0

(continues)

TABLE 22   *Continued*

Country/Area	Mammals	Birds	Reptiles	Amphibians	Fishes	Invertebrates
Seychelles	2	9	4	4	0	3
Sierra Leone	9	12	3	0	0	4
Singapore	6	9	1	0	1	1
Slovakia	8	4	0	0	7	20
Slovenia	10	3	0	1	5	38
Solomon Islands	20	18	4	0	0	5
Somalia	18	8	2	0	3	1
South Africa	33	16	19	9	27	101
South Georgia	0	1	0	0	0	0
South Korea	6	19	0	0	0	1
Spain	19	10	6	3	10	57
Sri Lanka	14	11	8	0	8	2
St. Helena	0	9	1	0	7	2
St. Kitts and Nevis	0	1	5	0	0	0
St. Lucia	0	3	6	0	0	0
St. Vincent	1	2	4	0	0	0
Sudan	21	9	3	0	0	1
Suriname	10	2	6	0	0	0
Swaziland	5	6	0	0	0	0
Sweden	5	4	0	0	1	13
Switzerland	6	4	0	0	4	25
Syria	4	7	3	0	0	3
Taiwan	10	13	3	0	6	1
Tajikistan	5	9	1	0	1	2
Tanzania	33	30	4	0	19	46
Thailand	34	45	16	0	14	1
Togo	8	1	3	0	0	0
Tokelau	0	1	2	0	0	0
Tonga	0	2	3	0	0	1
Trinidad and Tobago	1	3	5	0	0	0
Tunisia	11	6	2	0	0	5
Turkey	15	14	12	2	18	9
Turkmenistan	11	12	2	0	5	3
Turks and Caicos Islands	0	3	4	0	0	0
Tuvalu	0	1	2	0	0	1
Uganda	18	10	1	0	28	10
Ukraine	15	10	2	0	12	13
United Arab Emirates	3	4	2	0	1	0
United Kingdom	4	2	0	0	1	10
United States	35	50	28	24	123	594
Uruguay	5	11	0	0	0	1
U.S. Minor Pacific Islands	0	1	2	0	0	0
Uzbekistan	7	11	0	0	3	1

TABLE 22 *Continued*

Country/Area	Mammals	Birds	Reptiles	Amphibians	Fishes	Invertebrates
Vanuatu	3	6	3	0	0	1
Venezuela	24	22	14	0	5	1
Viet Nam	38	47	12	1	3	0
Virgin Islands (British)	1	2	6	1	0	0
Virgin Islands (U.S.)	0	2	5	0	0	0
Wallis and Futuna Islands	0	0	0	0	0	0
Western Sahara	7	1	2	0	0	0
Western Samoa	2	6	2	0	0	1
Yemen	5	13	2	0	0	2
Yugoslavia	12	8	1	0	13	19
Zaire	38	26	3	0	1	45
Zambia	11	10	0	0	0	6
Zimbabwe	9	9	0	0	0	2

Note: Data from the World Conservation Monitoring Centre, United Kingdom Document URL: http://www.wcmc.org.uk/species/animals/table3.html Includes Critically Endangered, Endangered, and Vulnerable Species.

Table 23. Countries with the Largest Number of Fish Species

Description

This table shows those countries with the largest number of fish species and the relative density in fish species per square kilometer of land area. Many large countries have substantial numbers of fish species, but the countries with the largest number of species per unit area may be smaller countries with moderate overall numbers of species (see next Table). Many of these countries are dependent on hydroelectricity for a substantial fraction of total electricity. Threats to biodiversity from dam development may be particularly severe in these regions.

Limitations

A listing of the countries with the largest number of fish species is only a simple measure of biodiversity, not an accurate measure of threats to those species.

Sources

Sea Wind, 1997, *Bulletin of Ocean Voice International*, Vol. 11, No. 3 (July–September), p. 32.

World Wide Fund for Nature, 1999, *A Place for Dams in the Twenty-first Century?* Gland, Switzerland.

TABLE 23   Countries with the Largest Number of Fish Species

Country	Number of Fish Species	Species per 1000 km^2
Brazil	3,000	0.355
Indonesia	1,300	0.718
China	1,010	0.108
Dem. Republic of Congo	962	0.424
Peru	855	0.668
United States	779	0.085
India	748	0.252
Thailand	690	1.351
Tanzania	682	0.770
Malaysia	600	1.826
Venezuela	512	0.580
Vietnam	450	1.383

Table 24. Countries with the Largest Number of Fish Species per Unit Area

Description

This table shows those countries with the greatest density of fish species based on the number of species per square kilometer of land area. Many large countries have substantial numbers of fish species, but the countries with the largest number of species per unit area may be smaller countries with moderate overall numbers of species (see previous table). Many of these countries are dependent on hydroelectricity for a substantial fraction of total electricity. Threats to biodiversity from dam development may be particularly severe in countries like Malaysia, Thailand, and Vietnam, which appear on both lists.

Limitations

A high density of fish species is only one measure of the potential threats to biodiversity. Others may be as or more important.

Sources

Sea Wind, 1997, *Bulletin of Ocean Voice International*, Vol. 11 No. 3, (July–September) p. 32.

World Wide Fund for Nature, 1999, *A Place for Dams in the Twenty-first Century?* Gland, Switzerland.

TABLE 24   Countries with the Largest Number of Fish Species per Unit Area

Country	Number of Fish Species	Species per 1000 km^2
Burundi	209	8.148
Malawi	361	3.837
Bangladesh	260	1.997
Malaysia	600	1.826
Sierra Leone	117	1.634
Cambodia	260	1.473
Vietnam	450	1.383
Thailand	690	1.351
Uganda	247	1.238
Laos PDR	262	1.135
Philippines	330	1.107
Ghana	224	0.984

Table 25. Water Units, Data Conversions, and Constants

Water experts, managers, scientists, and educators work with a bewildering array of different units and data. These vary with the field of work: engineers may use different water units than hydrologists; urban water agencies may use different units than reservoir operators; academics may use different units than water managers. But they also vary with regions: water agencies in England may use different units than water agencies in France or Africa; hydrologists in the eastern United States often use different units than hydrologists in the western United States And they vary over time: today's water agency in California may sell water by the acre-foot, but its predecessor a century ago may have sold miner's inches or some other now-arcane measure.

These differences are of more than academic interest. Unless a common "language" is used, or a dictionary of translations is available, errors can be made or misunderstandings can ensue. In some disciplines, unit errors can be more than embarrassing; they can be expensive, or deadly. In September 1999, the $125 million Mars Climate Orbiter spacecraft was sent crashing into the face of Mars instead of into its proper safe orbit above the surface because one of the computer programs controlling a portion of the navigational analysis used English units incompatible with the metric units used in all the other systems. The failure to translate English units into metric units was described in the findings of the preliminary investigation as the principal cause of mission failure.

This table is a comprehensive list of water units, data conversions, and constants related to water volumes, flows, pressures, and much more. Most of these units and conversions were compiled by Kent Anderson and initially published in P.H. Gleick, 1993, *Water in Crisis: A Guide to the World's Fresh Water Resources,* Oxford University Press, New York.

TABLE 25 Water Units, Data Conversions, and Constants

Prefix (Metric)	Abbreviation	Multiple	Prefix (Metric)	Abbreviation	Multiple
deka-	da	10	deci-	d	0.1
hecto-	h	100	centi-	c	0.01
kilo-	k	1000	milli-	m	0.001
mega-	M	10^6	micro-	µ	10^{-6}
giga-	G	10^9	nano-	n	10^{-9}
tera-	T	10^{12}	pico-	p	10^{-12}
peta-	P	10^{15}	femto-	f	10^{-15}
exa-	E	10^{18}	atto-	a	10^{-18}

LENGTH (L)

1 micron (µ)	$= 1 \times 10^{-3}$ mm	10 hectometers	= 1 kilometer
	$= 1 \times 10^{-6}$ m	1 mil	= 0.0254 mm
	$= 3.3937 \times 10^{-5}$ in		$= 1 \times 10^{-3}$ in
1 millimeter (mm)	= 0.1 cm	1 inch (in)	= 25.4 mm
	$= 1 \times 10^{-3}$ m		= 2.54 cm
	= 0.03937 in		= 0.08333 ft
1 centimeter (cm)	= 10 mm		= 0.0278 yd
	= 0.01 m	1 foot (ft)	= 30.48 cm
	$= 1 \times 10^{-5}$ km		= 0.3048 m
	= 0.3937 in		$= 3.048 \times 10^{-4}$ km
	= 0.03281 ft		= 12 in
	= 0.01094 yd		= 0.3333 yd
1 meter (m)	= 1000 mm		$= 1.89 \times 10^{-4}$ mi
	= 100 cm	1 yard (yd)	= 91.44 cm
	$= 1 \times 10^{-3}$ km		= 0.9144 m
	= 39.37 in		$= 9.144 \times 10^{-4}$ km
	= 3.281 ft		= 36 in
	= 1.094 yd		= 3 ft
	$= 6.21 \times 10^{-4}$ mi		$= 5.68 \times 10^{-4}$ mi
1 kilometer (km)	$= 1 \times 10^5$ cm	1 mile (mi)	= 1609.3 m
	= 1000 m		= 1.609 km
	= 3280.8 ft		= 5280 ft
	= 1093.6 yd		= 1760 yd
	= 0.621 mi	1 fathom (nautical)	= 6 ft
10 millimeters	= 1 centimeter	1 league (nautical)	= 5.556 km
10 centimeters	= 1 decimeter		= 3 nautical miles
10 decimeters (dm)	= 1 meter	1 league (land)	= 4.828 km
			= 5280 yd
10 meters	= 1 dekameter		= 3 mi
10 dekameters (dam)	= 1 hectometer	1 international nautical mile	= 1.852 km
			= 6076.1 ft
			= 1.151 mi

(continues)

TABLE 25 *Continued*

AREA (L²)

1 square centimeter (cm²)	$= 1 \times 10^{-4}\ m^2$	1 square foot (ft²)	$= 929.0\ cm^2$
	$= 0.1550\ in^2$		$= 0.0929\ m^2$
	$= 1.076 \times 10^{-3}\ ft^2$		$= 144\ in^2$
	$= 1.196 \times 10^{-4}\ yd^2$		$= 0.1111\ yd^2$
1 square meter (m²)	$= 1 \times 10^{-4}\ hectare$		$= 2.296 \times 10^{-5}\ acre$
	$= 1 \times 10^{-6}\ km^2$		$= 3.587 \times 10^{-8}\ mi^2$
	$= 1\ centare$ (French)	1 square yard (yd²)	$= 0.8361\ m^2$
	$= 0.01\ are$		$= 8.361 \times 10^{-5}$ hectare
	$= 1550.0\ in^2$		$= 1296\ in^2$
	$= 10.76\ ft^2$		$= 9\ ft^2$
	$= 1.196\ yd^2$		$= 2.066 \times 10^{-4}\ acres$
	$= 2.471 \times 10^{-4}\ acre$		$= 3.228 \times 10^{-7}\ mi^2$
1 are	$= 100\ m^2$	1 acre	$= 4046.9\ m^2$
1 hectare (ha)	$= 1 \times 10^4\ m^2$		$= 0.40469\ ha$
	$= 100\ are$		$= 4.0469 \times 10^{-3}\ km^2$
	$= 0.01\ km^2$		$= 43,560\ ft^2$
	$= 1.076 \times 10^5\ ft^2$		$= 4840\ yd^2$
	$= 1.196 \times 10^4\ yd^2$		$= 1.5625 \times 10^{-3}\ mi^2$
	$= 2.471\ acres$	1 square mile (mi²)	$= 2.590 \times 10^6\ m^2$
	$= 3.861 \times 10^{-3}\ mi^2$		$= 259.0\ hectares$
1 square kilometer (km²)	$= 1 \times 10^6\ m^2$		$= 2.590\ km^2$
	$= 100\ hectares$		$= 2.788 \times 10^7\ ft^2$
	$= 1.076 \times 10^7\ ft^2$		$= 3.098 \times 10^6\ yd^2$
	$= 1.196 \times 10^6\ yd^2$		$= 640\ acres$
	$= 247.1\ acres$		$= 1\ section$ (of land)
	$= 0.3861\ mi^2$		
1 square inch (in²)	$= 6.452\ cm^2$	1 feddan (Egyptian)	$= 4200\ m^2$
	$= 6.452 \times 10^{-4}\ m^2$		$= 0.42\ ha$
	$= 6.944 \times 10^{-3}\ ft^2$		$= 1.038\ acres$
	$= 7.716 \times 10^{-4}\ yd^2$		

TABLE 25 *Continued*

VOLUME (L³)

1 cubic centimeter (cm³)	= 1×10^{-3} liter = 1×10^{-6} m³ = 0.06102 in³ = 2.642×10^{-4} gal = 3.531×10^{-3} ft³	1 cubic foot (ft³)	= 2.832×10^4 cm³ = 28.32 liters = 0.02832 m³ = 1728 in³ = 7.481 gal = 0.03704 yd³
1 liter (l)	= 1000 cm³ = 1×10^{-3} m³ = 61.02 in³ = 0.2642 gal = 0.03531 ft³	1 cubic yard (yd³)	= 0.7646 m³ = 6.198×10^{-4} acre-ft = 46656 in³ = 27 ft³
1 cubic meter (m³)	= 1×10^6 cm³ = 1000 liter = 1×10^{-9} km³ = 264.2 gal = 35.31 ft³ = 6.29 bbl = 1.3078 yd³ = 8.107×10^{-4} acre-ft	1 acre-foot (acre-ft or AF)	= 1233.48 m³ = 3.259×10^5 gal = 43560 ft³
		1 Imperial gallon	= 4.546 liters = 277.4 in³ = 1.201 gal = 0.16055 ft³
1 cubic decameter (dam³)	= 1000 m³ = 1×10^6 liter = 1×10^{-6} km³ = 2.642×10^5 gal = 3.531×10^4 ft³ = 1.3078×10^3 yd³ = 0.8107 acre-ft	1 cfs-day	= 1.98 acre-feet = 0.0372 in-mi²
		1 inch-mi²	= 1.738×10^7 gal = 2.323×10^6 ft³ = 53.3 acre-ft = 26.9 cfs-days
1 cubic hectometer (ha³)	= 1×10^6 m³ = 1×10^3 dam³ = 1×10^9 liter = 2.642×10^8 gal = 3.531×10^7 ft³ = 1.3078×10^6 yd³ = 810.7 acre-ft	1 barrel (of oil) (bbl)	= 159 liter = 0.159 m³ = 42 gal = 5.6 ft³
		1 million gallons	= 3.069 acre-ft
1 cubic kilometer (km³)	= 1×10^{12} liter = 1×10^9 m³ = 1×10^6 dam³ = 1000 ha³ = 8.107×10^5 acre-ft = 0.24 mi³	1 pint (pt)	= 0.473 liter = 28.875 in³ = 0.5 qt = 16 fluid ounces = 32 tablespoons = 96 teaspoons
		1 quart (qt)	= 0.946 liter = 57.75 in³ = 2 pt = 0.25 gal
1 cubic inch (in³)	= 16.39 cm³ = 0.01639 liter = 4.329×10^{-3} gal = 5.787×10^{-4} ft²	1 morgen-foot (S. Africa)	= 2610.7 m³
		1 board-foot	= 2359.8 cm³ = 144 in³ = 0.0833 ft³
1 gallon (gal)	= 3.785 liters = 3.785×10^{-3} m³ = 231 in³ = 0.1337 ft³ = 4.951×10^{-3} yd³	1 cord	= 128 ft³ = 0.453 m³

(continues)

T A B L E 25   *Continued*

VOLUME/AREA (L³/L²)

| 1 inch of rain | = 5.610 gal/yd² | 1 box of rain | = 3,154.0 lesh |
| | = 2.715×10^4 gal/acre | | |

MASS (M)

1 gram (g or gm)	= 0.001 kg	1 ounce (oz)	= 28.35 g
	= 15.43 gr		= 437.5 gr
	= 0.03527 oz		= 0.0625 lb
	= 2.205×10^{-3} lb	1 pound (lb)	= 453.6 g
1 kilogram (kg)	= 1000 g		= 0.45359237 kg
	= 0.001 tonne		= 7000 gr
	= 35.27 oz		= 16 oz
	= 2.205 lb	1 short ton (ton)	= 907.2 kg
1 metric ton (tonne or te or MT)	= 1000 kg		= 0.9072 tonne
	= 2204.6 lb		= 2000 lb
	= 1.102 ton	1 long ton	= 1016.0 kg
	= 0.9842 long ton		= 1.016 tonne
1 dalton (atomic mass unit)	= 1.6604×10^{-24} g	1 long ton	= 2240 lb
			= 1.12 ton
1 grain (gr)	= 2.286×10^{-3} oz	1 stone (British)	= 6.35 kg
	= 1.429×10^{-4} lb		= 14 lb

TIME (T)

1 second (s or sec)	= 0.01667 min	1 day (d)	= 24 hr
	= 2.7778×10^{-4} hr		= 86400 s
1 minute (min)	= 60 s	1 year (yr or y)	= 365 d
	= 0.01667 hr		= 8760 hr
1 hour (hr or h)	= 60 min		= 3.15×10^7 s
	= 3600 s		

DENSITY (M/L³)

1 kilogram per cubic meter (kg/m³)	= 10^{-3} g/cm³	1 metric ton per cubic meter (te/m³)	= 1.0 specific gravity
	= 0.062 lb/ft³		= density of H_2O at 4°C
1 gram per cubic centimeter (g/cm³)	= 1000 kg/m³		= 8.35 lb/gal
	= 62.43 lb/ft³	1 pound per cubic foot (lb/ft³)	= 16.02 kg/m³

TABLE 25   *Continued*

VELOCITY (L/T)

1 meter per second (m/s)	= 3.6 km/hr = 2.237 mph = 3.28 ft/s	1 foot per second (ft/s)	= 0.68 mph = 0.3048 m/s
1 kilometer per hour (km/h or kph)	= 0.62 mph = 0.278 m/s	velocity of light in vacuum (c)	= 2.9979×10^8 m/s = 186,000 mi/s
1 mile per hour (mph or mi/h)	= 1.609 km/h = 0.45 m/s = 1.47 ft/s	1 knot	= 1.852 km/h = 1 nautical mile/hour = 1.151 mph = 1.688 ft/s

VELOCITY OF SOUND IN WATER AND SEA WATER
(assuming atmospheric pressure and sea water salinity of 35,000 ppm)

Temp, °C	Pure water, (meters/sec)	Sea water, (meters/sec)
0	1,400	1,445
10	1,445	1,485
20	1,480	1,520
30	1,505	1,545

FLOW RATE (L³/T)

1 liter per second (l/sec)	= 0.001 m³/sec = 86.4 m³/day = 15.9 gpm = 0.0228 mgd = 0.0353 cfs = 0.0700 AF/day	1 cubic decameters per day (dam³/day)	= 11.57 l/sec = 1.157×10^{-2} m³/sec
1 cubic meter per second (m³/sec)	= 1000 l/sec = 8.64×10^4 m³/day = 1.59×10^4 gpm = 22.8 mgd = 35.3 cfs = 70.0 AF/day	1 cubic decameters per day (dam³/day)	= 1000 m³/day = 1.83×10^6 gpm = 0.264 mgd = 0.409 cfs = 0.811 AF/day
1 cubic meter per day (m³/day)	= 0.01157 l/sec = 1.157×10^{-5} m³/sec = 0.183 gpm = 2.64×10^{-4} mgd = 4.09×10^{-4} cfs = 8.11×10^{-4} AF/day	1 gallon per minute (gpm)	= 0.0631 l/sec = 6.31×10^{-5} m³/sec = 1.44×10^{-3} mgd = 2.23×10^{-3} cfs = 4.42×10^{-3} AF/day
		1 million gallons per day (mgd)	= 43.8 l/sec = 0.0438 m³/sec = 3785 m³/day = 694 gpm = 1.55 cfs = 3.07 AF/day

(continues)

TABLE 25 *Continued*

FLOW RATE (L³/T) (*continued*)

1 cubic foot per second (cfs)	= 28.3 l/sec
	= 0.0283 m³/sec
	= 2447 m³/day
	= 449 gpm
		1 miner's inch	= 0.02 cfs (in Idaho,
			    Kansas, Nebraska,
			    New Mexico, North
			    Dakota, South Dakota,
			    and Utah)

1 cubic foot per second (cfs)	= 0.646 mgd = 1.98 AF/day		= 0.026 cfs (in Colorado)
1 acre-foot per day (AF/day)	= 14.3 l/sec = 0.0143 m³/sec = 1233.48 m³/day = 226 gpm = 0.326 mgd = 0.504 cfs		= 0.028 cfs (in British Columbia)
		1 garcia	= 0.02 weir
		1 quinaria (ancient Rome)	= 0.47–0.48 l/sec

1 miner's inch	= 0.025 cfs (in Arizona,
	    California, Montana,
	    and Oregon: flow of
	    water through 1 in²
	    aperture under
	    6-inch head)		

ACCELERATION (L/T²)

standard acceleration of gravity	= 9.8 m/s² = 32 ft/s²

FORCE (ML/T² = Mass × Acceleration)

1 newton (N)	= kg·m/s² = 10⁵ dynes = 0.1020 kg force = 0.2248 lb force	1 dyne	= g·cm/s² = 10^{-5} N
		1 pound force	= lb mass × acceler- ation of gravity = 4.448 N

TABLE 25   *Continued*

PRESSURE (M/L² = Force/Area)

1 pascal (Pa)	$= N/m^2$
1 bar	$= 1 \times 10^5$ Pa
	$= 1 \times 10^6$ dyne/cm²
	$= 1019.7$ g/cm²
	$= 10.197$ te/m²
	$= 0.9869$ atmosphere
	$= 14.50$ lb/in²
	$= 1000$ millibars
1 atmosphere (atm)	$=$ standard pressure
	$= 760$ mm of mercury at 0°C
	$= 1013.25$ millibars
	$= 1033$ g/cm²
	$= 1.033$ kg/cm²
	$= 14.7$ lb/in²
	$= 2116$ lb/ft²
	$= 33.95$ feet of water at 62°F
	$= 29.92$ inches of mercury at 32°F

1 kilogram per sq. centimeter (kg/cm²)	$= 14.22$ lb/in²
1 inch of water at 62°F	$= 0.0361$ lb/in²
	$= 5.196$ lb/ft³
1 inch of water at 62°F	$= 0.0735$ inch of mercury at 62°F
1 foot of water at 62°F	$= 0.433$ lb/in²
	$= 62.36$ lb/ft²
	$= 0.833$ inch of mercury at 62°F
	$= 2.950 \times 10^{-2}$ atmosphere
1 pound per sq. inch (psi or lb/in²)	$= 2.309$ feet of water at 62°F
	$= 2.036$ inches of mercury at 32°F
	$= 0.06804$ atmosphere
	$= 0.07031$ kg/cm²
1 inch of mercury at 32°F	$= 0.4192$ lb/in²
	$= 1.133$ feet of water at 32°F

TEMPERATURE

degrees Celsius or Centigrade (°C)	$= (°F - 32) \times 5/9$
	$= K - 273.16$
Kelvins (K)	$= 273.16 + °C$
	$= 273.16 + ((°F - 32) \times 5/9)$

degrees Fahrenheit (°F)	$= 32 + (°C \times 1.8)$
	$= 32 + ((°K - 273.16) \times 1.8)$

(*continues*)

TABLE 25   *Continued*

ENERGY (ML^2/T^2 = Force × Distance)

1 joule (J)	= 10^{7}ergs	1 kilowatt-hour	= 3.6 × 10^6 J
	= N·m	(kWh)	= 3412 Btu
	= W·s		= 859.1 kcal
	= kg·m^2/s^2	1 quad	= 10^{15} Btu
	= 0.239 calories		= 1.055 × 10^{18} J
	= 9.48 × 10^{-4} Btu		= 293 × 10^9 kWh
1 calorie (cal)	= 4.184 J		= 0.001 Q
	= 3.97 × 10^{-3} Btu		= 33.45 GWy
	(raises 1 g H$_2$O	1 Q	= 1000 quads
	1°C)		≈ 10^{21} J
1 British thermal	= 1055 J	1 foot-pound (ft-lb)	= 1.356 J
unit (Btu)	= 252 cal (raises		= 0.324 cal
	1 lb H$_2$O 1°F)	1 therm	= 10^5 Btu
	= 2.93 × 10^{-4} kWh	1 electron-volt (eV)	= 1.602 × 10^{-19} J
1 erg	= 10^{-7} J	1 kiloton of TNT	= 4.2 × 10^{12} J
	= g·cm^2/s^2	1 10^6 te oil equiv.	= 7.33 × 10^6 bbl oil
	= dyne·cm	(Mtoe)	= 45 × 10^{15} J
1 kilocalorie (kcal)	= 1000 cal		= 0.0425 quad
	= 1 Calorie (food)		

POWER (ML^2/T^3 = rate of flow of energy)

1 watt (W)	= J/s	1 horsepower	= 0.178 kcal/s
	= 3600 J/hr	(H.P. or hp)	= 6535 kWh/yr
	= 3.412 Btu/hr		= 33,000 ft-lb/min
1 TW	= 10^{12} W		= 550 ft-lb/sec
	= 31.5 × 10^{18} J		= 8760 H.P.-hr/yr
	= 30 quad/yr	H.P. input	= 1.34 × kW input
1 kilowatt (kW)	= 1000 W		to motor
	= 1.341 horsepower		= horsepower
	= 0.239 kcal/s		input to motor
	= 3412 Btu/hr	Water H.P.	= H.P. required to
10^6 bbl (oil)/day	≈ 2 quads/yr		lift water at a
(Mb/d)	≈ 70 GW		definite rate
1 quad/yr	= 33.45 GW		to a given dis-
	≈ 0.5 Mb/d		tance assum-
1 horsepower	= 745.7 W		ing 100%
(H.P. or hp)	= 0.7457 kW		efficiency
			= gpm × total head
			(in feet)/3960

Table 25 *Continued*

EXPRESSIONS OF HARDNESS[a]

1 grain per gallon	= 1 grain CaCO$_3$ per U.S. gallon	1 French degree	= 1 part CaCO$_3$ per 100,000 parts water
1 part per million	= 1 part CaCO$_3$ per 1,000,000 parts water	1 German degree	= 1 part CaO per 100,000 parts water
1 English, or Clark, degree	= 1 grain CaCO$_3$ per Imperial gallon		

CONVERSIONS OF HARDNESS

| 1 grain per U.S. gallon | = 17.1 ppm, as CaCO$_3$ | 1 French degree | = 10 ppm, as CaCO$_3$ |
| 1 English degree | = 14.3 ppm, as CaCO$_3$ | 1 German degree | = 17.9 ppm, as CaCO$_3$ |

WEIGHT OF WATER

1 cubic inch	= 0.0361 lb	1 imperial gallon	= 10.0 lb
1 cubic foot	= 62.4 lb	1 cubic meter	= 1 tonne
1 gallon	= 8.34 lb		

DENSITY OF WATER[a]

Temperature		Density
°C	°F	gm/cm^3
0	32	0.99987
1.667	35	0.99996
4.000	39.2	1.00000
4.444	40	0.99999
10.000	50	0.99975
15.556	60	0.99907
21.111	70	0.99802
26.667	80	0.99669
32.222	90	0.99510
37.778	100	0.99318
48.889	120	0.98870
60.000	140	0.98338
71.111	160	0.97729
82.222	180	0.97056
93.333	200	0.96333
100.000	212	0.95865

Note: Density of Sea Water: approximately 1.025 gm/cm^3 at 15°C.

[a]*Source:* F. van der Leeden, F.L. Troise, and D.K. Todd, 1990. *The Water Encyclopedia,* 2d edition. Lewis Publishers, Inc., Chelsea, Michigan.

Index